Combustion Efficiency
and Air Quality

Combustion Efficiency and Air Quality

Edited by

István Hargittai

*Budapest Technical University
and Hungarian Academy of Sciences
Budapest, Hungary*

and

Tamás Vidóczy

*Central Research Institute for Chemistry
of the Hungarian Academy of Sciences
Budapest, Hungary*

SPRINGER SCIENCE+BUSINESS MEDIA, LLC

Library of Congress Cataloging-in-Publication Data

On file

ISBN 978-1-4613-5739-1   ISBN 978-1-4615-1827-3 (eBook)
DOI 10.1007/978-1-4615-1827-3

© 1995 Springer Science+Business Media New York
Originally published by Plenum Press in 1995
Softcover reprint of the hardcover 1st edition 1995

10 9 8 7 6 5 4 3 2 1

Contributors

Hans Asbjørn Aaheim, Center for International Climate and Energy Research—Oslo (CICERO), University of Oslo, 0317 Oslo, Norway

Zoltán Adonyi, Department of Chemical Technology, Budapest Technical University, H-1521 Budapest, Hungary

Kristin Aunan, Center for International Climate and Energy Research—Oslo (CICERO), University of Oslo, 0317 Oslo, Norway

Paul Benedek, Department of Chemistry, Eötvös University, H-1518 Budapest, Hungary

Costantino Boni, Materials Division, CISE Tecnologie Innovative S.p.A., 20134 Milan, Italy

Grazia Maria Braga Marcazzan, Istituto di Fisica Generale ed Applicata, Università degli Studi di Milano, Milan, Italy

Thomas B. Brill, Department of Chemistry, University of Delaware, Newark, Delaware 19716

Ezio Cereda, Materials Division, CISE Tecnologie Innovative S.p.A., 20134 Milan, Italy

Istvan L. Gebefuegi, Institute for Ecological Chemistry, GSF Research Center, Neuherberg, D-85758 Oberschleissheim, Germany

Arthur Greenberg, Department of Chemistry, University of North Carolina at Charlotte, Charlotte, North Carolina 28223

István Hargittai, Institute of General and Analytical Chemistry, Budapest Technical University, H-1521 Budapest, Hungary and Structural Chemistry Research Group of the Hungarian Academy of Sciences, Eötvös University, H-1431 Budapest, Hungary

Sándor Kántor, Department of Chemical Technology, Budapest Technical University, H-1521 Budapest, Hungary

Thomas M. Klapötke, Institut für Anorganische und Analytische Chemie, Technische Universität Berlin, D-10623 Berlin, Germany

Ernő Kulcsár, Department of Chemistry, Eötvös University, H-1518 Budapest, Hungary

Jerzy Leszczyński, Department of Chemistry, Jackson State University, Jackson, Mississippi 39217

Alfred H. Lowrey, Laboratory for the Structure of Matter, Naval Research Laboratory, Washington, D.C. 20375

András Németh, Central Research Institute for Chemistry, H-1525 Budapest, Hungary

Fulvio Parmigiani, Materials Division, CISE Tecnologie Innovative S.p.A., 20134 Milan, Italy and Facoltà di Scienze, Università di Como, Como, Italy

Károly Reményi, Institute for Electric Power Research, H-1051 Budapest, Hungary

James L. Repace, Office of Research and Development, Exposure Assessment Division, U.S. Environmental Protection Agency, Washington, D.C. 20460

György Schultz, Structural Chemistry Research Group of the Hungarian Academy of Sciences, Eötvös University, H-1431 Budapest, Hungary

Hans Martin Seip, Center for International Climate and Energy Research—Oslo (CICERO), University of Oslo, 0317 Oslo, Norway

Inis C. Tornieporth-Oetting, Institut für Anorganische und Analytische Chemie, Technische Universität Berlin, D-10623 Berlin, Germany

Lance A. Wallace, Atmospheric Research and Exposure Assessment Laboratory, U.S. Environmental Protection Agency, Warrenton, Virginia 22186

Sol William Weller, Department of Chemical Engineering, State University of New York at Buffalo, Buffalo, New York 14260

Preface

There is a broad recognition in the scientific community that much-needed efficiencies in energy production will have to produce, at the same time, substantial benefits in environmental protection. Chemists and engineers understand that innovative approaches to combustion (still a major source of energy) will have to improve the conversion of fuel to clean byproducts and maximize the production of heat. It is satisfying that much of the fundamental research directed towards increasing combustion efficiency serves the dual purpose of protecting air quality. Innovative improvements in combustion efficiency must come from fundamental research. Fundamental research on molecular structure is now an important source of information for the thermodynamic understanding of reactants and atmospheric pollutants. The same physical experimental and computational techniques now contribute to the advances of our understanding of basic chemistry and combustion engineering.

The focus on basic research provides a link for the application of multidisciplinary approaches to the problems of combustion efficiency. Many techniques are being combined in innovative ways to make advances in our understanding of molecular structure and interactions with direct application to problems in combustion. This is a path to focus international efforts in many scientific areas on the pressing energy and environmental needs at both the global and local levels.

The Committee of Physical Chemistry and Inorganic Chemistry of the

Hungarian Academy of Sciences has followed the progress of various fields and encouraged interactions among its eight subcommittees covering much of physical chemistry and inorganic chemistry. Combustion efficiency and air quality have also involved analytical chemistry and chemical engineering. Thus an International Workshop was organized by the Committee, September 26–28, 1993, titled Combustion Efficiency and Air Quality at the Institute of General and Analytical Chemistry of the Budapest Technical University.

The following organizations gave generous support for the Workshop: the Hungarian Academy of Sciences and its Chemistry Division, the Budapest Technical University, the PHARE-ACCORD Program, and the MOL Hungarian Oil and Gas Industries. The organization of the meeting was carried out by a Workshop Committee, István Hargittai (Chair), Tamás Vidóczy (Secretary), Tibor Bérces, Jenő Fekete, József Kőmives, Ernő Mészáros, András J. Németh, József Pázmány, and Károly Reményi. The topics discussed during the Workshop included molecular structure and energetics, elementary reactions, kinetic modeling, combustion theory, combustion in practical systems, emission analysis and control, atmospheric chemistry, and waste incineration.

It was suggested during the Workshop to organize a volume along the lines of the topics of the Workshop. The purpose of this book is to demonstrate the contributions of the various fields of physical chemistry, analytical chemistry, and chemical engineering to the concerted handling of energy production through efficient combustion with simultaneous concern for the protection of the quality of the atmosphere. It was decided not to produce a proceedings volume but a monograph. Not all the workshop participants chose to contribute to this volume but further contributors have provided a more balanced presentation.

We would like to thank all the contributors for an efficient and pleasant cooperation and Plenum Press for bringing out this volume.

István Hargittai
Tamás Vidóczy

Budapest

Contents

Chapter 3

Covalent Inorganic Nonmetal Azides

Inis C. Tornieporth-Oetting and Thomas M. Klapötke

Chapter 4

Chemistry of a Burning Propellant Surface

Thomas B. Brill

Chapter 5

Competitive Reactions of Methyl Radicals in Partial Oxidation of Methane

Ernő Kulcsár, Paul Benedek, and András Németh

Chapter 9

Incineration of Waste Solvents Containing Chlorinated Hydrocarbons: Some Critical Remarks

Zoltán Adonyi and Sándor Kántor

Chapter 10

Concentrations of Combustion Particulates in Outdoor and Indoor Environments

Alfred H. Lowrey, Lance A. Wallace, Sándor Kántor,
   and James L. Repace

Chapter 11

Bulk and Surface Studies of Fly Ash Particles

Costantino Boni, Ezio Cereda, Grazia Maria Braga Marcazzan,
   and Fulvio Parmigiani

Chapter 1

Models for Environmental and Chemical Systems

Paul Benedek

At the beginning were the prophets. Heaven told them the future by hotline.

Later arrived the shamans. Their prophecies were read from the carcasses of sacrificed animals,

Fortune tellers told the future by cards.

The great stargazers with *great dreams* emerged in the 19th century: Jules Verne, Theodor Herzl, V. I. Lenin. They expected from the perfect technology the prosperity of mankind.

Recently, the Club of Rome (CR), members of which are honorable gentlemen, has been dealing with the new problems of the post-industrial society. At the beginning of the 1970s, the activity of the Club of Rome awoke high interest.

1. CLUB OF ROME

The CR guided the attention of the intellectuals to the fact, already known to them, that though increasing production serves the public welfare by providing

PAUL BENEDEK • Department of Chemistry, Eötvös University, H-1518 Budapest, Hungary.

Combustion Efficiency and Air Quality, edited by István Hargittai and Tamás Vidóczy. Plenum Press, New York, 1995.

the population in the developed countries with an increasing volume of products, it increases the exploitation of natural resources and the pollution of the environment.

The CR did not say anything new, but the way of saying it was different and so was the justification. It was different because the CR used the global world model of Forrester and later Meadows and based its picture of the future on calculations performed on the world model. Let us come to the point.

The world is a closed system with the following characteristic variables:

- Population of the world
- Stock of nonrenewable raw materials
- Volume of industrial production
- Pollution of the environment
- Quality of life

For each of the variables, a differential equation (DE) may be written. Let us assume that the rate of change of the world population is equal to the birth rate minus the death rate (Fig. 1). Both are proportional to the population. A key problem in the building of a model is the determination of the constants of proportionality, k. Applying the data for the seven decades between 1900 and 1970, the coefficients k were elaborated not only for this DE, but also for the other four DEs. Each of these k coefficients is a multivariate function of the above-mentioned five variables and of other internal variables as well (Fig. 2).

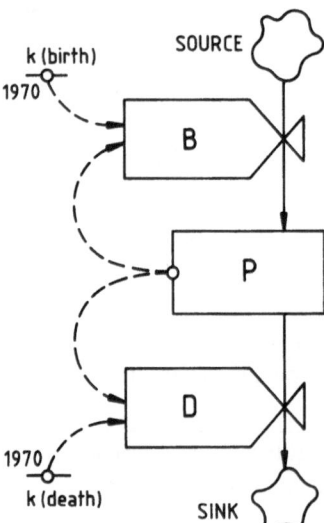

Figure 1. Population control loop in the world model. B, Birth; P, population; D, death.

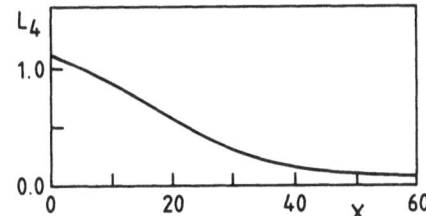

Figure 2. Pollution is contribution no. 4 to the quality of life. X, Pollution; L_4, factor no. 4 for quality of life.

We obtain a DE system (DES) with nonconstant coefficients. Its numerical solution does not present any difficulties. The DES is solved with the assumption that the k functions are valid for the next 70 years. This assumption means, however, that mankind will live in the next decades without research and development.

The solution of such a DES can be seen in Fig. 3. This indicates that an ecological catastrophe will occur in the thirties of the next century.

The concept of *zero growth* derives from this result. This is the suggestion that mankind should freeze the model variables at the level of 1970, that is, people should not be propagated (birth control), and mining of nonrenewable natural resources and the industrial production should not increase. This guarantees that the level of environmental pollution, and consequently the quality of life, will be kept at approximately the level of the 1980s (Fig. 4).

The courage with which Forrester and Meadows applied these few highly aggregated variables is really fascinating, and the prophecy of the calculation makes one shudder.

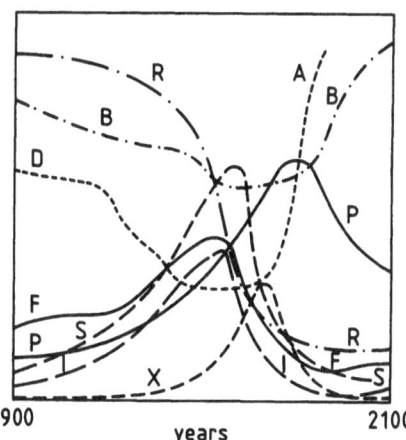

Figure 3. Solution of the basic world model. P, Population; R, nonrenewable resources; D, death; B, birth; F, food; X, pollution; S, service; I, quality of life.

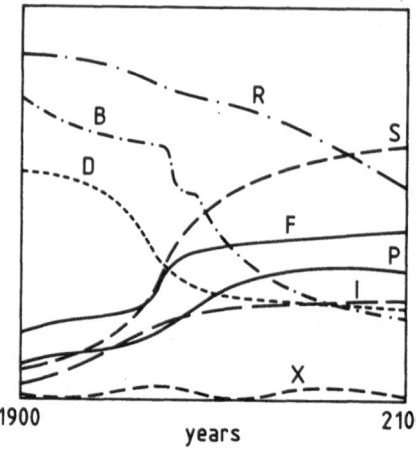

1900
years
2100

Figure 4. Solution of the modified world model, based on limits on growth. For definitions of abbreviations, see caption to Fig. 3.

2. TWO EXAMPLES

However, more than 20 years have elapsed since the concept of zero growth was proposed. One has to realize, on the one hand, that mankind—if it considered at all the recommendations of the CR—took good decisions organized at most national, regional, and later world conferences. Such a good decision was that CO_2 emissions should not surpass the level of 1987. On the other hand, the degree of aggregation of the model variables conceals what should be actually done to avoid the global ecological catastrophe.

The problem is not so trivial. I present two examples. I just mentioned the agreement about CO_2 emission, which deals with national CO_2 emission (note that this is a severely aggregated variable). Hungary engaged in this agreement and was able to comply with the international agreement not to increase national CO_2 emission. However, no technical measures were taken. The explanation is very simple: during the past five years, industrial production declined in Hungary. However, while the national CO_2 emission did not increase, let us look at the local situation. Figure 5, which shows the local CO_2 emissions on the map of Hungary, speaks for itself.

Let us now consider another example. Two of my colleagues [Tibor Tóth (TT) and Ferenc Garai (FG)] were collecting an air sample in the street at the Petőfi Bridge in Buda. This is a location with extremely heavy traffic. The sample was analyzed by gas–liquid chromatography (GLC). More precisely, a chromatogram was constructed by computer simulation (Fig. 6). It is exactly as if TT and FG had worked with the same chromatograph as Martin employed in 1952, when he was awarded a Nobel prize for his work on chromatography. The peak in the neighborhood of toluene is so diffuse that one cannot know how many and what type of

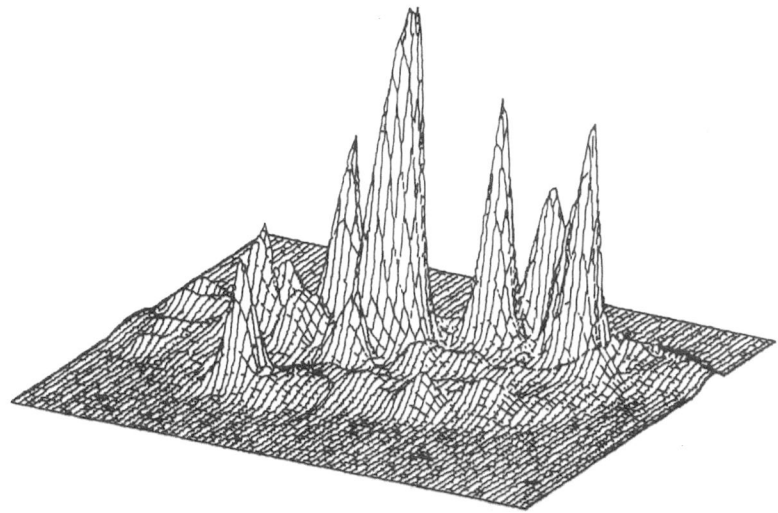

Figure 5. CO_2 emissions on the map of Hungary.

components are covered by the peak. Figure 7 similarly shows the result of a computer simulation. I would like to tell you now a story about the development of chromatography.

3. MARTIN AND GOLAY

At the Second Symposium on Gas Chromatography in Amsterdam in 1958, somebody presented a paper about a catharometer made from a sparking plug. In the discussion following this paper, Martin rose to speak and told his *dream* that a

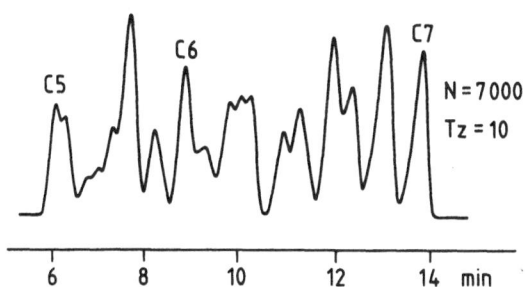

Figure 6. Simulated chromatogram of the air sample. Separation number = 10.

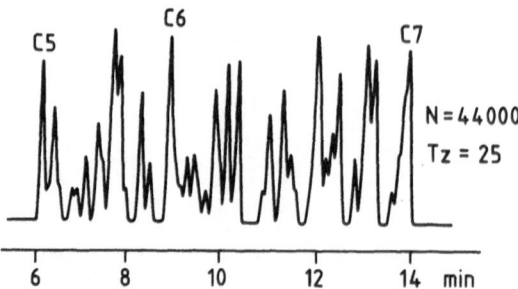

Figure 7.   Simulated chromatogram of the air sample. Separation number = 25.

chemist would be able to build a chromatograph in his laboratory by such simple means.

At the same symposium, Golay presented an absolutely theoretical conception of the open-tube column. His presentation was perfectly theoretical because a chromatograph with an open-tube column could not be built. The resolution power of the classical catharometer was not sensitive enough for the open-tube column. Golay also had a dream that somebody would construct a detector that could be used with his column.

And here is the wonder—the flame ionization detector was developed, with exactly the sensitivity that corresponded to the resolution performance of the Golay column. The marriage took place very soon, within the American instrument industry. This instrument could not be constructed by a chemist in his own laboratory. So not Martin's but Golay's dream came true.

The chromatogram in Fig. 7 is just as if L. S. Ettre had obtained it with the first Golay column. The separation number became the double as compared to that of the first. The separation number became double again on the next chromatogram (Fig. 8), as if it had been obtained with the first glass capillary column. This means

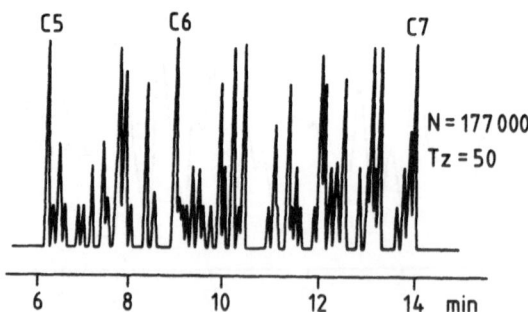

Figure 8.   Simulated chromatogram of the air sample. Separation number = 50.

that the instrument industry continuously improved the resolution power of the bench-scale chromatograph for us chemists, and consequently we obtained increasingly more detailed and more specific data on flue gas pollution.

4. TOTH AND GARAI

The chromatograms in Fig. 9 are not the result of simulation. At the top is the chromatogram of the air sample collected by Toth and Garai and at bottom is the chromatogram of the gasoline taken from Toth's car. Now it is already visible that the resolution of severely aggregated variables has sense. The point is that the flue gas contains generally unburned hydrocarbons, but among them the aromatic compounds play a considerable role.

A vast literature deals with the health-impairing effects of aromatics, and it is well known that, although the antiknocking properties of aromatics are good, their burning velocity is slow as compared to that of paraffinic hydrocarbons. We also know that the majority of the aromatic components of gasoline come not from the depths of the earth but rather arise from the use of aromatization in order to increase the octane number or are blended voluntarily into the gasoline.

It appears clear that the concept of zero growth is not acceptable. People living in modern society cannot be threatened by ecological catastrophe or persuaded not to use their cars. Electric car belongs to the future. The environment, however, has to be protected now, with technical solutions built on the existing knowledge basis. For example, 20 years ago the task was the introduction of lead-free gasoline, and the petroleum industry achieved this task by blending

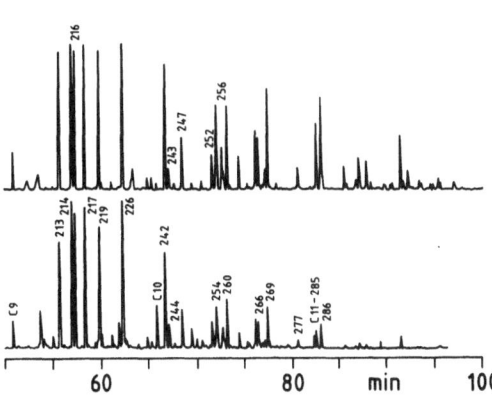

Figure 9. Original chromatograms of the air sample and a gasoline sample. *Top*: Hydrocarbon components of air. *Bottom*: hydrocarbon components of gasoline.

aromatic components into the gasoline. One decade ago, the task had already become the production of nonaromatic gasoline. This has been solved too. As a matter of fact, the direction and the speed of the spread of nonaromatic gasoline is merely an economical question in Europe.

5. A PLAY WITH THREE PARTICIPANTS

In the meantime, the speed of growth has been so great that the pollution of the environment has evoked the interest of the society as a whole. As a result, a play with three participants has begun. The interests of the society are defined by the first participant—the public authorities—for instance, by limiting and controlling emissions. Each of the individually defined limits means individually defined R&D tasks for the second participant—the chemist. The latter has to find the best solution for a given task. As we have seen, the third participant is the instrument industry. This participant stays between the public authorities and the chemists in the struggle for environmental protection. This fact saved the instrument industry of the United States from the consequences of the recent economic recession. One of the causes of the recession was the decline of the defense industry.

In my report were five star-gazers. Permit me as the sixth to tell you *my dream* too. If the economic power of the defense industry were to have been taken channeled toward protection of the environment, it would have piloted the economy to a crisis-free world.

Chapter 2

Structural Investigation of Molecules of Energetic Materials in the Gas Phase

György Schultz and István Hargittai

1. INTRODUCTION: AIMS AND SCOPE

Recently, Gilardi and Karle (1991) published a review on structural investigations of energetic materials in the crystalline state. The purpose of the study was to promote the understanding of the relationship between structure and function and to facilitate the synthesis of new materials. This study was based on the Cambridge Structural Database (Allen *et al.*, 1983) and also on the Energetic Materials Structural Database of the (U.S.) Naval Research Laboratory (NRL) with over 300 relevant structures (Gilardi *et al.*, 1992).

As discussed by Gilardi and Karle (1991), density is closely related with the behavior of energetic materials. The dependency of the detonation pressure on the density is greater than quadratic. Thus, it is appropriate to study these materials in the solid phase. In addition to density, molecular strain is also an important source

GYÖRGY SCHULTZ • Structural Chemistry Research Group of the Hungarian Academy of Sciences, Eötvös University, H-1431 Budapest, Hungary.   ISTVÁN HARGITTAI • Institute of General and Analytical Chemistry, Budapest Technical University, H-1521 Budapest, Hungary and Structural Chemistry Research Group of the Hungarian Academy of Sciences, Eötvös University, H-1431 Budapest, Hungary.

Combustion Efficiency and Air Quality, edited by István Hargittai and Tamás Vidóczy. Plenum Press, New York, 1995.

for higher energy release. The synthetic goal is to obtain highly strained molecules that pack densely and contain a large number of energetic groups. The overwhelming majority of molecules discussed by Gilardi and Karle (1991) and/or reported in the NRL Database (Gilardi *et al.*, 1992) contain the NO_2 fragment as an energetic group.

The energetic properties of NO_2-containing hydrocarbon molecules are closely related to the oxidation of carbon and hydrogen by oxygen of the NO_2 group(s) of the same molecule and by a sudden increase of the pressure due to the creation of gaseous nitrogen, carbon dioxide, and water, followed by a volume increase.

In view of the above, we would like to rationalize our interest in the gaseous molecular structures of compounds whose importance is intimately related to the solid state. Our main argument is that in order to evaluate and understand the interactions stemming from the fact that the molecules are part of a crystal structure/packing arrangement in the solid, the structure of the isolated molecule itself must be known in the first place. The structure of the isolated/free molecule can then serve as reference structure in the evaluation of the consequences of intermolecular interactions in the crystal.

Interest in gas/solid molecular structure comparisons has grown steadily during the past years (see, e.g., I. Hargittai and M. Hargittai, 1987; M. Hargittai and I. Hargittai, 1987; Domenicano and Hargittai, 1993; Burns and Leopold, 1993). This is due to the importance of the observed structural differences and also to the increased accuracies that have become attainable in structure determinations and have made such comparisons meaningful (Domenicano and Hargittai, 1992).

We would also like to comment upon the choice of compound class selected for discussion in this review. While nitro compounds are an important class of energetic materials, there are other important classes. The present choice was dictated by the availability of Gilardi and Karle's (1991) review, on the one hand, and by space limitations, on the other.

Another class of compounds of conspicuous recent interest as energetic materials is the fullerenes and related compounds such as the polyynes, for example, whose investigation played a pivotal role in the discovery of buckminsterfullerene itself (Kroto and Walton, 1993).

Incidentally, soon after buckminsterfullerene (Fig. 1) had become available in sufficient amounts, its molecular structure was determined by various experimental techniques, including its gas-phase structure by electron diffraction, as well as by theoretical calculations. The results are summarized in Table I.

We mention, in passing, another class of compounds of interest, the polycyclic hydrocarbons, which cover the broadest range of unstable (e.g., tetrahedrane) to extremely stable (e.g., adamantane) compounds. A recent review of cubane and derivatives is a case in point (Eaton, 1992).

That the subject matter of the present review is of high interest is charac-

Figure 1. Truncated icosahedral structure of buckminsterfullerene (see, e.g., Kroto, 1989).

terized by the fact that two of the substances related to it have recently been named "molecule of the year" in *Science*. One is buckminsterfullerene, in December 1991, and the other is nitrogen oxide, in December 1992.

The present review is restricted to the gas-phase molecular geometries of NO_2-containing molecules determined from experimental data. In selected cases, they will be compared with the corresponding structures in the solid phase. Molecular geometry will be characterized by shape, symmetry, and internal coordinates, i.e., bond lengths, bond angles, and angles of torsion.

The primary techniques whose results will be presented are the two principal physical methods of gaseous molecular structure determination, viz., electron diffraction (see, e.g., Hargittai and Hargittai, 1988) and microwave spectroscopy (see, e.g., van Eijck, 1992). Table II presents some merits of various techniques of molecular structure determination as well as some difficulties that are encountered in the application.

In modern work much emphasis is placed on the exact physical meaning of the structural information originating from different techniques. This gains importance beyond a certain accuracy level. References will be made to the physical meaning of the parameters in our compilation. However, in view of the relatively large experimental errors, our discussion will be staying at a level where differences in the physical meaning of the parameters hardly matter. We issue this caveat though, because future and more detailed comparisons and discussions will have to take such considerations into account when data of the necessary accuracy level become available in increasing amounts.

Table I. Bond Lengths (Å) in Buckminsterfullerene

Technique	Phase	Temperature (K)	Distance type[a]	C5/C6	C6/C6	Reference
Electron diffraction	Gas	1000	r_g	1.458(6)	1.401(10)	Hedberg et al., 1991
Neutron diffraction	Crystal	5	r_α	1.455(12)	1.391(18)	David et al., 1991
X-ray diffraction	Crystal	110	r_α	1.445(5)	1.399(7)	Bürgi et al., 1992[b]
NMR spectroscopy	Solid	77	r_α	1.45(15)	1.40(15)	Yannoni et al., 1991
Ab initio (MP2 triple-zeta)	Free molecule	0	r_e	1.446	1.406	Hasser et al., 1991

[a]See, e.g., Domenicano and Hargittai (1992).
[b]A reanalysis of previous data (Liu et al., 1991) and further work eliminated the considerable discrepancy between the X-ray diffraction results and the electron diffraction results.

Table II. Merits and Difficulties of Five Techniques for Molecular Structure Determination

Technique	Merits	Difficulties
X-ray crystallography	Vast amount of structural data accumulated and accessible from data banks Accurate bond angles Differences between bond lengths may often be determinded accurately	Intermolecular interactions may influence structure Bond lengths are not very accurate and do not correspond to internuclear distance
Neutron crystallography	Accurate bond angles and internuclear distances, including those involving hydrogen atoms	Internuclear interactions may influence structure
Electron diffraction	Yields structure of free molecules Increased accuracy for symmetrical molecules Gives also information on intramolecular motion	Accuracy diminishes with increasing molecular complexity
Microwave spectroscopy	Yields structure of free molecules Very accurate geometrical parameters from substitution coordinates	Poor for determining position of atoms near center of gravity Not suitable for highly symmetrical molecules
Quantum-chemical calculations	Predict directions of even small structural changes May give information also on experimentally inaccessible molecules May help to interpret structural peculiarities	There may be difficulties in assessing the uncertainty of the parameters determined Basis set dependence, correlations, other factors must be carefully examined

A special comment is made here concerning the temperature dependence of structural parameters. One of the few molecules that has been studied in great detail is nitrous oxide, N_2O (Fink and Kohl, 1988). Figure 2 presents the internuclear distances, the vibrational amplitudes, and the so-called shrinkage effect as a function of temperature. The shrinkage, $[r_g(N-N) + r_g(N-O)] - r_g(N\cdots O)$, is the shortening of the average $N\cdots O$ nonbonded distance as compared to the sum of the two bond lengths of the N_2O molecule, which is linear in its equilibrium structure. Due to the bending vibrations, there is a shortening of the average $N\cdots O$ distance. Obviously, the mean vibrational amplitudes and the

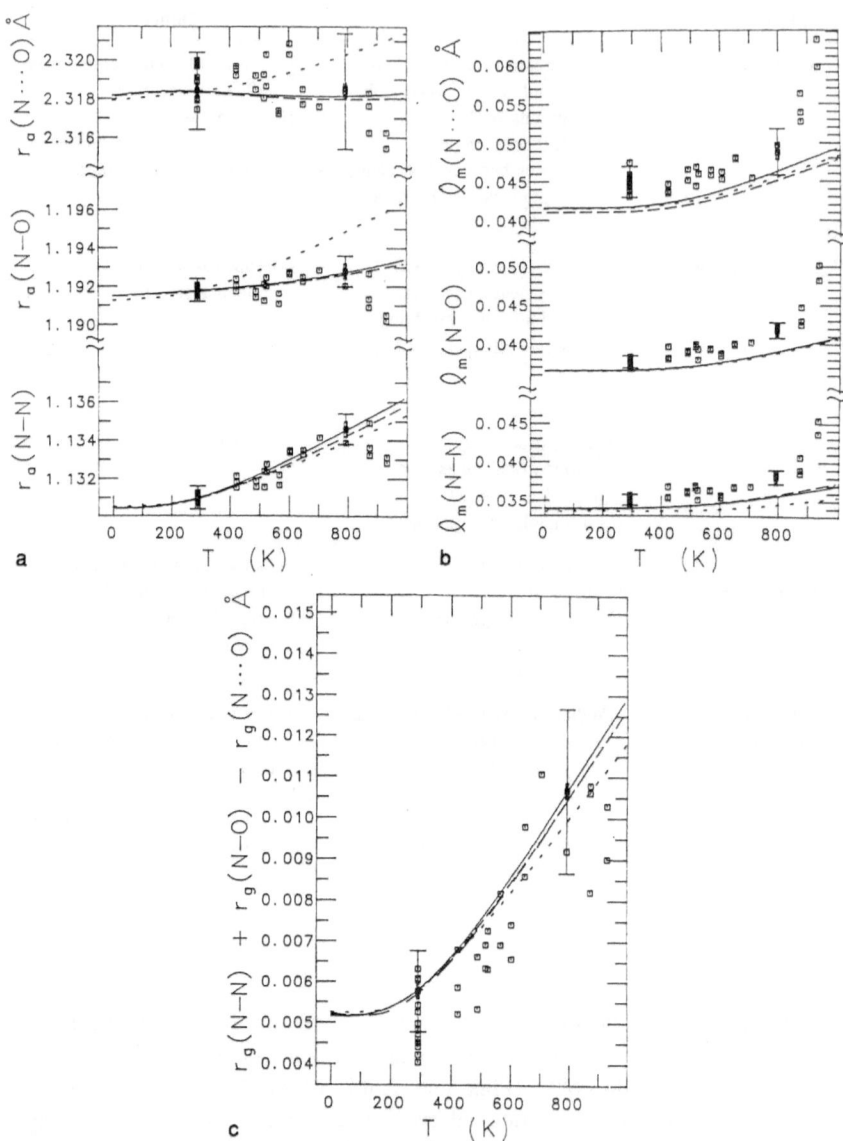

Figure 2. Internuclear distances r_a (a), mean vibrational amplitudes (b), and (c) the shrinkage effect $r_g(N–N) + r_g(N–O) - r_g(N\cdots O)$ of N_2O as a function of temperature. After Fink and Kohl (1988); for details of the calculations, see the original publication.

shrinkage effect are considerably more sensitive to temperature changes than the internuclear distances.

We would like to issue yet another caveat prior to embarking on the detailed presentation of our discussion. The NO_2 group is usually depicted as containing two N=O bonds. These should not be taken to be double bonds but bonds that are probably somewhat stronger than single bonds. In employing this representation, we are following general usage rather than implying rigorous bonding considerations.

Our work is based on the primary literature with very few exceptions. This we find important especially in error evaluation. Another consideration is the lack of a uniform approach in evaluating the consequences of various assumptions in electron diffraction structure analysis, which, again, makes consultation with the original work useful. In quoting the parameters and their uncertainties, we follow the original reports, with very few exceptions.

The three Landolt–Börnstein volumes (Callomon et al., 1976, 1987; Kuchitsu, 1992) on gas-phase molecular structures report structural parameters of about 80 molecules containing the NO_2 group. Only a fraction of these molecules have also been studied in the solid phase. Because of the limited space available, we have not generally considered theoretical calculations although their growing importance is well recognized.

Structural studies of azides, another important class of energetic materials, are discussed by Tornieporth-Oetting and Klapötke in Chapter 3 of this volume.

2. OXIDES OF NITROGEN

The following oxides of nitrogen containing the NO_2 moiety have been investigated in the gas phase: nitrogen dioxide, NO_2 (Hedberg, 1966; Bird et al., 1964); nitrogen trioxide radical, NO_3 (Ishiwata et al., 1985); dinitrogen trioxide, N_2O_3 (Brittain et al., 1969); dinitrogen tetroxide, N_2O_4 (McClelland et al., 1972); and dinitrogen pentoxide, N_2O_5 (McClelland et al., 1983).

The structure of the NO_2 molecule was studied both by gas-phase electron diffraction (ED) and microwave spectroscopy (MW). According to ED, r_a(N=O) 1.203(3) Å and $\langle$O=N=O 134.0(13)° (Hedberg, 1966), and according to MW, r_s(N=O) 1.1934(10) Å and $\langle_s$O=N=O 134.07(10)° (Bird et al., 1964).

From an infrared diode laser spectroscopic study of the gaseous NO_3 radical, its D_{3h} symmetry was deduced. The N–O bond length, calculated from the B_0 rotational constant, is 1.240 Å (Ishiwata et al., 1985).

Dinitrogen trioxide is planar according to MW (Brittain et al., 1969). Its geometrical parameters are given in Fig. 3a.

Dinitrogen tetroxide is also planar (D_{2h} symmetry) with r_a(N–N) 1.782(8) Å, r_a(N=O) 1.190(2) Å, $\langle_a$O=N=O 135.4(6)° [ED by McClelland et al. (1972)].

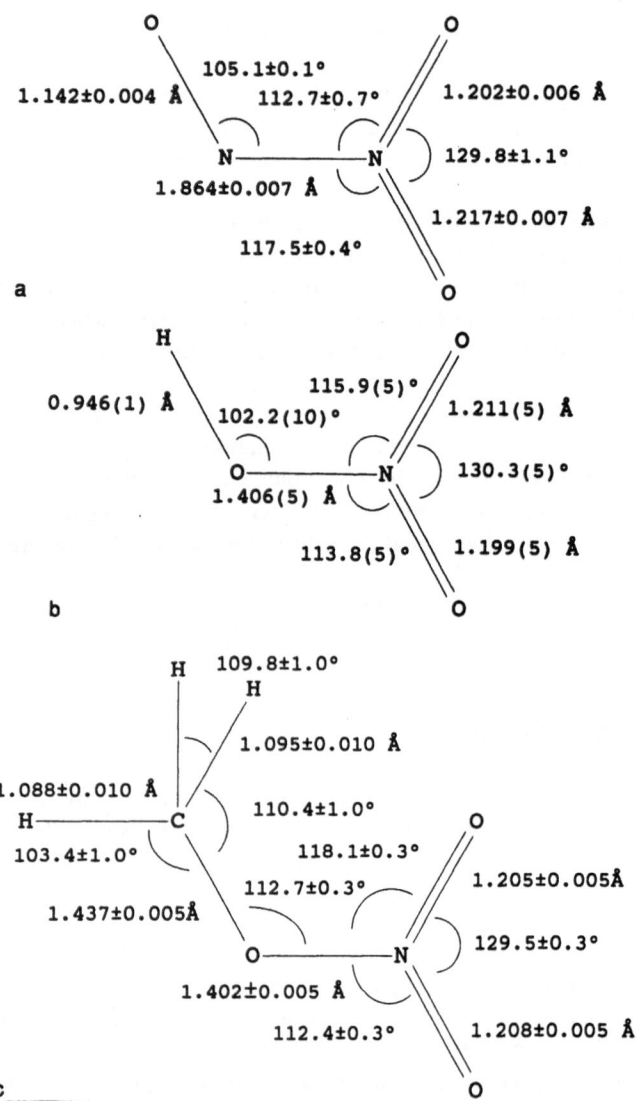

Figure 3. (a) Structure (r_s) of dinitrogen trioxide from MW (Brittain *et al.*, 1969); (b) structure (r_s) of nitric acid from MW (Cox and Riveros, 1965); (c) structure (r_s) of methyl nitrate from MW (Cox and Waring, 1971); (d) structure (r_g, $\langle_\alpha\rangle$) of the *anti* form of ethyl nitrate from ED (Shishkov *et al.*, 1992); (e) structure (r_a, $\langle_\alpha\rangle$) of chlorine nitrate from a combined ED/MW analysis (Casper *et al.*, 1993); (f) structure (r_a, $\langle_\alpha\rangle$) of bromine nitrate from ED (Casper *et al.*, 1993).

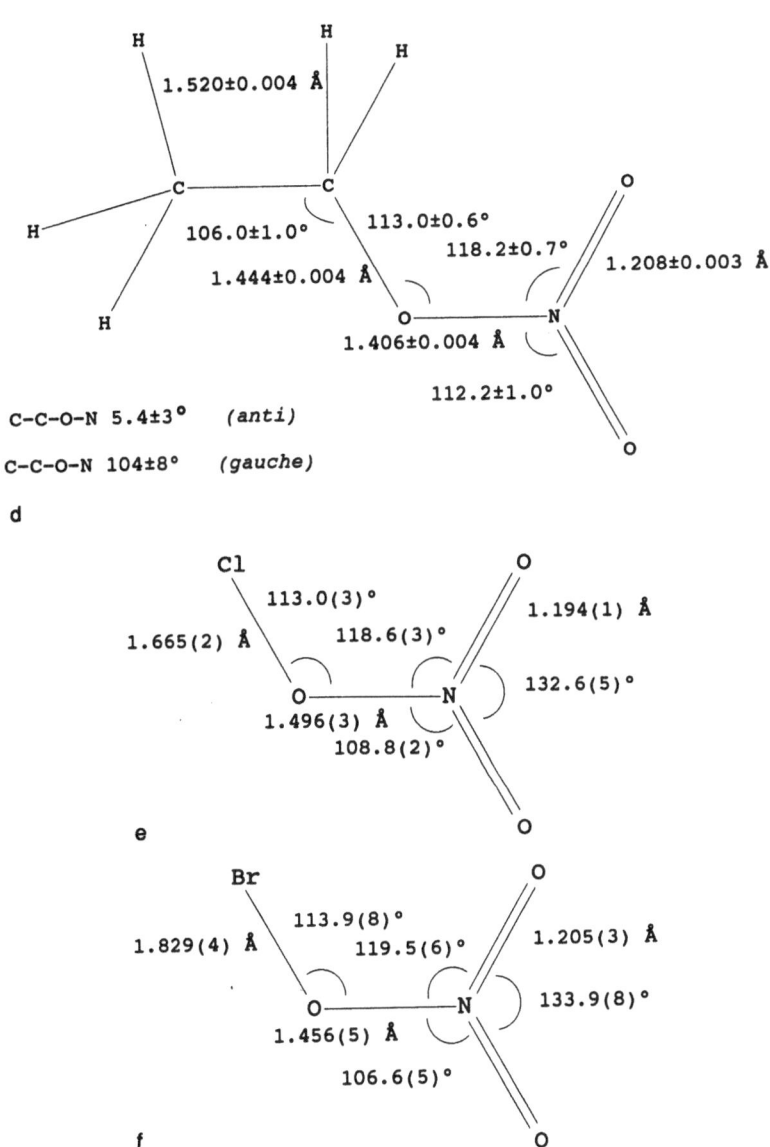

C-C-O-N 5.4±3°   *(anti)*

C-C-O-N 104±8°   *(gauche)*

d

e

f

Figure 3. *(Continued)*

The structures of both N_2O_3 and N_2O_4 are unusual in that the N–N bonds are extremely long. For comparison, the N–N bond length is 1.417(6) Å in 1,2-dimethylhydrazine (r_z; Nakata et al., 1981) and 1.447(2) Å in hydrazine (r_a; Kohata et al., 1982). The long N–N bonds in dinitrogen trioxide and dinitrogen tetroxide are part of a planar molecular skeleton that may be a source of considerable strain. The molecular geometry is known in the crystal as well (Groth, 1963; Obermeyer et al., 1991). Two modifications, viz., monoclinic and cubic, have been studied, and the molecule was found to be planar in both. Here we quote the geometrical parameters (mean values) from the more recent study on the cubic modification: N=O 1.179(1) Å, N–N 1.758(2) Å, O=N–N 112.9(1)°, O=N=O 134.3(1)°. Noteworthy is, again, the very long N–N bond, merely 0.02 Å shorter than in the gas phase, the difference being hardly significant in view of the relatively large experimental error of the gas-phase study.

Molecular orbital calculations suggest a σ type N–N bond in N_2O_4. Its unusual length may be due to the delocalization of the electron pair over the whole molecule with strong repulsions between doubly occupied molecular orbitals of the NO_2 groups. The coplanarity results from a balance of forces favoring skew and planar forms. The barrier to internal rotation about the N–N bond was estimated to be 9.6 kJ/mol in N_2O_4 (Cotton and Wilkinson, 1988).

The ED study of N_2O_5 (McClelland et al., 1983) indicated that the two NO_2 groups are joined through the fifth oxygen atom by noncollinear bonds. The NO_2 groups perform large-amplitude torsional motion about the minimum energy position, corresponding to C_2 symmetry of the molecule. The dihedral angle is 30° between each of the NO_2 groups and the N–O–N plane. Assuming C_{2v} local symmetry for the ONO_2 groups, the following bond lengths (r_g) and bond angles ($\langle_\alpha$) were obtained: N=O 1.188(2) Å, N–O 1.498(2) Å, O=N=O 133.2(6)°, N–O–N 111.8(16)°.

N_2O_5 has an altogether different structure in the solid phase: linear nitronium (NO_2^+) and planar nitrate (NO_3^-) ions are arranged in layers with the axes of the former perpendicular to the planes of the latter (Grison et al., 1950).

The gas-phase structures of NO_2 groups in oxides of nitrogen are rather similar in spite of their different environments. The N=O distances vary between 1.188 and 1.202 Å, and the O=N=O angles between 129.8° and 135.4°. The only exception is the NO_3 radical (vide supra).

3. NITRIC ACID AND NITRATES

According to MW, nitric acid, $HONO_2$, takes a planar conformation with the NO_2 group tilted 2° away from the hydrogen atom (Cox and Riveros, 1965). The cis N=O bond is slightly longer than the trans N=O bond. The structural parameters are given in Fig. 3b. The error estimates are taken from Callomon et al. (1976).

The MW study of peroxynitric acid, $HOONO_2$ (Suenram et al., 1986), discovered in the reaction of nitrogen dioxide with the HOO radical (Niki et al., 1977), provided only partial structural information. The heavy-atom skeleton is coplanar, and the hydrogen atom is oriented at a rotational angle of 106° with respect to the syn form.

The structure of methyl nitrate, CH_3ONO_2, given in Fig. 3c, was determined by MW (Cox and Waring, 1971). It has a planar skeleton, and one of the C–H bonds is anti to the O–N bond. The methyl group and the nitro group are tilted away from each other.

The early MW studies of ethyl nitrate, $CH_3CH_2ONO_2$, showed the existence of two rotational isomers, anti and a less stable gauche, with respect to the C–O bond (Scroggin et al., 1974). Infrared (IR) spectroscopy (Durig and Lindsay, 1990) has confirmed the existence of the two conformers. ED (Shishkov et al., 1992) provided structural data on the more abundant anti form, shown in Fig. 3d. Differences between the parameters of the anti and gauche forms were assumed from ab initio calculations in the ED analysis. The most important anti/gauche parameter difference is in the C–C–O angle, being some 7° smaller in the anti form than in the gauche form. Durig and Sheehan (1990) noted this difference in their Raman spectroscopic study and ascribed the decrease of the C–C–O angle in the anti form to the attractive interactions between the oxygen lone pairs and two of the methyl hydrogens. The torsional angle, C–C–O–N, characterizing the gauche form, is 104 ± 8°. In both methyl nitrate and ethyl nitrate, the NO_2 group is tilted toward the lone pair of the third oxygen atom. The energy difference between the gauche and anti forms of ethyl nitrate, estimated from their relative abundance, is 2.5 ± 0.8 kJ/mol according to the ED study. This result is in agreement with the results of MW (Scroggin et al., 1974) and IR (Durig and Lindsay, 1990) studies. The nitrogen bond configurations are similar in nitric acid, methyl nitrate, and ethyl nitrate.

An early ED study of fluorine nitrate, $FONO_2$, reported a nonplanar structure with the O–F bond perpendicular to the ONO_2 plane (Pauling and Brockway, 1937). Later, the infrared spectra were interpreted in terms of a planar structure (Arvia et al., 1963).

A recent ED/MW combined analysis of chlorine nitrate, $ClONO_2$, and an ED analysis of bromine nitrate, $BrONO_2$, found both modules to be planar (Casper et al., 1993). The parameters are presented in Figs. 3e and 3f.

The structure of $ClONO_2$ has also been studied in the solid state by X-ray diffraction (XD) (Obermeyer et al., 1994). As compared with the gaseous phase, the most important differences are in the N–O and O–Cl bond lengths: the former is shorter by 0.026 Å and the latter is longer by 0.019 Å in the crystal. Casper et al. (1993) interpreted these differences by a higher ionic character, $Cl^+NO_2^-$, in the solid than in the gas.

Comparison of the structures of nitric acid with those of various nitrates (cf.

Figs. 3b–3f) indicates little change in the NO_2 geometries. However, the tilt of the nitro group relative to the N–O bond increases with increasing substituent size. The length of the N–O single bond varies considerably in these molecules; it is the same in $HONO_2$, $MeONO_2$, and $EtONO_2$ and increases in halogen nitrates with increasing substituent electronegativity.

A number of metal nitrates have been investigated by ED. Some of these studies were carried out in the early days by the visual technique or, owing to technical difficulties, provided intensity data in a narrow range only. Thus, these results are of limited accuracy.

A visual ED study of copper(II) nitrate, $Cu(ONO_2)_2$, indicated a bidentate structure with the copper occupying the center of inversion (Scheme 1; LaVilla

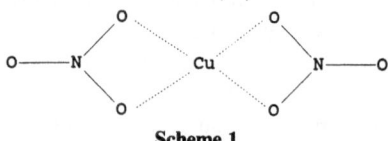

Scheme 1

and Bauer, 1963). No information was obtained about the relative orientation of the nitrate groups. In the crystalline state, copper(II) nitrate was found to consist of parallel chains of NO_3 groups bridged by copper atoms (Wallwork, 1959).

The structures of sodium nitrate, $NaNO_3$, and lithium nitrate, $LiNO_3$, reported in another visual ED study (Khodchenkov et al., 1965) are analogous to those of nitric acid and halogen nitrates (vide supra). The ED study of two other alkali nitrates, $RbNO_3$ (Kulikov et al., 1981a) and $CsNO_3$ (Kulikov et al., 1981b), as well as that of thallium(I) nitrate, $TlNO_3$ (Kulikov et al., 1981c), indicated these molecules to be planar with C_{2v} symmetry (Scheme 2). The structural parameters

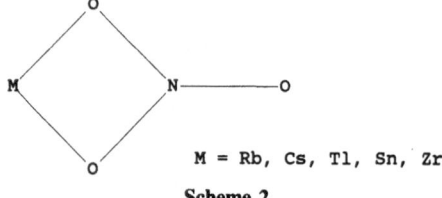

M = Rb, Cs, Tl, Sn, Zr

Scheme 2

were determined with large uncertainties. Reinvestigation of these structures with modern techniques is warranted.

The ED data on tin tetranitrate, $Sn(NO_3)_4$ (Touseev et al., 1976), and zirconium tetranitrate, $Zr(NO_3)_4$ (Spiridonov et al., 1975), were consistent with a model of S_4 symmetry and planar four-membered ring fragments similar to the structure shown in Scheme 2. The geometrical parameters are given in Table III.

The ED data on μ_4-oxohexa-μ-nitratotetraberyllium, $Be_4O(NO_3)_6$, were

Table III. Geometrical Parameters (r_a) of $Sn(NO_3)_4$ and $Zr(NO_3)_4$ Molecules from ED[a]

Parameter[b]	$Sn(NO_3)_4$[c] (M = Sn)	$Zr(NO_3)_4$[d] (M = Zr)
M–O	2.172(5)	2.219(6)
M–N	2.597(11)	2.650(14)
O–N$_{ring}$	1.281(5)	1.284(7)
O–N$_{ext}$	1.179(8)	1.184(10)
N–M–N	134(2)	131
τ[e]	9(2)	12

[a]Distances are given in angstroms, and angles in degrees.
[b]See also Scheme 2.
[c]Touseev et al., 1976.
[d]Spiridonov et al., 1975.
[e]Rotation angle of the NO_3 group around the M–N–O axis.

found consistent with a T symmetry model (Touseev et al., 1984). The molecule consists of six six-membered rings (Scheme 3) which are partially fused. The O1

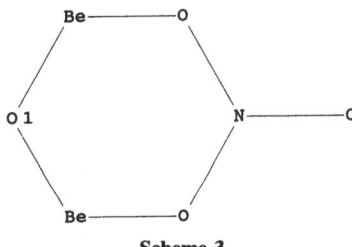

Scheme 3

atom is common to all six rings, each Be atom belongs to three rings, and each ONO moiety belongs to one ring only. The principal geometrical parameters are: Be–O1 1.665(21), Be–O 1.620(8), N=O 1.185(5), N–O 1.298(4) Å (r_g bond lengths), and O–N–O 117.0(9)°. The dihedral angle between the O1Be$_2$ and NO$_2$ planes is 25(2)°.

A model of C_2 symmetry was used to interpret the ED data on chromyl nitrate, $CrO_2(NO_3)_2$ (Marsden et al., 1991). In addition to the two nominal double bonds, Cr=O, chromium forms two nominal single bonds, Cr–O(N), and two long and weak bonds, Cr$\cdots$O(N) (Scheme 4). The following bond lengths (r_g) and bond angles were reported: Cr=O 1.586(2), Cr–O 1.975(5), Cr$\cdots$O 2.254(20), N–O 1.341(4), N=O(Cr) 1.254(4), N=O 1.193(4) Å, O=Cr=O 112.6(35), O–Cr–O 140.4(33), O=Cr–O4 97.2(18), O=Cr–O8 104.5(9), O=Cr$\cdots$O 83.7(34), Cr–O–N

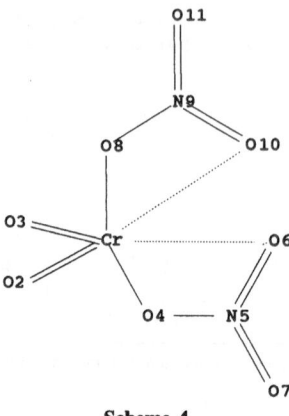

Scheme 4

97.5(5), O–N=O(Cr) 112.2(1), O–N=O 119.7(40), O=N=O 128.1(36)°. The two NO$_2$ groups are turned toward chromium, indicating some attractive interaction between the Cr and O atoms.

The ED study (Kulikov *et al.*, 1981d) of rubidium nitrite, RbNO$_2$, and cesium nitrite, CsNO$_2$, has suggested a planar four-membered cyclic structure of C_{2v} symmetry for these molecules. The bond lengths and bond angles are collected in Table IV.

MW studies of nitryl fluoride, FNO$_2$ (Legon and Millen, 1968), and nitryl chloride, ClNO$_2$ (Endo, 1979), showed these molecules to be planar with the geometrical parameters given in Table V.

Nitromethane, H$_3$CNO$_2$, has also been studied by MW (Cox and Waring, 1972), and its geometrical parameters are also given in Table V. The trend of increasing O=N=O angle and decreasing N=O distance with increasing ligand electronegativity is in agreement with the predictions of the valence-shell electron-pair repulsion (VSEPR) model (Gillespie and Hargittai, 1991). According to XD

Table IV. Bond Lengths (r_g) and Bond Angles of RbNO$_2$ and CsNO$_2$ Molecules from ED[a,b]

Parameter	RbNO$_2$ (M = Rb)	CsNO$_2$ (M = Cs)
N–O	1.252(5)	1.256(2)
M–O	2.64(2)	2.79(2)
O–N–O	116(3)	118(3)

[a]Kulikov *et al.*, 1981d.
[b]Distances are given in angstroms, and angles in degrees.

Table V. Geometrical Parameters (Partial Substitution Structures) of Nitryl Fluoride, Nitryl Chloride, and Nitromethane from MW[a]

Parameter	FNO_2[b] (X = F)	$ClNO_2$[c] (X = Cl)	CH_3NO_2[d] (X = C)
X–N	1.467 ± 0.015	1.843(3)	1.489(5)
N=O	1.1798 ± 0.0035	1.198(2)	1.224(5)
C–H	—	—	1.088[e]
O=N=O	136.0 ± 1.5	130.9(5)	125.3(3)
N–C–H	—	—	107.2(10)

[a]Distances are given in argstroms, and angles in degrees.
[b]Legon and Millen, 1968.
[c]Endo, 1979.
[d]Cox and Waring, 1972.
[e]Assumed.

crystallography studies of nitromethane (Trevino *et al.*, 1980; Bagryanskaya and Gatilov, 1983) the O=N=O angle is about 1° smaller in the crystal than in the gas.

A recent high-level (MP2/6-31G*) *ab initio* calculation reported a geometry of nitromethane remarkably close to the experimental structure (Lammertsma and Prasad, 1993). The nitro⇌*aci*-nitro tautomerism was also investigated, and the geometrical parameters of the *aci*-nitro tautomer were found to be markedly different. Figure 4a presents some of the results. It is seen that not only the tautomerized N–O bond and the O–N–O angle but even the N–C bond undergoes substantial changes. In spite of these interesting structural changes, there is no experimental evidence for them because the tautomerization in question has a prohibitively high barrier. The closest substance that could be studied experimentally was obtained by the dimerization of *aci*-nitrodiphenylmethane (Scheme 5).

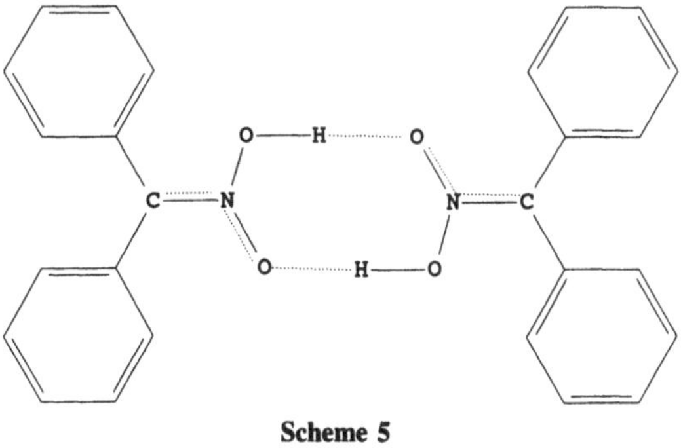

Scheme 5

According to an X-ray crystallographic study (Bock *et al.*, 1993), the geometrical parameters for this species are: C–N 1.30, N=O 1.30, N–O 1.38 Å, C–N=O 127,

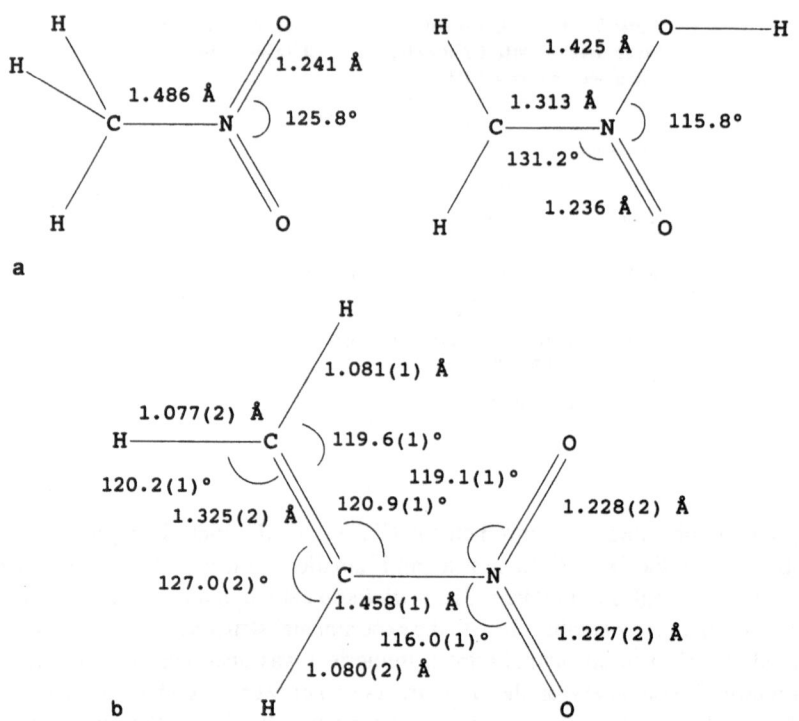

Figure 4. (a) Geometrical parameters of nitromethane and *aci*-nitromethane from MP2/6-31G* *ab initio* calculations by Lammertsma and Prasad (1993); (b) structure (r_s) of nitroethylene from MW (Nösberger *et al.*, 1983).

C–N–O 118, O–N≡O 115°. There is then experimental evidence here for the drastic shortening of the C–N bond, lengthening of the N–O bond, and closing of the O–N≡O angle in this hydrogen-bonded dimer.

The following noncyclic hydrocarbon nitro derivatives with one nitro group have been studied in the gas phase in addition to nitromethane: nitroehtylene, $CH_2=CH–NO_2$ (Nösberger *et al.*, 1983), dimethylnitromethane, $(CH_3)_2CHNO_2$ (Shishkov *et al.*, 1983a), and trimethylnitromethane, $(CH_3)_3CNO_2$ (Shishkov *et al.*, 1983a; Langridge-Smith *et al.*, 1980). The nitroethylene molecule is planar; its geometrical parameters are presented in Fig. 4b. The structure of the nitro group is similar to that in nitromethane, but it is twisted with respect to the C–N bond by 1.5°. The C–N bond is shorter than that of nitromethane by about 0.03 Å.

The ED studies of dimethylnitromethane and trimethylnitromethane (Shishkov *et al.*, 1983a) allowed only the determination of mean values of the C–C and C–N bond lengths within each molecule. It is expected that H-to-methyl substitution would increase the N–O bond length and decrease the O≡N≡O bond angle,

Table VI. Comparison of N=O Bond Lengths and
O=N=O Bond Angles in Nitromethane,
Dimethylnitromethane, and Trimethylnitromethane[a]

Parameter	CH_3NO_2[b]	$(CH_3)_2HCNO_2$[c]	$(CH_3)_3CNO_2$[c]
N=O	1.244(5)	1.226(2)	1.240(2)
O=N=O	125.3(3)	125.4(3)	122.2(6)

[a]Distances are given in angstroms, and angles are in degrees.
[b]r_s structure; Cox and Waring, 1972.
[c]r_g structure; Shishkov et al., 1983a.

due to the σ-electron-releasing character of the CH_3 group. This effect is best seen in trimethylnitromethane, according to the data of Table VI. The O=N=O angle in this molecule has the lowest observed value among the gaseous aliphatic nitro compounds studied so far. The MW investigation of $(CH_3)_3CNO_2$ (Langridge-Smith et al., 1980) yielded a larger C–N bond length (1.53 ± 0.02 Å) than that in H_3CNO_2. Curiously, this bond lengthening is accompanied by an increase of the V_6 potential barrier.

The structures of the following halogenated nitromethanes with one nitro group have been studied by ED: nitrotrifluoromethane, F_3CNO_2 (Karle and Karle, 1962); nitrochloromethane, H_2ClCNO_2 (Sadova et al., 1972); nitrotrichloro-methane, Cl_3CNO_2 (Knudsen et al., 1966); and nitrotribromomethane, Br_3CNO_2 (Karle and Karle, 1962). The geometrical parameters are given in Table VII. The CNO_2 group is nonplanar in the trihalogenated derivatives. Nitrotrifluoromethane exists as a mixture of staggered and eclipsed forms, while nitrotrichloromethane

Table VII. Geometrical Parameters of F_3CNO_2, H_2ClCNO_2, Cl_3CNO_2, and Br_3CNO_2[a]

Parameter	F_3CNO_2[b] (X = F)	H_2ClCNO_2[c] (X = Cl)	Cl_3CNO_2[d] (X = Cl)	Br_3CNO_2[b] (X = Br)
C–X	1.325 ± 0.005	1.765 ± 0.009	1.726 ± 0.005	1.920 ± 0.007
C–N	1.56 ± 0.02	1.493 ± 0.010	1.594 ± 0.020	1.59 ± 0.02
N=O	1.21 ± 0.01	1.230 ± 0.002	1.190 ± 0.006	1.22 ± 0.01
C–H	—	1.095[e]	—	—
O=N=O	132	128 ± 2	131.7 ± 2.6	134
O=N–C	112	116 ± 1	113.2 ± 2.1	110.7
N–C–X	109	114 ± 1	106.0 ± 1.1	108.3
X–C–X	110	—	111.8 ± 0.7	110.3

[a]r_a Structures from ED; distances are given in angstroms, and angles in degrees.
[b]Karle and Karle, 1962.
[c]Sadova et al., 1972.
[d]Knudsen et al., 1966.
[e]Assumed.

and nitrotribromomethane have a staggered conformation, and there is free, or nearly free, rotation about the C–N bond in H_2ClCNO_2. The C–N bond is very long in the trihalogenated nitromethanes. The thermal instability of these molecules seems to be consistent with the lengthening of the C–N bond.

The geometrical parameters of gaseous nitromethanes with more than one nitro group are presented in Table VIII. In dimethyldinitromethane, $(CH_3)_2C(NO_2)_2$, the two nitro groups are not equivalent with respect to the carbon backbone; they are found in perpendicular planes (Shishkov et al., 1983b). For dichlorodinitromethane, $Cl_2C(NO_2)_2$, a model of C_2 symmetry with two equal angles of torsion of the two NO_2 groups with respect to the ClCCl plane approximated best the experimental distributions (Sadova et al., 1977a). The ED data for $HC(NO_2)_3$, $ClC(NO_2)_3$, and $BrC(NO_2)_3$ are consistent with a molecular model of C_3 symmetry, characterized by one angle of torsion about the C–N axes (Sadova et al., 1976a). The ED data for $C(NO_2)_4$ are consistent with a model of S_4 symmetry; the angle of torsion of the NO_2 groups is 47° (Sadova et al., 1976b). Comparison of H_3CNO_2, $HC(NO_2)_3$, and $C(NO_2)_4$ indicates a tendency of C–N bond elongation accompanied by an opening of the O=N=O angle with increasing number of nitro groups (Table VIII).

4. NITRAMINES

The most intriguing question of the nitramine structures is the dependence of the N–N bond length and amide pyramidality on the nature of the substituents.

The structure of nitramine, NH_2NO_2, was studied by MW (Tyler, 1963). The

Table VIII.   Geometrical Parameters of Nitromethanes with Two, Three, or Four Nitro Groups from ED[a]

Parameter	$(CH_3)_2C(NO_2)_2$[b] $(X = C)$	$Cl_2C(NO_2)_2$[c] $(X = Cl)$	$HC(NO_2)_3$[d] $(X = H)$	$ClC(NO_2)_3$[d] $(X = Cl)$	$BrC(NO_2)_3$[d] $(X = Br)$	$C(NO_2)_4$[e]
N=O	1.227(2)	1.224(4)	1.219(3)	1.213(3)	1.214(3)	1.218(2)
C–N	} 1.517(15)[f]	1.506(12)	1.505(5)	1.513(3)	1.514(6)	1.526(6)
C–X		1.757(8)	1.13[g]	1.712(4)	1.885(9)	—
O=N=O	124.5(6)	124.9(11)	128.6(10)	128.3(6)	132.5(20)	129.3(10)
N–C–N	111.6(12)	109.2(20)	110.7(10)	106.7(6)	107.9(10)	109.5[g]
X–C–X	113.1(23)	108.9(19)	—	—	—	—

[a]Distances are given in angstroms, and angles in degrees.
[b]$r_g$structure; Shishkov et al., 1983b
[c]$r_a$structure; Sadova et al., 1977a
[d]$r_a$structure; Sadova et al., 1976a.
[e]$r_a$structure; Sadova et al., 1976b.
[f]Mean value.
[g]Assumed.

bond lengths and bond angles are given in Fig. 5a. The heavy-atom skeleton is planar. C_s molecular symmetry of the molecule was assumed in deriving the r_0 structure.

The bond lengths and bond angles of N-methylnitramine, CH_3HNNO_2, from ED (Sadova et al., 1977b) are given in Fig. 5b. The heavy-atom skeleton is nonplanar. The HNN and HNC angles could not be determined.

The heavy-atom skeleton of N-chloro-N-methylnitramine, CH_3ClNNO_2, was also found to be nonplanar by ED (Sadova et al., 1977b). The bond lengths and bond angles are given in Fig. 5c.

The structural parameters of N,N-dimethylnitramine, $(CH_3)_2NNO_2$, obtained by ED (Stølevik and Rademacher, 1969) and XD crystallography (Krebs et al., 1979), are given in Table IX along with those of methyl(chloromethyl)-nitramine, $CH_3(CH_2Cl)NNO_2$ (Shishkov et al., 1982). A planar heavy-atom skeleton was found in both phases of $(CH_3)_2NNO_2$. However, there were differences in bond lengths and bond angles. The longer N–N bond and shorter N=O bonds in the gas, compared with the crystal, are conspicuous. The O=N=O and C–N–C angles are also considerably larger in the gas than in the crystal. These changes of the bond angles and bond lengths appear to be self-consistent according to the VSEPR model (Gillespie and Hargittai, 1991), which predicts angular openings to accompany stronger/shorter bonds.

The bond configuration of the amine nitrogen is also planar in methyl(chloro-methyl)nitramine. However, the NO_2 plane is rotated by 33(3)° with respect to the CNC plane, in variance with the principal features of N,N-dimethylnitramine (vide supra). The ClCN plane is perpendicular to the C_2N_2 plane.

The geometrical parameters of methyldinitramine, $CH_3N(NO_2)_2$, are presented in Fig. 5d from ED (Tarasenko et al., 1977). The configuration of the amine nitrogen is nonplanar, and the two nitro groups are coplanar.

The structural parameters of N-methyl-N-nitrovinylamine, $CH_3(H_2CCH)$-NNO_2, from ED (Batyukhnova et al., 1984a) are given in Fig. 5e. The bond configuration of the amine nitrogen is planar, and the nitro group and the N–N–C=C chain are coplanar. On the basis of comparison of experimental and calculated dipole moments, anti orientation of the N–N and C=C bonds was suggested.

Table X contains the ED structural parameters of the cyclic N-nitroamines N-nitropyrrolidine, $(CH_2)_4NNO_2$ (NP), 1,3-dinitro-1,3-diazacyclopentane, CH_2-$(CH_2NNO_2)_2$ (DDCP), and 1,3,5-trinitro-1,3,5-triazacyclohexane, $(CH_2NNO_2)_3$ (TCH, usually referred to as RDX) from Shishkov et al. (1991). The greater deviation from planarity of the amine nitrogen bond configuration in NP and DDCP than in TCH is the consequence of the confinement of nitrogen in a five-membered ring. The ring of NP takes an envelope conformation with the N–N bond in the equatorial orientation. The ring of DDCP takes a half-chair form of C_2 symmetry with the N–N bonds in the equatorial orientation. The TCH molecule

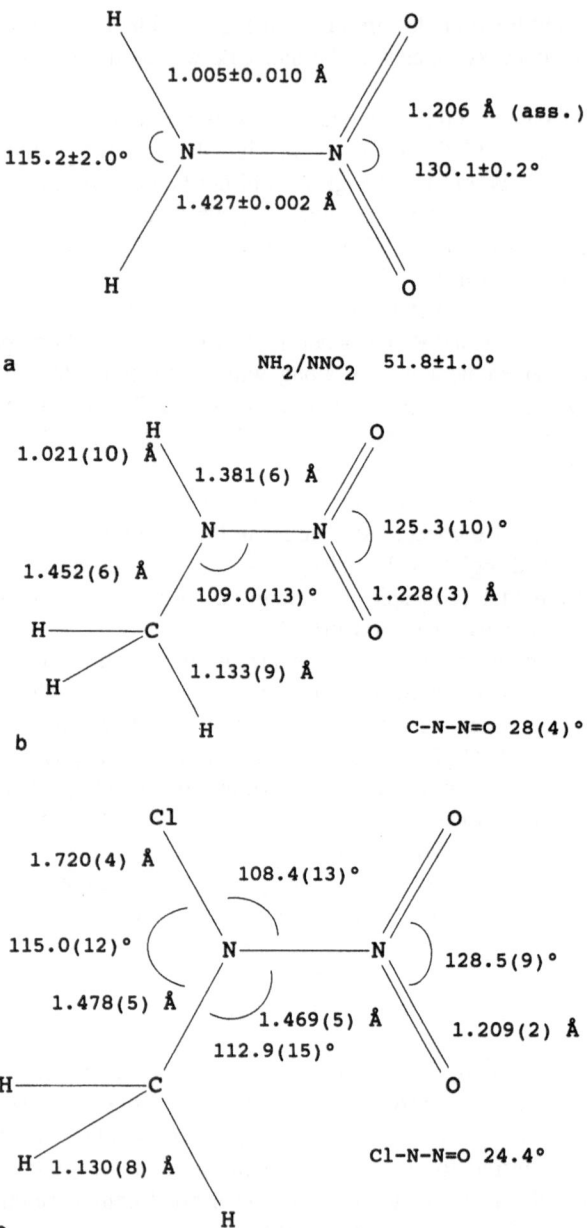

Figure 5. (a) Structure (r_0) of nitramine from MW (Tyler, 1963); (b) Structure (r_a) of *N*-methylnitramine from ED (Sadova *et al.*, 1977b); (c) structure (r_a) of *N*-chloro-*N*-methylnitramine from ED (Sadova *et al.*, 1977b); (d) structure (r_a) of methyldinitramine molecule from ED (Tarasenko *et al.*, 1977); (e) structure (r_g bond lengths) of *N*-methyl-*N*-nitrovinylamine from ED (Batyukhnova *et al.*, 1984a).

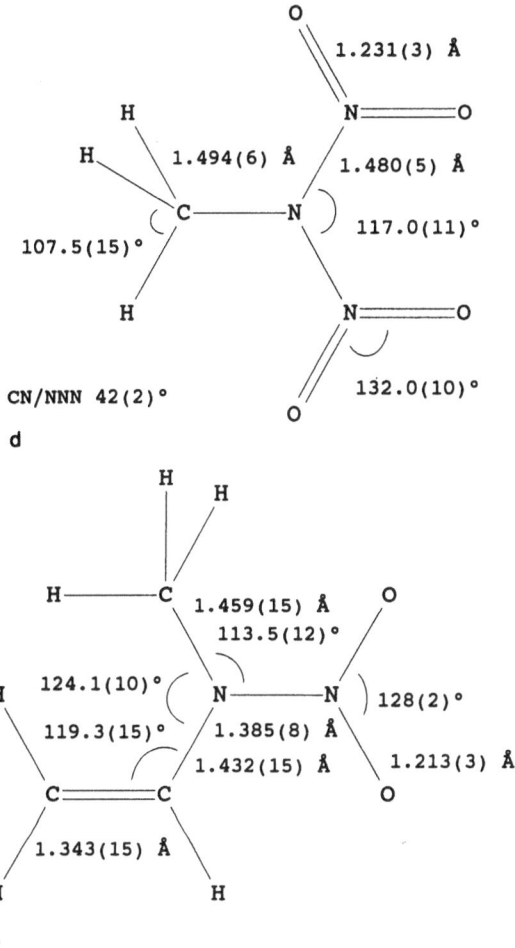

Figure 5. (*Continued*)

is of C_3 symmetry, and its six-membered ring has a chair form, with the N–N bonds in the axial orientation.

Table X gives also the structural parameters of crystalline TCH in a polymorphic form from neutron diffraction (ND) (Choi and Prince, 1972), represented by mean values of bond lengths and bond angles. The molecule possesses a plane of approximate mirror symmetry perpendicular to the plane of the three carbon atoms of the ring in the solid state. The three N–NO$_2$ groups are not equivalent. There is a slight decrease of mean N–N bond length and a considerable decrease of C–N–C bond angle in the solid phase as compared with the gas, and thus the six-

Table IX.   Geometrical Parameters of
N,N-Dimethylnitramine in the Gas Phase and in the
Crystal and of Methyl(chloromethyl)nitramine in the Gas
Phase[a]

Parameter	$(CH_3)_2NNO_2$ ED,[b] r_g	XD[c]	$CH_3(CH_2Cl)NNO_2$ ED,[d] r_g
C–N	1.460(3)	1.456(5)	1.476(9)
N–N	1.382(3)	1.342(4)	1.424(5)
N=O	1.223(2)	1.239(4)	1.212(2)
C–Cl	—	—	1.809(4)
O=N=O	130.4(13)	124.3(2)	127.5(10)
O=N–N	114.8(7)	117.8(2)	
C–N–C	127.6(6)	124.3(3)	127.0(9)
N–C–Cl	—	—	107.9(9)
CNC/ONO			33(3)
Cl–C–N–N	—	—	86(2)

[a]Distances are given in angstroms, and angles in degrees.
[b]Stølevik and Rademacher, 1969.
[c]Krebs et al., 1979.
[d]Shishkov et al., 1992.

Table X.   Structural Parameters (r_g Bond Lengths) of Cyclic
N-Nitramines: N-Nitropyrrolidine (NP) from ED,
1,2-Dinitro-1,3-diazacyclopentane (DDCP) from ED, and
1,3,5-Trinitrotriazacyclohexane (TCH) from ED and from a
Single-Crystal Neutron Diffraction Study[a]

Parameter	NP[b]	DDCP[b]	TCH ED[b]	ND[c]
C–N	1.477(8)	1.483(8)	1.464(6)	1.454(4)
N–N	1.363(4)	1.393(8)	1.413(5)	1.380(3)
N=O	1.225(2)	1.226(2)	1.213(2)	1.210(5)
C–N–C	110.3(14)	109.7(17)	123.7(6)	114.8
N–N–C	116.0(11)	114.8(8)	116.3(5)	117.9
Σ N[d]	342.3	339.3	356.3	347.4
N–C–N	—	—	109.4(6)	109.3
N–C–C	96.6(4)	100.2(24)	—	—
O=N=O	126.3(19)	128.2(18)	125.5(10)	125.4
N–N/CNC	40	43	19	

[a]Distances are in angstroms, and angles are in degrees.
[b]Shishkov et al., 1992.
[c]Mean values are given; Choi and Prince, 1972.
[d]Sum of the bond angles of the amine nitrogen.

membered ring is more puckered in the solid phase than in the gas. The bond configuration around the endocyclic N atom is pyramidal in both phases. This pyramidality is most pronounced for the N of nonplanar $N-NO_2$ groups in the crystal.

The gas-phase structure of DDCP may be compared with the crystal structure of 2,4,6,8-tetranitro-2,4,6,8-tetraazabicyclo[3.3.0]octane (bicyclo-DDCP) from XD (Coon, 1988). This molecule consists of two five-membered rings, fused at the C–C bond. These five-membered rings are less puckered than that in DDCP; the sum of the endocyclic angles is 525.9° in the latter compound, while it is 536.1° and 536.6° in the two rings of bicyclo-DDCP. The dihedral angle between the two five-membered rings is about 113°. The bond configurations of the endocyclic N atoms are flattened in the crystal as compared with the gas phase. The N–N bonds of one ring point away from the other ring in the fused system.

5. NONAROMATIC CYCLIC SYSTEMS

An MW study of nitrocyclopropane, $C_3H_5NO_2$, indicated that the NO_2 plane is perpendicular to and bisects the cyclopropyl ring (Mochel et al., 1973). The r_0 structure based on a single set of rotational constants is given in Fig. 6a.

The ED study of 2-methyl-4-trinitromethyl-1,2,3-triazole (Belyakov et al., 1992) provided information on the influence of the electronegative triazole ring on the geometry of the trinitromethyl fragment. The structural parameters of the latter are given in Fig. 6b.

The NO_2 group of 1-nitrocyclohexene (Scheme 6) is coplanar with the double bond of the twisted ring according to an MW study by González et al. (1989) in which only a single set of rotational constants was available for structure analysis.

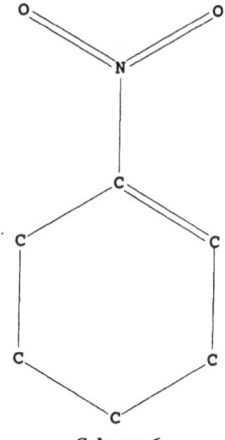

Scheme 6

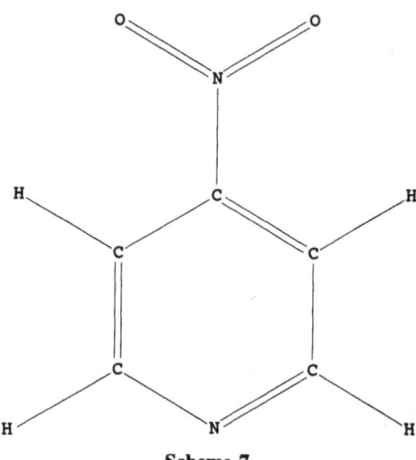

Scheme 7

The MW study of 4-nitropyridine (Scheme 7) by Mulas *et al.* (1989) was also based on a single set of rotational constants. The small inertial defect indicates the planarity of the molecule.

The geometrical parameters of 4-nitropyridine-*N*-oxide in the crystal, given in Table XI, were determined by XD (Wang *et al.*, 1976) and ND (Coppens and Lehmann, 1976). The ED data of Chiang and Song (1982) have large uncertainties and are at variance with the XD and ND results. It is especially puzzling that the ED value of $C-N(O_2)$ is some 0.17 Å longer than the XD and ND values. The electron diffraction analysis must have suffered from a strong correlation between the parameters, and the results may thus be suspect.

6. BENZENE DERIVATIVES

In accordance with the σ-electron-withdrawing properties of the NO_2 group, in nitro derivatives of benzene, an increase of the $C-C(NO_2)-C$ endocyclic angle is expected (Domenicano *et al.*, 1975).

The structure of nitrobenzene was studied by MW (Høg, 1971) and by ED (Shishkov *et al.*, 1984; Domenicano *et al.*, 1989) in the gas phase. In these and other studies on nitro derivatives of benzene, special care was taken to determine the deformation of the benzene ring under the influence of the NO_2 group. The structural parameters of gaseous nitrobenzene are given in Table XII along with the latest crystal-phase data (Boese *et al.*, 1992). The observed differences between the ED and MW parameters may be due to the different meaning of r_g and r_s structures and to the limitations of the MW method in obtaining the coordinates of atoms near the center of mass of the molecule. It is noteworthy that the C–N bond appears to be shorter and the O=N=O and C–N(N)–C bond angles to be

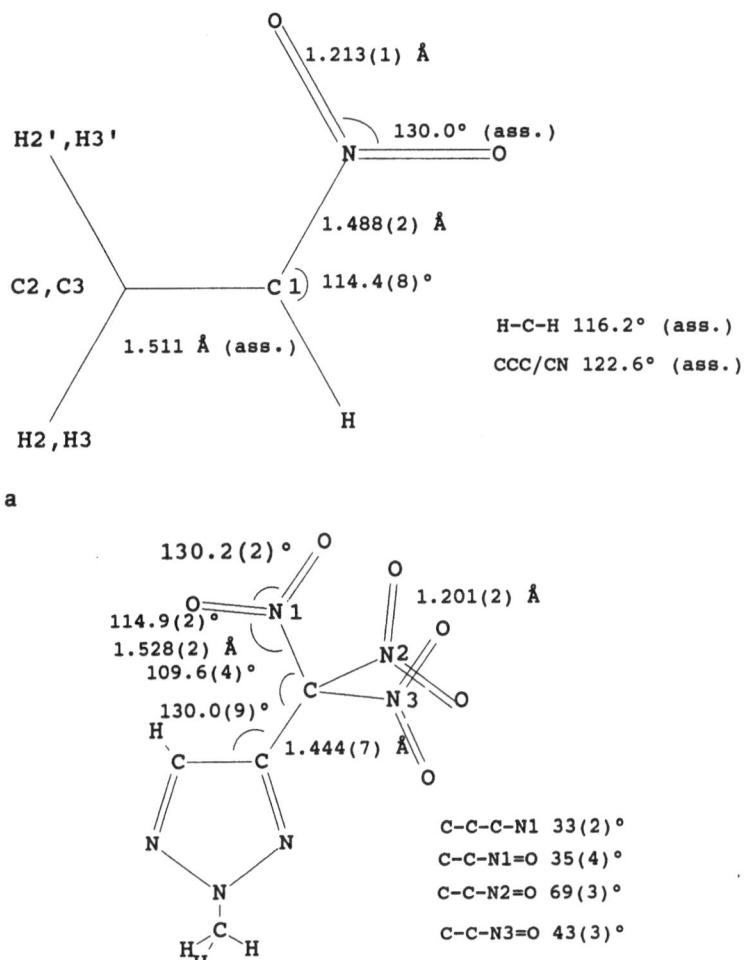

Figure 6. (a) Structure (r_0) of nitrocyclopropane from MW (Mochel *et al.*, 1973); (b) structure $(r_a, \langle_\alpha\rangle)$ of the trinitromethyl fragment of 2-methyl-4-trinitromethyl-1,2,3-triazole from ED (Belyakov *et al.*, 1992).

smaller in the crystal than in the gas phase. These changes, if real, might be interpreted as a consequence of increased conjugation between the ring and the NO_2 substituent in the crystal (Domenicano *et al.*, 1989).

Table XIII gives the structural parameters of the following *para*-substituted benzene derivatives with one NO_2 group, studied in the gas phase by ED: *p*-nitroaniline, *p*-$NH_2C_6H_4NO_2$ (Sadova *et al.*, 1976c), along with the X-ray crystallo-

Table XI. Geometrical Parameters of
4-Nitropyridine-N-oxide in the Crystal
from XD and ND[a]

Parameter	XD[b]	ND[c]
N1–C2	1.369(1)	1.369(2)
C2–C3	1.377(2)	1.377(3)
C3–C4	1.388(2)	1.387(3)
N1–O8	1.297(2)	1.291(2)
C4–N7	1.455(1)	1.458(2)
N7=O	1.235(1)	1.227(2)
C2–N1–C6	120.2[d]	120.0[d]
C3–C4–C5	121.4[d]	120.0[d]
O=N=O	123.5[d]	124.6[d]

[a]Distances are given in angstroms and angles in
degrees.
[b]Mean values; Wang et al., 1976.
[c]Mean values; Coppens and Lehmann, 1976.
[d]No error estimation available.

graphic results (Colapietro et al., 1981), p-chloronitrobenzene, p-ClC$_6$H$_4$NO$_2$ (Sadova et al., 1976d), p-bromonitrobenzene, p-BrC$_6$H$_4$NO$_2$ (Almenningen et al., 1984), and p-iodonitrobenzene, p-IC$_6$H$_4$NO$_2$ (Brunvoll et al., 1990). The effect of through-conjugation was invoked in the interpretation of the structural changes observed in p-nitroaniline as compared with aniline and nitrobenzene. The

Table XII. Comparison of the Geometry of Nitrobenzene in
the Gas Phase and in the Crystal[a]

Parameter	MW[b]	ED[c]	XD[d]
(C–C)$_{mean}$	1.3911(5)	1.399 ± 0.003	1.390(1)
C–N	1.4916(17)	1.486 ± 0.004	1.471(1)
N=O	1.2272(2)	1.223 ± 0.003	1.231(1)
C–C(N)–C	124.99(13)	123.4 ± 0.3	122.7(1)
C(N)–C–C	117.11(7)	117.6 ± 0.3	118.2(1)
C(N)C–C–C	120.30(2)	120.6 ± 0.2	120.2(1)
C(N)CC–C–C	120.18(1)	120.18[e]	120.5(1)
O=N–O	124.35(1)	125.3 ± 0.2	123.2(1)
Φ[f]		13.2 ± 1.0	

[a]Distances are given in angstroms, and angles in degrees.
[b]r_s structure; Høg, 1971.
[c]r_g structure; Domenicano et al., 1989.
[d]Mean values; Boese et al., 1992.
[e]Assumed.
[f]Effective twist angle of the NO$_2$ group.

Table XIII. Geometrical Parameters of p-Nitroaniline, p-Chloronitrobenzene, p-Bromonitrobenzene, and p-Iodonitrobenzene from ED, and of p-Nitroaniline in the Crystal from XD[a]

	p-NH$_2$C$_6$H$_4$NO$_2$ (X = N)		p-ClC$_6$H$_4$NO$_2$[d] (X = Cl)	p-BrC$_6$H$_4$NO$_2$[e] (X = Br)	p-IC$_6$H$_4$NO$_2$[f] (X = I)
Parameter	ED[b]	XD[c]			
(C–C)mean	1.402(4)	1.388	1.388(4)	1.399(3)	1.396(2)
C–N(O$_2$)	1.474(14)	1.437	1.469(9)	1.454(4)	1.458(15)
N=O	1.225(3)	1.231	1.233(4)	1.239(2)	1.228(3)
C–X	1.362(25)	1.356	1.710(7)	1.896(2)	2.102(8)
C–C(NO$_2$)–C	123.1(20)	120.9	123.2(16)	121.6(2)	122.2(5)
C–C(X)–C	119.8(22)	118.5	120.4(10)	122.6(2)	122.2(5)
O=N=O	125.4(11)	121.8	122.6(12)	125.0(7)	124.2
Φ[g]	<33		21(7)	18.8(24)	15.6(4)

[a]Distances are given in angstroms, and angles in degrees.
[b]Sadova *et al.*, 1976a.
[c]Colapietro *et al.*, 1981; no uncertainties given.
[d]Sadova *et al.*, 1976b.
[e]Almenningen *et al.*, 1984.
[f]Brunvoll *et al.*, 1990.
[g]Effective twist angle of the nitro group.

C–N(H$_2$) bond is shorter than in aniline, and the C–N(O$_2$) bond is shorter than in nitrobenzene, both by about 0.04 Å. The N=O bonds are longer and the O=N=O angle is smaller than in nitrobenzene. The amino group makes an angle of 10° with the benzene plane, and the corresponding angle is 37° in aniline. The effect of through-conjugation is less expressed in the gas phase.

The structural parameters of p-dinitrobenzene, p-(NO$_2$)$_2$C$_6$H$_4$, from ED (Penionzhkevich *et al.*, 1979) and XD crystallography (DiRienzo *et al.*, 1980) are presented in Table XIV. The results of the two studies are in reasonable agreement, considering the many potential sources of uncertainty. The comparison of ED data from p-dinitrobenzene and nitrobenzene indicates a shortening of all bonds, except the N=O bond, and an increase of the O=N=O and C–C(N)–N angles in p-dinitrobenzene, as compared with nitrobenzene.

The geometrical parameters of o-chloronitrobenzene (Batyukhnova *et al.*, 1985), o-bromonitrobenzene (Batyukhnova *et al.*, 1988), o-iodonitrobenzene (Samdal *et al.*, 1992), and o-dinitrobenzene (Penionzhkevich *et al.*, 1979), all determined by ED, are presented in Table XV. The results of XD on o-(NO$_2$)$_2$C$_6$H$_4$ (Herbstein and Kapon, 1990) are also given in this table. Mean values are listed wherever applicable. There is generally satisfactory agreement between the ED and XD data.

The possibility of O$\cdots$Se and O$\cdots$S nonbonded interactions was investigated in an ED study of o-nitrophenylselenylbromide, o-NO$_2$C$_6$H$_4$SeBr (Zaripov *et al.*,

Table XIV. Structural Parameters of
p-Dinitrobenzene from ED (r_g Values)
and X-Ray Crystallography[a]

Parameter	ED[b]	XD[c]
(C–C)$_{mean}$	1.392(2)	1.380(2)
C–N	1.463(10)	1.478(2)
N=O	1.221(2)	1.219(2)
C–C(N)–C	122.9(9)	123.4(2)
O=N=O	125.8(5)	124.5(2)
Φ[d]	18(4)	10.2

[a]Distances are given in angstroms, and angles in
degrees.
[b]Penionzhkevich et al., 1979.
[c]DiRienzo et al., 1980.
[d]Effective twist angle of the NO_2 group.

1984), 2-nitrobenzenesulfenyl chloride, o-$NO_2C_6H_4SCl$ (Schultz et al., 1984), and
2-methyl nitrophenyl sulfide, o-$NO_2C_6H_4SCH_3$ (Schultz et al., 1987). There is
some indication of such interactions in the structural results. The Se···O and S···O
distances in these molecules are markedly shorter than the sum of van der Waals
radii of the two atoms, 3.30 Å and 3.25 Å, respectively. In the XD study of nitro-

Table XV. Geometrical Parameters of o-Chloronitrobenzene, o-Bromonitrobenzene,
o-Iodonitrobenzene, and o-Dinitrobenzene[a]

Parameter	o-$ClC_6H_4NO_2$ (X = Cl) ED,[b] r_g	o-$BrC_6H_4NO_2$ (X = Br) ED,[c] r_g	o-$IC_6H_4NO_2$ (X = I) ED,[d] r_a, ⟨α	o-$(NO_2)_2C_6H_4$ (X = N) ED,[e] r_g	o-$(NO_2)_2C_6H_4$ (X = N) XD[f]
(C–C)$_{mean}$	1.387(2)	1.386(3)	1.399(3)	1.397(2)	1.379
C–N	1.462(12)	1.494(14)	1.463(16)	1.475(12)	1.470(2)
N=O	1.226(2)	1.218(3)	1.237(3)	1.224(2)	1.222(2)
C–X	1.721(3)	1.894(6)	2.092(7)		
C–C(N)–C	121.4(12)	120.4(24)	122.2	120.4(12)	120.3
C–C(X)–C	121.7(7)	119.6(12)	119.9(15)		120.3
C(X)–C(H)–C(H)	117.2(10)	120.6(24)	119.3(15)		119.2
C(X)–C(N)–N	123.5(10)	125.2(21)	122.6(14)		121.6
C(N)–C(X)–X	120.8(11)	121.1(21)	123.8(9)		121.6
O=N=O	123.6(10)	128.6(15)	121.4(6)	124.9(5)	125.3
Φ[g]	33.9(4)	43.3(27)	70 (4)	31.1(18)	

[a]Distances are given in angstroms, and angles are in degrees.
[b]Batyukhnova et al., 1985.
[c]Batyukhnova et al., 1988.
[d]Samdal et al., 1992.
[e]Penionzhkevich et al., 1979.
[f]Herbstein and Kapon, 1990.
[g]Effective twist angle of the NO_2 group.

benzenesulfenyl chloride (Kucsman *et al.*, 1989), two kinds of molecules were observed in the crystal, both with short S···O distances: 2.379 and 2.408 Å. Similarly to the gas-phase structures, the XD structure shows a nearly linear Cl–S···O sequence. Two different intermolecular S···Cl close contacts were observed in the crystal at 3.447 and 3.452 Å. The presence of the two kinds of molecules is consistent with the difference in these interactions.

The geometrical parameters of *m*-chloronitrobenzene (Batyukhnova *et al.*, 1983), *m*-bromonitrobenzene (Batyukhnova *et al.*, 1988), *m*-iodonitrobenzene (Samdal *et al.*, 1992), and *m*-dinitrobenzene (Batyukhnova *et al.*, 1983) from ED are given in Table XVI, along with the XD results on *m*-dinitrobenzene (Trotter and Williston, 1966). Here, again, mean values of bond lengths and bond angles are quoted. According to the XD results, the benzene ring and the nitrogen atoms are in the same plane in the crystal.

The structure of 1,3,5-trinitrobenzene from ED (Penionzhkevich *et al.*, 1979) and ND (Choi and Abel, 1972) is given in Table XVII. Two different molecules were found in an asymmetric unit in the crystal: one with a planar and the other with a nonplanar benzene ring. The NO_2 groups are either twisted or bent out of the benzene ring plane. The oxygen atoms undergo large-amplitude motion. The short intermolecular and intramolecular O···H distances indicate the presence of C–H···O type hydrogen bonding.

The geometrical parameters of 2,6-dinitrochlorobenzene (Batyukhnova *et*

Table XVI. **Geometrical Parameters of *m*-Chloronitrobenzene, *m*-Bromonitrobenzene, *m*-Iodonitrobenzene, and *m*-Dinitrobenzene**[a]

Parameter	m-$ClC_6H_4NO_2$ (X = Cl) ED,[b] r_g	m-$BrC_6H_4NO_2$ (X = Br) ED,[c] r_g	m-$IC_6H_4NO_2$ (X = I) ED,[d] r_a, $\langle\alpha$	m-$(NO_2)_2C_6H_4$ (X = N) ED,[b] r_g	XD[e]
(C–C)mean	1.388(3)	1.394(3)	1.391(2)	1.382(2)	1.387(6)
C–N	1.442(10)	1.448(14)	1.494(10)	1.461(6)	1.493(6)
N=O	1.243(3)	1.238(8)	1.225(2)	1.225(2)	1.248(6)
C–X	1.746(6)	1.865(8)	2.102(6)		
O=N=O	122.6(10)	121.8(14)	123.6(4)	125.3(7)	125.8(4)
C–C(N)–C	123.0(15)	124.4(15)	123.9	121.8(10)	123.6(4)
C–C(X)–C	121.5(2)	121.4(10)	121.7(7)	121.8(10)	123.6(4)
C(X)–C–C(N)	118.3(15)	118.7(18)	116.6(8)	118.5(15)	115.7(4)
Φ[f]	13(6)	25.0(51)	14(4)	23(3)	

[a]Distances are given in angstroms, and angles in degrees.
[b]Batyukhnova *et al.*, 1983.
[c]Batyukhnova *et al.*, 1988.
[d]Samdal *et al.*, 1992.
[e]Trotter and Williston, 1966.
[f]Effective twist angle of the NO_2 group; given only where available.

Table XVII. Geometrical Parameters of
1,3,5-$(NO_2)_3C_6H_3$ from ED and ND[a]

Parameter	ED,[b] r_g	ND (crystal)[c]
$(C–C)_{mean}$	1.388(2)	1.379
C–N	1.475(5)	1.477
N=O	1.224(2)	1.207
O=N=O	125.7(5)	125.0
$C–C_N–C$	123.2(9)	123.4
Φ[d]	21.0(23)	

[a]Distances are given in angstroms, and angles in
degrees.
[b]Penionzhkevich et al., 1979.
[c]Choi and Abel, 1972; mean values, no uncertainties
given.
[d]Effective twist angle of the NO_2 group.

al., 1984b) and 2,6-dinitrobromobenzene (Batyukhnova et al., 1988) are presented in Table XVIII. The ED results indicate considerable torsion of the NO_2
groups around the C–N axis in both molecules.

Formation of intramolecular hydrogen bonds between the nitro group oxy-
gens and the hydroxy hydrogens was observed in 2-nitroresorcinol (Borisenko and
Hargittai, 1993) and 2-nitrophenol (Borisenko et al., 1994) from ED. The resonance forms assisting intramolecular hydrogen bonding are shown in Schemes 8
and 9. The O···O nonbonded distances between the nitro and hydroxy groups are
much shorter, 2.56(1) and 2.58(1) Å, than twice the oxygen van der Walls radius.
Selected geometrical parameters of the two molecules are given in Table XIX. The

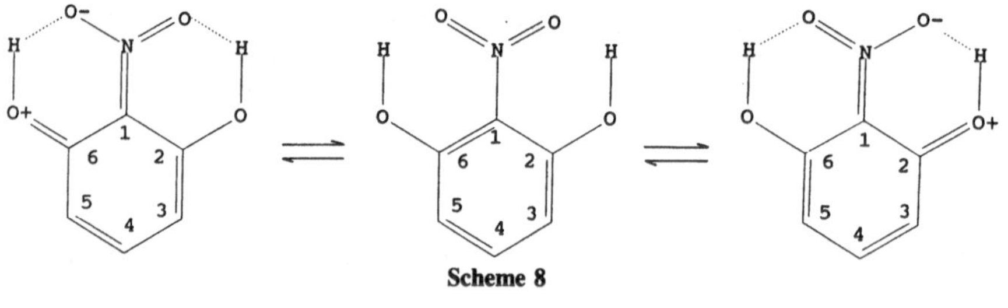

Scheme 8

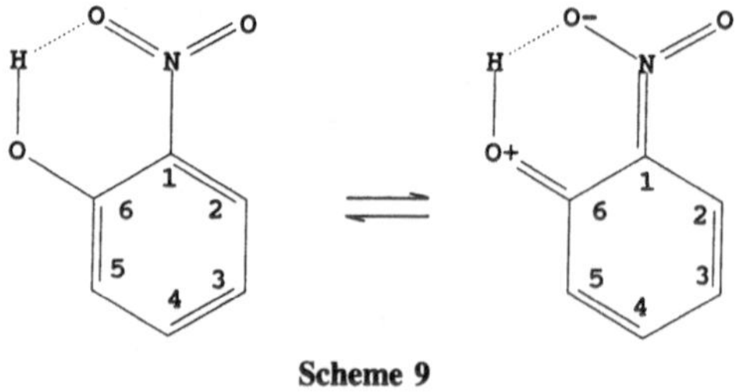

Scheme 9

Table XVIII. Geometrical Parameters of 2,6-Dinitrochlorobenzene and 2,6-Dinitrobromobenzene[a]

Parameter	$2,6\text{-}(NO_2)_2ClC_6H_3$ (X = Cl) ED,[b] r_g	$2,6\text{-}(NO_2)_2BrC_6H_3$ (X = Br) ED,[c] r_g
$(C\text{-}C)_{mean}$	1.395(3)	1.393(3)
C–N	1.447(4)	1.469(9)
N=O	1.229(2)	1.229(3)
C–X	1.712(5)	1.899(9)
O=N=O	123.0(3)	125.6(9)
C–C(X)–C	118.9(6)	117.1(12)
C–C(N)–C	121.6(4)	122.7(9)
C(X)–C(N)–N	122.8(4)	121.2(7)
Φ^d	54.0(10)	57.9(19)

[a]Distances are in angstroms, and angles in degrees.
[b]Batyukhnova et al., 1984b.
[c]Batyukhnova et al., 1988
[d]Effective twist angle of the NO_2 group.

Table XIX. Geometrical Parameters of 2-Nitroresorcinol and 2-Nitrophenol from ED[a]

Parameter	$O_2NC_6H_3(OH)_2$[b]	$O_2NC_6H_4OH$[c]
$(C\text{-}C)_{mean}$	1.404 ± 0.003	1.399 ± 0.003
$(C\text{-}H)_{mean}$	1.090 ± 0.015	1.089 ± 0.007
N–C	1.449 ± 0.007	1.464 ± 0.005
$(N=O)_{mean}$	1.239 ± 0.003	1.233 ± 0.003
C–O	1.354 ± 0.004	1.359 ± 0.009
O···O	2.56 ± 0.01	2.58 ± 0.01
O=N=O	121.4 ± 0.5	123.3 ± 0.4
$(O=N\text{-}C)_{mean}$	119.3 ± 0.3	118.4 ± 0.3
N–C1–C2	120.5 ± 0.4	120.8 ± 0.7
C1–C2–O	122.8 ± 0.7	123.9 ± 0.8
C6–C1–C2	119.1 ± 0.7	121.4 ± 0.5
C1–C2–C3	120.4 ± 0.5	119.4 ± 0.8
C2–C3–C4	118.3 ± 0.5	118.1 ± 1.6
C3–C4–C5	123.6 ± 0.6	122.9 ± 0.9
C4–C5–C6	118.3 ± 0.5	119.3 ± 0.8
C1–C6–C5	120.4 ± 0.5	119.0 ± 0.8
Φ^d	0.0[e]	7.3 ± 5.7

[a]Distances (r_g) are given in angstroms, and angles in degrees.
[b]Borisenko and Hargittai, 1993. The numbering of atoms corresponds to that in Scheme 8.
[c]Borisenko et al., 1994. The numbering of atoms corresponds to that in Scheme 9.
[d]Effective angle of torsion of the NO_2 group.
[e]Assumed; very small when refined.

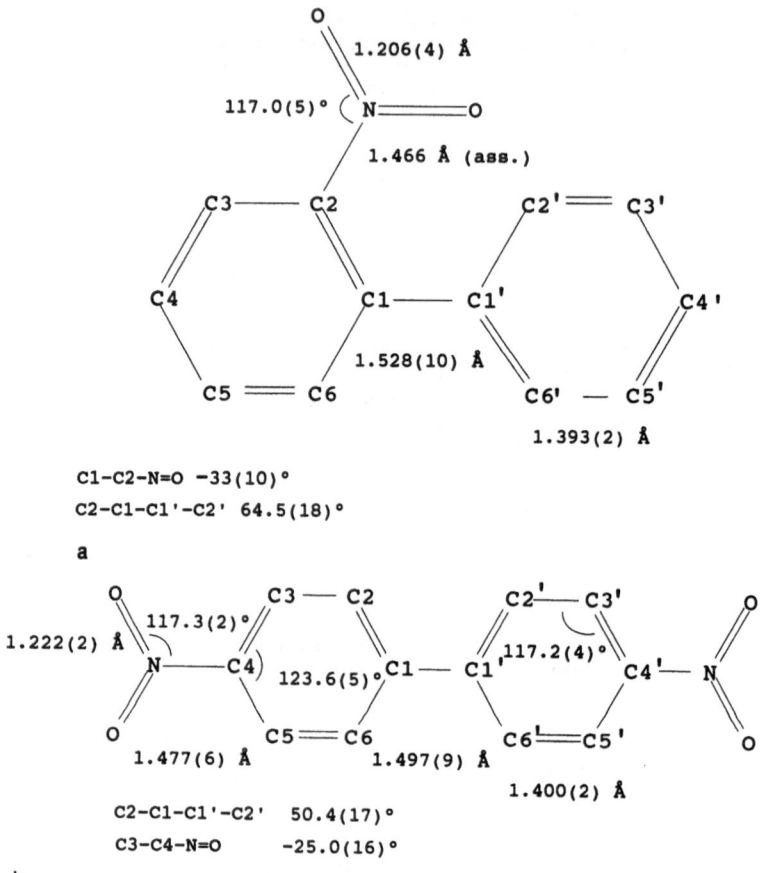

Figure 7. Selected structural parameters of 2-nitrodiphenyl (Levit *et al.*, 1991) (a), 4,4′-dinitro-diphenyl (Levit *et al.*, 1992) (b), and 6,6′-dinitro-2,2′-diphenic acid (Novikov *et al.*, 1979) (c), from ED.

ED data are consistent with the coplanarity of the benzene ring and the NO$_2$ group. This is probably due to the presence of hydrogen bonds as nitro group torsion is hindered by a higher torsional barrier in these molecules than in nitrobenzene.

Three ED studies were published on the structure of nitro derivatives of biphenyls: 2-nitrobiphenyl (Levit *et al.*, 1991), 4,4′-dinitrobiphenyl (Levit *et al.*, 1992), and 6,6′-dinitro-2,2′-diphenic acid (Novikov *et al.*, 1979). The most intriguing question was the determination of the relative orientation of the two benzene rings in these studies. Some results are presented in Fig. 7. The

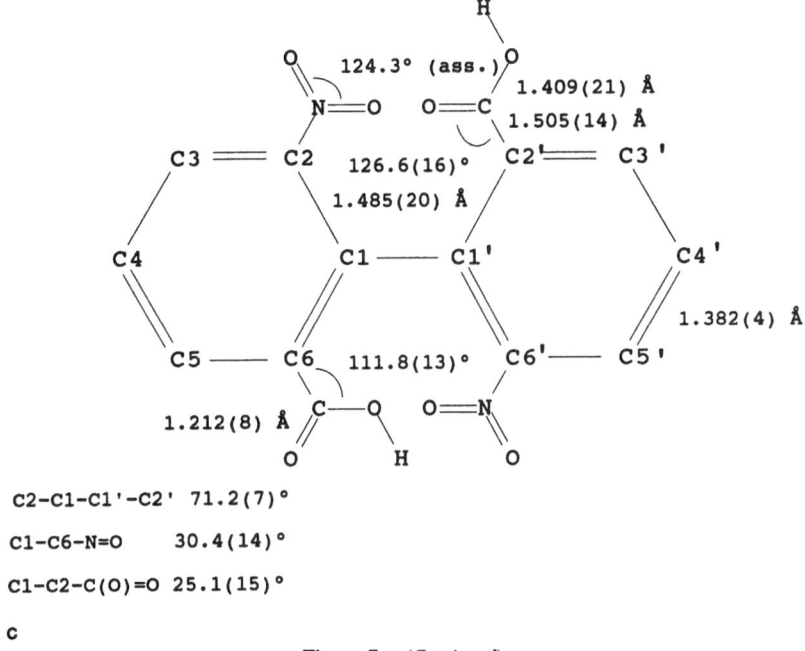

C2-C1-C1'-C2' 71.2(7)°

C1-C6-N=O   30.4(14)°

C1-C2-C(O)=O 25.1(15)°

c

Figure 7. (*Continued*)

conformation of 2-nitrobiphenyl is probably stabilized by hydrogen bonding between one of the oxygen atoms of the NO_2 group and the ortho hydrogen atom of the adjacent ring. Earlier crystallographic studies (Boonstra, 1963) reported smaller rotational angles, 32° and 10°, respectively, than those in the gas phase. Crystal-packing effects may be responsible for the stabilization of the biphenyl conformation with smaller twist angles (Brock and Minton, 1989). The distances between the oxygen atoms of nitro groups and oxygen atoms of hydroxyl groups, 2.69(3) Å, correspond to intramolecular hydrogen bonding. On the other hand, Popova *et al.* (1989) stressed the importance of intermolecular hydrogen bonds in the crystal on the basis of an XD study and concluded that intramolecular hydrogen bonding may not be of importance, or present at all, in the gas phase either.

7. CONCLUSIONS

The gas-phase studies on the geometry of nitro compounds have been reviewed in this chapter. One of the questions raised in the study of the geometries

of nitro compounds is the mutual influence of the NO_2 group and the rest of the
molecule on each other's geometries.

Figure 8 attempts to give an answer to the question of the influence of the rest
of the molecule on the NO_2 geometry. The O=N=O angle is plotted against the
N=O distance in Fig. 8 for the nitro compounds studied in the gas phase. There is
an overall tendency of opening O=N=O angle and shortening N=O bonds with
increasing electronegativity of the X ligand in XNO_2, in complete agreement with
the predictions of the VSEPR model (Gillespie and Hargittai, 1991). There is also a
constancy in the bulk of the data. The O···O nonbonded distance is about 2.17 Å.
A similar constancy has been observed for the O···O distance of the SO_2 groups of
sulfone molecules at 2.48 Å (Hargittai, 1985). This constancy has been discussed
in terms of balancing electron-pair repulsions and nonbonded interactions.

As regards the influence of the NO_2 group on the structure of the rest of the
molecule, it is expected that substitution of a less electronegative ligand by NO_2
shifts some electron density along the $X-NO_2$ bond toward NO_2, and this, in turn,
gives rise to an increase of the neighboring bond angles, in accordance with the
VSEPR model (Gillespie and Hargittai, 1991).

Concerning gas/solid structural differences, the changes are commensurable
with the experimental errors; thus, no conclusion can be made, except that these
changes seem to be very small. Thus, the geometry of the nitro group and that of
the adjacent parts of the molecule demonstrate a remarkable structural stability.
The geometries hardly change upon the molecules entering the crystal structure,
and this is noteworthy as these are high-density solids with presumably very
efficient close packing. The intermolecular interactions in the crystal though do
not seem to influence strongly the molecular geometry. A case in point is the

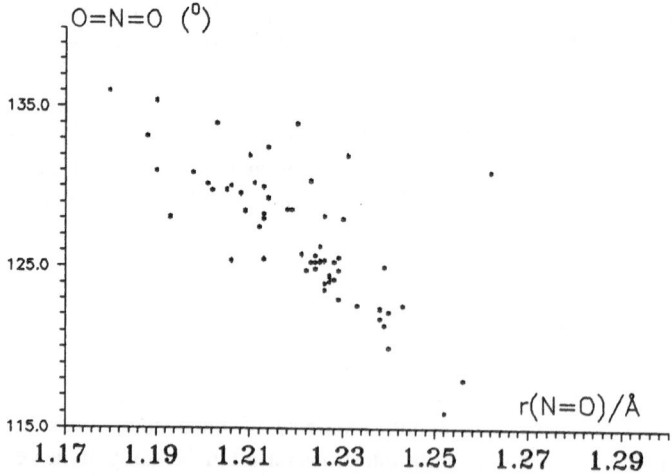

Figure 8.   The O=N=O angle plotted against the N=O distance in gaseous nitro compounds.

structure of nitrobenzene, which has been studied recently both in the gase phase and in the crystal (cf. Table XII). Boese *et al.* (1992) established the presence of some intermolecular interactions, evidenced by the relatively short $O \cdots H$ distances and supported by X–X density maps. Yet the structural changes are relatively small. This is in contrast with the drastic structural changes calculated for the tautomerization of nitromethane (cf. Fig. 4a), which is, of course, an extreme case.

Another remarkable example of structural stability of the nitro compounds is the case of dinitrogen tetroxide, again studied both in the gas and in the crystal. The N_2O_4 molecule has a conspicuously long N–N bond, which may be interpreted as an indication of a rather weak bond. For other relatively weak bonds, the gas-to-crystal change is accompanied by a bond shortening of a large extent by any measure. Thus, for example, in $HCN-BF_3$ a shortening of 0.84 Å was reported for the N–B bond (Burns and Leopold, 1993), and in fluorosilatrane a shortening of 0.28 Å was observed for the N–Si bond (Forgács *et al.*, 1990). The shortening of the N–N bond of N_2O_4 is 0.02 Å only, upon the gas-to-solid transition.

The structural chemistry of nitro compounds offers a host of interesting features and questions. We hope that our review will call attention to their further investigation.

ACKNOWLEDGMENTS. We are grateful to Dr. Alfred H. Lowrey (Washington, D.C.) for information on the NRL Energetic Materials Structural Database and to Dr. Mátyás Czugler (Budapest) for assistance in using the Cambridge Crystallographic Database. Partial support was provided by the National Scientific Research Foundation (OTKA, No. 2103).

REFERENCES

Allen, F. H., Kennard, O., and Taylor, R., 1983, Systematic analysis of the structural data as a research technique in organic chemistry, *Acc. Chem. Res.* **16**:146–153.

Almenningen, A., Brunvoll, J., Popik, M. V., Vilkov, L. V., and Samdal, S., 1984, The molecular structure and barrier to internal rotation of *p*-bromonitrobenzene, determined by gas-phase electron diffraction, *J. Mol. Struct.* **118**:37–45.

Arvia, A. J., Cafferata, F. R., and Schumacher, H. J., 1963, Das Infrarotspektrum und die Struktur von NO_3Cl und NO_3F, *Chem. Ber.* **96**:1187–1194.

Bagryanskaya, I. Yu., and Gatilov, Yu. V., 1983, Crystal structure of nitromethane, *Zh. Strukt. Khim.* **24**(1):158–160.

Batyukhnova, O. G., Sadova, N. I., Vilkov, L. V., and Pankrushev, Yu. A., 1983, Electron diffraction study of meta-dinitrobenzene and meta-chloronitrobenzene in the gas phase, *J. Mol. Struct.* **97**:153–163 [in Russian].

Batyukhnova, O. G., Sadova, N. I., Vilkov, L. V., Ivshin, V. P., and Pankrushev, Yu. A., 1984a, Electron diffraction study of the structure of the *N*-methyl-*N*-nitrovinylamine molecule in the gas phase, *Zh. Strukt. Khim.* **25**(6):47–54 [in Russian].

Batyukhnova, O. G., Vilkov, L. V., Sadova, N. I., and Pankrushev, Yu. A., 1984b, Electron diffraction study of the structure of the 2,6-dinitrochlorobenzene molecule in the gas phase, *Zh. Strukt. Khim.* **25**(3):166–168 [in Russian].

Batyukhnova, O. G., Sadova, N. I., Vilkov, L. V., and Pankrushev, Yu. A., 1985, Electron diffraction study of the structure of the chloronitrobenzene molecule in the gas phase, *Zh. Strukt. Khim.* **26**(5):175–178.

Batyukhnova, O. G., Sadova, N. I., Syshchikov, Yu. I., Vilkov, L. V., and Pankrushev, Yu. A., 1988, Electron diffraction study of the structure of *o*-bromonitrobenzene, *m*-bromonitrobenzene, and 2,6-dinitrobromobenzene in the gas phase, *Zh. Strukt. Khim.* **29**(4):53–63 [in Russian].

Belyakov, A. V., Levit, P. B., Tselinskii, I. V., Shlyapochnikov, V. A., Ladyzhnikova, T. D., Altukhov, K. V., and Manuel, D. V., 1992, Electron diffraction analysis of the molecular structure of gaseous 2-methyl-4-trinitromethyl-1,2,3-triazol, $HC=N-N(CH_3)-N=C-(NO_2)_3$, *J. Mol. Struct.* **265**:337–345 [in Russian].

Bird, G. R., Baird, J. C., Jache, A. W., Hodgeson, J. A., Curl, R. F., Kunkle, A. C., Bransford, J. W., Rastrup-Andersen, J., and Rosenthal, J., 1964, Microwave spectrum of NO_2: Fine structure and magnetic coupling. *J. Chem. Phys.* **40**:3378–3390.

Bock, H., Dienelt, R., Schödel, H., Havlas, Z., Herdtweck, E., and Herrmann, W. A., 1993, Aci-Nitrodiphenylmethan: ein über Wasserstoffbrücken vernüpftes Dimer, *Angew. Chem.* **105**:1826–1828.

Boese, R., Bläser, D., Nussbaumer, M., and Krygowski, T. M., 1992, Low temperature crystal and molecular structure of nitrobenzene, *Struct. Chem.* **3**:363–368.

Boonstra, E. G., 1963, The crystal and molecular structure of 4,4'-dinitrodiphenyl, *Acta Crystallogr.* **16**:816–823.

Borisenko, K. B., and Hargittai, I., 1993, Intramolecular hydrogen bonding and molecular structure of 2-nitroresorcinol from gas-phase electron diffraction, *J. Phys. Chem.* **97**:4080–4084.

Borisenko, K. B., Bock, C. W., and Hargittai, I., 1994, Intramolecular hydrogen bonding and molecular geometry of 2-nitrophenol from a joint gas-phase electron diffraction and *ab initio* molecular orbital investigation, *J. Phys. Chem.* **98**:1442–1448.

Brittain, A. H., Cox, A. P., and Kuczkowski, R. L., 1969, Microwave spectrum, low frequency vibrations, dipole moment and quadrupole coupling constants of dinitrogen trioxide, *Trans. Faraday Soc.* **65**:1963–1974.

Brock, C. P., and Minton, R. P., 1989, Systematic effects of crystal-packing forces: Biphenyl fragments with H atoms in all four ortho positions, *J. Am. Chem. Soc.* **111**:4586–4593.

Brunvoll, J., Samdal, S., Thomassen, H., Vilkov, L. V., and Volden, H. V., 1990, The molecular structure of iodobenzene and *p*-iodonitrobenzene in the gaseous state, *Acta Chem. Scand.* **44**:23–30.

Bürgi, H.-B., Blance, E., Schwarzenbach, D., Liu, S., Lu, Y.-J., Kappes, M. M., and Ibers, J. A., 1992, The structure of C_{60}: Orientational disorder in the low-temperature modification of C_{60}, *Angew. Chem. Int. Ed. Engl.* **31**:640–643.

Burns, W. A., and Leopold, K. R., 1993, Unusually large gas–solid structure differences: a Crystallographic study of $HCN-BF_3$, *J. Am. Chem. Soc.* **115**:11622–11623.

Callomon, J. H., Hirota, E., Kuchitsu, K., Lafferty, W. J., Maki, A. G., and Pote, C. S., 1976, *Structure Data of Free Polyatomic Molecules, Landolt-Börnstein Numerical Data and Functional Relationships in Science and Technology* (New Series), Group II, Vol. 7, Springer-Verlag, Berlin.

Callomon, J. H., Hirota, E., Iijima, T., Kuchitsu, K., and Lafferty, W. J., 1987, *Structure Data of Free Polyatomic Molecules, Landolt-Börnstein Numerical Data and Functional Relationships in Science and Technology* (New Series), Group II, Vol. 15, Springer-Verlag, Berlin.

Casper, B., Lambotte, P., Minkwitz, R., and Oberhammer, H., 1993, Gas-phase structures of chlorine nitrate and bromine nitrate ($ClONO_2$ and $BrONO_2$), *J. Phys. Chem.* **97**:9992–9995.

Chiang, J. F., and Song, J. J., 1982, Molecular structure of 4-nitro-, 4-methyl- and 4-chloro-pyridine-*N*-oxides, *J. Mol. Struct.* **96**:151–162.

Choi, C. S., and Abel, J. E., 1972, The crystal structure of 1,3,5-trinitrobenzene by neutron diffraction, *Acta Crystallogr., Sect. B* **28**:193–201.

Choi, C. S., and Prince, E., 1972, The crystal structure of cyclotrimethylene-trinitramine. *Acta Crystallogr., Sect. B* **28**:2857–2862.

Colapietro, M., Domenicano, A., Marciante, C., and Portalone, G., 1981, *p*-Nitroaniline revisited, *Acta Crystallogr., Sect. A* **37**:C199.

Coon, C., 1988, cited in Gilardi *et al.* (1992).

Coppens, P., and Lehmann, M. S., 1976, Charge density studies below liquid nitrogen temperature. II. Neutron analysis of *p*-nitropyridine *N*-oxide at 30K and comparison with X-ray results, *Acta Crystallogr., Sect. B* **32**:1777–1784.

Cotton, F. A., and Wilkinson, G., 1988, *Advanced Inorganic Chemistry*, 5th ed., Wiley-Interscience, New York, p. 324.

Cox, A. P., and Riveros, J. M., 1965, Microwave spectrum and structure of nitric acid, *J. Chem. Phys.* **42**:3106–3112.

Cox, A. P., and Waring, S., 1971, Microwave spectrum, structure and dipole moment of methyl nitrate, *J. Chem. Soc., Faraday Trans. 2* **1971**:3441–3450.

Cox, A. P., and Waring, S., 1972, Microwave spectrum and structure of nitromethane, *J. Chem. Soc., Faraday Trans. 2* **68**:1060–1071.

David, W. I. F., Ibberson, R. M., Matthewman, J. C., Prassides, K., Dennis, T. J. S., Hare, J. P., Kroto, H. W., Taylor, R., and Walton, D. R. M., 1991, Crystal structure and bonding of ordered C_{60}, *Nature (London)* **353**:147–149.

DiRienzo, F., Domenicano, A., and Riva di Sanseverino, L., 1980, Structural studies of benzene derivatives. VIII. Refinement of the crystal structure of *p*-dinitrobenzene, *Acta Crystallogr., Sect. B* **36**:586–591.

Domenicano, A., and Hargittai, I. (eds.), 1992, *Accurate Molecular Structures. Their Determination and Importance*, Oxford University Press, Oxford.

Domenicano, A., and Hargittai, I., 1993, Gas/crystal structural differences in aromatic molecules, *ACH—Models in Chemistry* **130**:347–362.

Domenicano, A., Vaciago, A., and Coulson, C. A., 1975, Molecular geometry of substituted benzene derivatives. I. On the nature of the ring deformations induced by substitution, *Acta Crystallogr., Sect. B.* **31**:221–234.

Domenicano, A., Schultz, G., Hargittai, I., Colapietro, M., Portalone, G., George, P., and Bock, C. W., 1989, Molecular structure of nitrobenzene in the planar and orthogonal conformations: A concerted study by electron diffraction, X-ray crystallography, and molecular orbital calculations, *Struct. Chem.* **1**:107–122.

Durig, J. R., and Lindsay, N. E., 1990, Far-infrared spectra and barriers to internal rotation of ethyl nitrate, *Spectrochim. Acta, Part A* **46**:1125–1135.

Durig, J. R., and Sheehan, T. G., 1990, Raman spectra, vibrational assignment, structural parameters and *ab initio* calculations for ethyl nitrate, *J. Raman Spectrosc.* **21**:635–644.

Eaton, P. E., 1992, Cubanes: Starting materials for the chemistry of the 1990s and the new century, *Angew. Chem. Int. Ed. Engl.* **31**:1421–1436.

Endo, K., 1979, Microwave spectrum and structure of nitryl chloride, *J. Chem. Soc. Jpn.* **1979**:1129–1132 [in Japanese].

Fink, M., and Kohl, D. A., 1988, Temperature dependence of electron diffraction structural parameters: Theory and experiment, in *Stereochemical Applications of Gas-Phase Electron Diffraction*, Part A (I. Hargittai and M. Hargittai, eds.), pp. 139–190, VCH Publishers, New York.

Forgács, G., Kolonits, M., and Hargittai, I., 1990, The gas phase structure of 1-fluorosilatrane from electron diffraction, *Struct. Chem.* **1**:245–250.

Gilardi, R. D., and Karle, J., 1991, The structural investigation of energetic materials, in *Chemistry of Energetic Materials* (G. A. Olah and D. R. Squire, eds.), pp. 1–26, Academic Press, New York.

Gilardi, R., George, C., and Flippen-Anderson, J. L., 1992, A Pictorial Index to the NRL Energetic Materials Structural Database as of Oct 1992 Analyzed via X-Ray Single-Crystal Diffraction Analyses at the Laboratory for the Structure of Matter, Naval Research Laboratory, Washington, D.C., p. 36.

Gillespie, R. J., and Hargittai, I., 1991, *The VSEPR Model of Molecular Geometry*, Allyn and Bacon, Boston.

González, S. R., Mulas, R., and Alonso, J. L., 1989, Rotational spectrum of 1-nitrocyclohexene, *J. Mol. Spectrosc.* **133**:413–422.

Grison, P. E., Eriks, K., and deVries, J. L., 1950, Structure cristalline de l'anhydride azotique N_2O_5, *Acta Crystallogr.* **3**:290–293.

Groth, P., 1963, Crystal structure of an unstable monoclinic form of dinitrogen tetroxide, *Acta Chem. Scand.* **17**:2419–2422.

Hargittai, I., 1985, *The Structure of Volatile Sulphur Compounds*, Reidel, Dordrecht.

Hargittai, I., and Hargittai, M., 1987, The importance of small structural differences, in *Molecular Structure and Energetics*, Vol. 2 (J. F. Liebman and A. Greenberg, eds.), pp. 1–35, VCH Publishers, New York.

Hargittai, I., and Hargittai, M., 1988, *Stereochemical Applications of Gas-Phase Electron Diffraction*, Parts A and B, VCH Publishers, New York.

Hargittai, M., and Hargittai, I., 1987, Gas–solid molecular structure differences, *Phys. Chem. Miner.* **17**:413–425.

Hasser, M., Almlöf, J., and Scuseria, G. E., 1991, University of Minnesota Supercomputer Institute Research Report, UMSI91/142, May 1991.

Hedberg, K., 1966, The determination of vibrational potential functions using mean amplitudes from electron diffraction: Some recent results, *Trans. Am. Crystallogr. Assoc.* **2**:79–89.

Hedberg, K., Hedberg, L., Bethunde, D. S., Brown, C. A., Dorn, H. C., Johnson, R. D., and De Vries, M. 1991, Bond lengths in free molecules of buckminsterfullerene, C_{60}, from gas-phase electron diffraction, *Science* **254**:410–412.

Herbstein, F. H., and Kapon, M., 1990, Structure of 1,2-dinitrobenzene, *Acta Crystallogr., Sect. B* **46**:567–572.

Høg, J. H., 1971, A study of nitrobenzene, Thesis, University of Copenhagen, p. 129.

Ishiwata, T., Tanaka, I., Kavaguchi, K., and Hirota, E., 1985, Infrared diode laser spectroscopy of the NO_3 ν_3 band, *J. Chem. Phys.* **82**:2196–2205.

Karle, I. L., and Karle, J., 1962, Determination of the molecular structure of CF_3NO_2 and CBr_3NO_2 by electron diffraction, *J. Chem. Phys.* **36**:1969–1973.

Khodchenkov, A. N., Spiridonov, V. P., and Akishin, P. A., 1965, Electron diffraction study of the structure of nitrates of lithium and sodium in the gas-phase, *Zh. Strukt. Khim.* **6**(5):765–766 [in Russian].

Knudsen, R. E., George, C. F., and Karle, J., 1966, Molecular structure of CCl_3NO_2 by electron diffraction, *J. Chem. Phys.* **44**:2334–2337.

Kohata, K., Fukuyama, T., and Kuchitsu, K., 1982, Molecular structure of hydrazine as studied by gas electron diffraction, *J. Phys. Chem.* **86**:602–606.

Krebs, B., Mandt, J., Cobbledick, R. E., and Small, R. W. H., 1979, The structure of *N,N*-dimethylnitramine, *Acta Crystallogr., Sect. B* **35**:402–404.

Kroto, H. W., 1989, C_{60}[B] Buckminsterfullerene, other fullerenes and the icospiral shell, in *Symmetry 2, Unifying Human Understanding* (I. Hargittai, ed.), Pergamon Press, Oxford, pp. 417–423.

Kroto, H. W., and Walton, D. R. M., 1993, Polyynes and the formation of fullerenes, *Philos. Trans. R. Soc. London, Ser. A* **343**:103–112.

Kuchitsu, K. (ed.), 1992, *Structure Data of Free Polyatomic Molecules, Landolt-Börnstein Numerical Data and Functional Relationships in Science and Technology* (New Series), Group II, Vol. 21, Springer-Verlag, Berlin.

Kucsman, A., Kapovits, I., Czugler, M., Párkányi, L., and Kálmán, A., 1989, Intramolecular sulphur–oxygen interaction in organosulphur compounds with different sulphur valence state: An X-ray study of methyl-2-nitrobenzene-sulphenate, -sulphinate, -sulphonate and 2-nitrobenzene-sulphenyl chloride, *J. Mol. Struct.* **198**:339–353.

Kulikov, V. A., Ugarov, V. V., and Rambidi, N. G., 1981a, Structure of the RbNO$_3$ molecule, *Zh. Strukt. Khim.* **22**(2):196–198 [in Russian].

Kulikov, V. A., Ugarov, V. V., and Rambidi, N. G., 1981b, Structure of the CsNO$_3$ molecule, *Zh. Strukt. Khim.* **22**(3):168–171 [in Russian].

Kulikov, V. A., Ugarov, V. V., and Rambidi, N. G., 1981c, Structure of the TlNO$_3$ molecule, *Zh. Strukt. Khim.* **22**(3):166–168 [in Russian].

Kulikov, V. A., Ugarov, V. V., and Rambidi, N. G., 1981d, Structure of the RbNO$_2$ and CSNO$_2$ molecules, *Zh. Strukt. Khim.* **22**(5):183–185 [in Russian].

Lammertsma, K., and Prasad, B. V., 1993, Nitro *aci*-nitro tautomerism, *J. Am. Chem. Soc.* **115**:2348–2351.

Langridge-Smith, P. R. R., Stevens, R., and Cox, A. P., 1980, Microwave spectrum and barrier to internal rotation of 2-methyl-2-nitropropane, Me$_3$CNO$_2$, *J. Chem. Soc., Faraday Trans. 2* **76**:330–338.

LaVilla, R. E., and Bauer, S. H., 1963, The structure of gaseous copper(II) nitrate as determined by electron diffraction, *J. Am. Chem. Soc.* **85**:3597–3600.

Legon, A. C., and Millen, D. J., 1968, The microwave spectrum, structure and dipole moment of nitryl fluoride, *J. Chem. Soc. A* **1968**:1736–1740.

Levit, P. B., Belyakov, A. V., Tselinskii, I. V., Golubinskii, A. V., Vilkov, L. V., and Shlyapochnikov, V. A., 1991, Molecular structure of gaseous 2-nitrodiphenyl based on electron diffraction data, *Zh. Fiz. Khim.* **65**:1946–1948 [in Russian].

Levit, P. B., Belyakov, A. V., Tselinskii, I. V., Vilkov, L. V., and Shlyapochnikov, V. A., 1992, Molecular structure of 4,4′-dinitrobiphenyl in the gaseous state, *J. Mol. Struct.* **265**:329–336.

Liu, S., Lu, Y.-J., Kappes, M. M., and Ibers, J. A., 1991, The structure of the C$_{60}$ molecule: X-ray crystal structure determination of twin at 110 K, *Science* **254**:408–410.

Marsden, C. J., Hedberg, K., Ludwig, M. M., and Gard, G. L., 1991, Molecular structure of CrO$_2$(NO$_3$)$_2$ in the gas phase: A novel form of coordination for chromium?, *Inorg. Chem.* **30**:4761–4766.

McClelland, B. W., Gundersen, G., and Hedberg, K., 1972, Reinvestigation of the structure of dinitrogen tetroxide N$_2$O$_4$, by gaseous electron diffraction, *J. Chem. Phys.* **56**:4541–4545.

McClelland, B. W., Hedberg, L., Hedberg, K., and Hagen, K., 1983, Molecular structure of N$_2$O$_5$ in the gas phase. Large amplitude motion in a system of coupled rotors, *J. Am. Chem. Soc.* **105**:3789–3793.

Mochel, A. R., Britt, C. O., and Boggs, J. E., 1973, Microwave spectrum of nitrocyclopropane, *J. Chem. Phys.* **58**:3221–3229.

Mulas, R., González, S. R., and Alonso, J. L., 1989, Torsional frequency, barrier to internal rotation of 4-nitropyridine from microwave spectra, *J. Mol. Struct.* **213**:77–82.

Nakata, M., Takeo, H., Matsumura, C., Yamanouchi, K., Kuchitsu, K., and Fukuyama, T., 1981, Structures of 1,2-dimethylhydrazine conformers as determined by microwave spectroscopy and gas electron diffraction, *Chem. Phys. Lett.* **83**:246–248.

Niki, H., Maker, P. D., Savage, C. M., and Breitenbach, L. P., 1977, Fourier transform IR spectroscopic observation of pernitric acid formed via HOO + NO$_2$ → HOONO$_2$, *Chem. Phys. Lett.* **45**:564–566.

Nösberger, P., Bauder, A., and Günthard, H. H., 1983, The substitution structure of nitroethylene, *Chem. Phys.* **8**:245–251.

Novikov, V. P., Popik, M. V., Vilkov, L. V., Migachev, G. I., and Dyumaev, K. M., 1979, Gas phase electron diffraction study of the molecular structure of 6,6′-dinitro-2,2′-diphenic acid, *J. Mol. Struct.* **53**:211–218.

Obermeyer, A., Borrmann, H., and Simon, A., 1991, Die Phasenzusammenhänge zwischen kubischen und monoklinen N_2O_4, *Z. Kristallogr.* **196**:129–135.

Obermeyer, A., Borrmann, H., and Simon, A., to be published, as quoted by Casper *et al.* (1993).

Pauling, L., and Brockway, L. O., 1937, The adjacent charge rule and the structure of methyl azide, methyl nitrate, and fluorine nitrate, *J. Am. Chem. Soc.* **59**:13–20.

Penionzhkevich, N. P., Sadova, N. I., Popik, N. I., Vilkov, L. V., and Pankrushev, Yu. A., 1979, Electron diffraction study of the structure of the *o*-dinitrobenzen, *p*-dinitrobenzene, and 1,3,5-trinitrobenzene molecules in gas phase, *Zh. Strukt. Khim.* **20**(4):603–611 [in Russian].

Popova, E. G., Chyotkina, L. A., Sobolev, A. N., Bel'skii, V. K., Andrievskii, A. M., Poplavskii, A. N., and Dyumaev, K. M., 1989, Structure of the 6,6'-dinitro-2,2-dicarboxybiphenyl molecule in the crystal, *Dokl Akad, Nauk SSSR* **304**:127–130 [in Russian].

Sadova, N. I., Vilkov, L. V., and Anfimova, T. M., 1972, Electron diffraction study of the structure of chloronitromethane, *Zh. Strukt. Khim.* **13**(5):763–767 [in Russian].

Sadova, N. I., Popik, N. I., and Vilkov, L. V., 1976a, Electron diffraction study of the structure of the $HC(NO_2)_3$, $ClC(NO_2)_3$, and $BrC(NO_2)_3$ molecules in the gas phase, *Zh. Strukt. Khim.* **17**(2):298–303 [in Russian].

Sadova, N. I., Popik, N. I., and Vilkov, L. V., 1976b, Electron diffraction study of tetranitromethane, *J. Mol. Struct.* **31**:399–402.

Sadova, N. I., Penionzhkevich, N. P., and Vilkov, L. V., 1976c, Study of the structure of the *p*-nitroaniline *p*-$C_6H_4(NO_2)(NH_2)$ molecule by gas electron diffraction method, *Zh. Strukt. Khim.* **17**:1122–1123 [in Russian].

Sadova, N. I., Penionzhkevich, N. P., and Vilkov, L. V., 1976d, Electron diffraction study of the structure of *p*-chloronitrobenzene *p*-$C_6H_4ClNO_2$ in the gas phase, *Zh. Strukt. Khim.* **17**:753–754 [in Russian].

Sadova, N. I., Slepnev, G. E., and Vilkov, L. V., 1977a, Electron diffraction study of dichlorodinitromethane in the gas phase, *Zh. Strukt. Khim.* **18**(2):382–384 [in Russian].

Sadova, N. I., Slepnev, G. E., Tarasenko, N. A., Zenkin, A. A., Vilkov, L. V., Shishkov, I. F., and Pankrushev, Yu. A., 1977b, Geometrical structure of the *N*-methylnitramine and *N*-chloro-*N*-methylnitramine molecules in the gas phase, *Zh. Strukt. Khim.* **18**(5):865–872 [in Russian].

Samdal, S., Vilkov, L. V., and Volden, H. V., 1992, The molecular structure of *ortho*- and *meta*-iodonitrobenzene in the gaseous state as determined by the electron diffraction method. General trends in the bond lengths and STO-3G* *ab initio* calculations on iodonitrobenzenes, *Acta Chem. Scand.* **46**:712–719.

Schultz, G., Hargittai, I., Kapovits, I., and Kucsman, A., 1984, Molecular structure of 2-nitrobenzenesulphenyl chloride, an electron-diffraction study, *J. Chem. Soc., Faraday Trans. 2* **80**:1273–1279.

Schultz, G., Hargittai, I., Kapovits, I., and Kucsman, A., 1987, Molecular structure of methyl-2-nitrophenyl sulphide, an electron-diffraction study, *J. Chem. Soc., Faraday Trans. 2* **83**:2113–2121.

Scroggin, D. G., Riveros, J. M., and Wilson, E. B., 1974, Microwave spectrum and rotational isomerism of ethyl nitrate, *J. Chem. Phys.* **60**:1376–1385.

Shishkov, I. F., Sadova, N. I., Vilkov, L. V., and Ivshin, V. P., 1982, Electron diffraction study of the structure of the methyl(chloromethyl)nitramine molecule in the gas phase, *Zh. Strukt. Khim.* **23**(4):73–78 [in Russian].

Shishkov, I. F., Sadova, N. I., Vilkov, L. V., and Pankrushev, Yu. A., 1983a, Geometrical structure of the dimethylnitromethane and trimethylnitromethane molecules in the gas-phase, *Zh. Strukt. Khim.* **24**(2):27–30 [in Russian].

Shishkov, I. F., Sadova, N. I., Vilkov, L. V., and Pankrushev, Yu. A., 1983b, Electron diffraction study of the structure of dimethyldinitromethane in the gas-phase, *Zh. Strukt. Khim.* **24**:173–176 [in Russian].

Shishkov, I. F., Sadova, N. I., Novikov, V. P., and Vilkov, L. V., 1984, Electron diffraction study of the structure of the nitrobenzene molecule in the gas phase, *Zh. Strukt. Khim.* **25**(2):98–102 [in Russian].

Shishkov, I. F., Vilkov, L. V., Kolonits, M., and Rozsondai, B., 1991, The molecular geometries of some cyclic nitramines in the gas phase, *Struct. Chem.* **2**:57–64.

Shishkov, I. F., Vilkov, L. V., Bock, C. W., and Hargittai, I., 1992, Molecular structure of ethyl nitrate from gas-phase electron diffraction and *ab initio* MO calculations, *Chem. Phys. Lett.* **197**: 489–494.

Spiridonov, V. P., Zasorin, E. Z., Ischenko, A. A., and Touseev, N. I., 1975, Paper presented at the Xth International Congress of Crystallography, Amsterdam, The Netherlands.

Stølevik, R., and Rademacher, P., 1969, Elektronenbeugungsuntersuchung der Struktur des Dimethylnitramins, $(CH_3)_2NNO_2$, *Acta Chem. Scand.* **23**:672–682.

Suenram, R. D., Lovas, F. J., and Pickett, H. M., 1986, The microwave spectrum and molecular conformation of peroxynitric acid ($HOONO_2$), *J. Mol. Spectrosc.* **116**:406–421.

Tarasenko, N. A., Vilkov, L. V., Slepnev, G. E., and Pankrushev, Yu. A., 1977, Electron diffraction study of the structure of the methyldinitramine molecule, *Zh. Strukt. Khim.* **18**(5):953–954 [in Russian].

Touseev, N. I., Golubinsky, A. V., Zasorin, E. Z., and Spiridonov, V. P., 1976, The molecular structure of tin tetra-nitrate in the gas phase. An electron diffraction study, Paper presented at the Sixth Austin Symposium on Gas Phase Molecular Structure, The University of Texas at Austin, pp. 107–108.

Touseev, N. I., Sipachev, V. A., Galimzyanov, R. F., Golubinsky, A. V., Zasorin, E. Z., and Spiridonov, V. P., 1984, Gas-phase electron diffraction study of μ_4-oxohexa-μ-nitratotetraberyllium, *J. Mol. Struct.* **125**:277–286.

Trevino, S. F., Prince, E., and Hubbard, C. R., 1980, Refinement of the structure of solid nitromethane, *J. Chem. Phys.* **73**:2996–3006.

Trotter, J., and Williston, C. S., 1966, Bond lengths and thermal vibrations in *m*-dinitrobenzene, *Acta Crystallogr.* **21**:285–288.

Tyler, J. K., 1963, Microwave spectrum of nitramide, *J. Mol. Spectrosc.* **11**:39–46.

van Eijck, B. P., 1992, Reliability of structure determinations by microwave spectroscopy, in *Accurate Molecular Structures. Their Determination and Importance* (A. Domenicano and I. Hargittai, eds.), pp. 47–64, Oxford University Press, Oxford.

Wallwork, S. C., 1959, The crystal structure of anhydrous cupric nitrate, *Proc. Chem. Soc.* **1959**: 311–312.

Wang, Y., Blessing, R. H., Ross, F. K., and Coppens, P., 1976, Charge density studies below liquid nitrogen temperature: X-ray analysis of *p*-nitropyridine N-oxide at 30 K, *Acta Crystallogr., Sec. B* **32**:572–578.

Yannoni, C. S., Bernier, P. P., Bethunde, D. S., Meijer, G., and Salem, J. R., 1991, NMR determination of the bond lengths in C_{60}, *J. Am. Chem. Soc.* **113**:3190–3192.

Zaripov, N. M., Golubinskii, A. V., Sokolkov, S. V., Vilkov, L. V., and Mannafov, T. G., 1984, Molecular structure of *o*-nitrophenylselenyl bromide, *Dokl. Akad. Nauk SSSR* **278**:664–666 [in Russian].

Chapter 3

Covalent Inorganic Nonmetal Azides

Inis C. Tornieporth-Oetting and Thomas M. Klapötke

1. INTRODUCTION

Inorganic azides can be classified into (i) ionic salts (e.g., NaN_3), (ii) heavy-metal azides (e.g., AgN_3, PbN_3), and (iii) covalently bound nonmetal azides (XN_3: X = H, R_2B, R_3Si, NO, NO_2, R_2P, halogen; R = alkyl, aryl) (Jones, 1973). Whereas the ionic salts are reasonably stable materials and sodium azide is prepared on a commercial scale, the major use of the heavy-metal azides depends on their explosive nature (Greenwood and Earnshaw, 1984). Particularly, lead azide is used for detonators because of its reliability under a variety of adverse, especially damp conditions (Köhler and Meyer, 1991). Although the covalent azides have been known since the beginning of this century, it is only in recent years that these azides have found use in preparative chemistry and that their structures have been elucidated (Tornieporth-Oetting and Klapötke, 1993a). The characterization and usage of the covalent nonmetal azides have obviously been hampered by their thermodynamic and kinetic instability. HN_3 and all of the halogen azides are very hazardous. Especially dangerous are the pure halogen

Inis C. Tornieporth-Oetting and Thomas M. Klapötke • Institut für Anorganische und Analytische Chemie, Technische Universität Berlin, D-10623 Berlin, Germany.

Combustion Efficiency and Air Quality, edited by István Hargittai and Tamás Vidóczy. Plenum Press, New York, 1995.

azides in the condensed phase (Dehnicke, 1983). Violent explosions can also occur in the gaseous state upon sudden variations of the pressure (Dehnicke, 1983).

In this chapter, we direct our attention primarily to the structure, bonding, and energetics of covalent nonmetal azides in which the N_3 group is coordinated to a univalent atom (H, halogen) or a univalent group (−NO, −NO$_2$). We intend to establish perspective with respect to earlier work and to contemporary research, to evaluate the present state of the subject, and to cast a glance to the future.

2. PREPARATION

2.1. Hydrazoic Acid, HN$_3$

Hydrogen azide (hydrazoic acid, HN_3; mp = $-80°C$, bp = $36°C$) is best prepared by the action of molten stearic acid [$CH_3(CH_2)_{16}COOH$; mp = $69-70°C$] on sodium azide at $110-130°C$ (Krakow *et al.*, 1968; Christe *et al.*, 1993):

$$RCOOH + NaN_3 \xrightarrow{\text{neat, melt, } 110-130°C} RCO_2Na + HN_3 \qquad (1)$$

Anhydrous ether solutions of HN_3 can also be prepared by dropwise addition of concentrated sulfuric acid to a mixture of ether and aqueous NaN_3, during which time the bulk of the ether and HN_3 distills off (Jones, 1973).

2.2. Halogen Azides, XN$_3$ (X = F, Cl, Br, I)

The chemistry of the halogen azides has been reviewed by Dehnicke (1976, 1979). FN_3 (mp = $-154°C$, bp = $-82°C$) is best prepared from dry hydrazoic acid and fluorine (diluted with N_2) in the gas phase (Bauer, 1947):

$$4HN_3 + 2F_2 \rightarrow 3FN_3 + N_2 + NH_4F \qquad (2)$$

For preparative purposes the most convenient method to prepare dry ClN_3 (mp ≈ $-100°C$, bp ≈ $-15°C$) is the reaction of chlorine gas (diluted with N_2) with an aqueous solution of sodium azide (Eq. 3), followed by desiccation of the gaseous ClN_3 over P_4O_{10} prior to use (Brauer, 1975).

$$NaN_3 + Cl_2 \xrightarrow{0°C, H_2O} ClN_3 + NaCl \qquad (3)$$

Pure BrN_3 (mp ≈ $-45°C$) can be obtained as a very explosive red liquid by passing a stream of gaseous bromine (diluted with N_2) over dry sodium azide (Eq. 4) in a very slow reaction rate (Dehnicke, 1983).

$$NaN_3(s) + Br_2(g) \rightarrow NaBr(s) + BrN_3(l) \qquad (4)$$

The best way to prepare IN_3 ($T_{subl.}$ ≈ $4°C$, 0.25 bar) in nearly quantitative yield in a 0.5-g scale is the reaction of freshly prepared silver azide and iodine in $CFCl_3$

solution (Eq. 5). After separation of the IN_3 solution by filtration followed by slow evaporation of the solvent, pure, bright yellow IN_3 can be isolated. Sublimation under reduced pressure yields very pure, crystalline iodine azide (Buzek *et al.*, 1993).

$$I_2 + AgN_3 \xrightarrow{\text{0°C, CFCl}_3} IN_3 + AgI \qquad (5)$$

2.3. Nitrosyl Azide and Nitryl Azide (ON–N_3, O_2N–N_3)

Nitrosyl azide, ON_4, is best prepared by condensing gaseous NOCl at $-95°C$ onto an excess of anhydrous, activated NaN_3 (Eq. 6). The product, which is stable at low temperatures only, can be isolated as a pale yellow solid by low-temperature sublimation ($-55/-196°C$) under reduced pressure (Schulz *et al.*, 1993).

$$\text{Excess } NaN_3 + NOCl \xrightarrow{\text{neat, } -196/-55°C} NaCl + ON-N_3 \qquad (6)$$

In addition, there is some evidence for nitryl azide, O_2N-N_3 (from the reaction of NO_2Cl and NaN_3) (Doyle *et al.*, 1973). However, this species has yet to be prepared as a pure compound and to be characterized by low-temperature techniques.

3. STRUCTURE AND BONDING

Whereas the ionic azides NaN_3, KN_3, etc., are colorless crystalline salts and contain a symmetrical linear N_3^- group (Greenwood and Earnshaw, 1984), the corresponding heavy-metal azides are far less ionic and have more complex structures (Greenwood and Earnshaw, 1984).

The covalent X–N_3 azides (X = H, halogen, NO, etc.) in any case feature an unsymmetrical, slightly bent N_3 group with two significantly different N–N distances (Fig. 1 and Table I; Tornieporth-Oetting and Klapötke, 1993a). In the gas phase they all possess monomer trans-bent structures with C_s geometry (Tornieporth-Oetting and Klapötke, 1993a; Otto *et al.*, 1992).

The covalent XN_3 species feature a N1N2N3 bond angle of about 172°, essentially independent of the nature of X. In the halogen azides the XN1N2 angle increases regularly from FN_3 (104°) to IN_3 (110°), following the trend that substituents with higher electronegativity change the hybridization toward more *p* character, thus yielding smaller bond angles (Otto *et al.*, 1992). In general, the agreement between computed and experimentally observed structural parameters is satisfactory with the exception of the N–I bond distance in solid iodine azide (Fig. 2) (Buzek *et al.*, 1993). This compound, however, possesses a polymeric chain structure with two-coordinated iodine atoms in the polymeric chain (i.e., the iodine atom is a bridging atom). Quite recently, the gas-phase structure of highly

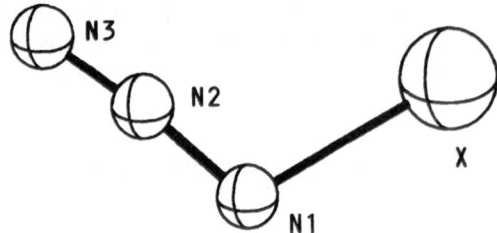

Figure 1. Schematic representation of the geometry of covalent $X-N_3$ molecules (X = H, F, Cl, Br, I, NO, etc.).

explosive iodine azide has been determined by electron diffraction (Hargittai et al., 1994). The experimental data are nicely in agreement with *ab initio* computed structural parameters (Table I).

In agreement with the experimental Raman data for nitrosyl azide, N_4O, the *ab initio* computation also shows a (planar, C_s) chainlike N_4O structure (Schulz et al., 1993). The optimized geometry of N_4O as well as the most likely Lewis representation is shown in Fig. 3. As indicated by natural bond orbital (NBO) analysis, there is a significant interaction of one of the lone pairs on oxygen with the unoccupied, antibonding σ^* orbital of the N1–N4 bond. This LP(O) $\rightarrow$ σ^*(N1–N4) (negative) hyperconjugation (Fig. 4) accounts for the rather long N1–N4 bond of 1.48 Å [cf. N_2F_4, d(N–N) = 1.49 Å]. Moreover, the ON–N_3 bond has a polarity $ON^{\delta+}-^{\delta-}N_3$ (cf. Fig. 3) that resembles the situation of the polarized covalent azides BrN_3 and IN_3. The NO distance of 1.20 Å is 0.07 Å longer than that in NOCl (1.13 Å), corresponding to a bond order (b) slightly less than two (usual values for d(N–O): b = 1, 1.36 Å; b = 2, 1.16 Å).

The cyclic, aromatic 6π isomer of ON–N_3 (isoelectronic to N_5^-; Fig. 5) was computed to be 13 kcal/mol more favorable (!) than the observed chainlike species [MP2(FC)/6-31+G*] (Schulz et al., 1993). The formation of the trans-chain form can easily be explained by the reaction mechanism. The trans-isomer which is formed in the first step could easily convert into the cis-isomer (1 kcal/mol less stable); however, the cyclization requires a high activation energy (49 kcal/mol) because of the necessary bending of the azide unit.

4. ENERGETICS

4.1. Calculations

In recent quantum-mechanical *ab initio* computations, the calculated total energies were used to predict theoretically the bond energies and thermodynamic stabilities of the halogen azides, which are (for obvious reasons) difficult to

Table I. Structural Parameters of Covalent Azides XN_3[a]

Compound	Method (state)[b]	$d(X-N1)$ (Å) Exptl	Calcd	$d(N1-N2)$ (Å) Exptl	Calcd	$d(N2-N3)$ (Å) Exptl	Calcd	<(N1N2N3)(°) Exptl	Calcd	<(XN1N2)(°) Exptl	Calcd
HN_3[c]	MW (g)	1.015	1.018	1.243	1.250	1.134	1.158	171.3	171.2	108.8	109.7
FN_3[d]	MW (g)	1.444	1.431	1.253	1.280	1.132	1.152	170.9	171.7	103.8	103.8
ClN_3[e]	MW (g)	1.745	1.753	1.252	1.265	1.133	1.157	171.9	171.3	108.6	109.3
BrN_3[f]	ED (g)	1.90	1.923	1.23	1.262	1.13	1.160	171	171.4	110	108.5
IN_3[g]	ED (g)	2.14	2.133	1.26	1.258	1.14	1.167	168	171.4	107	110.2
	X (s)	2.30	—	1.28	—	1.09	—	174	—	114	—
$ON-N_3$[h]	—	—	1.484	—	1.270	—	1.154	—	174.3	—	107.1

[a] Ab initio data: MP2/6-31+G* (Otto et al., 1992).
[b] MW, Microwave; ED, electron diffraction; X, X-ray.
[c] Winnewisser, 1980.
[d] Christen et al., 1988.
[e] Cook and Gerry, 1970.
[f] Hargittai et al., 1993.
[g] Buzek et al., 1993; Hargittai et al., 1994.
[h] Schulz et al., 1993.

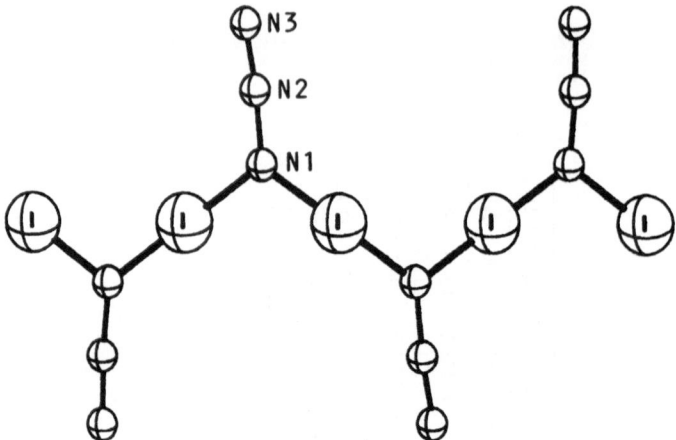

Figure 2. Solid-state structure of polymeric iodine azide.

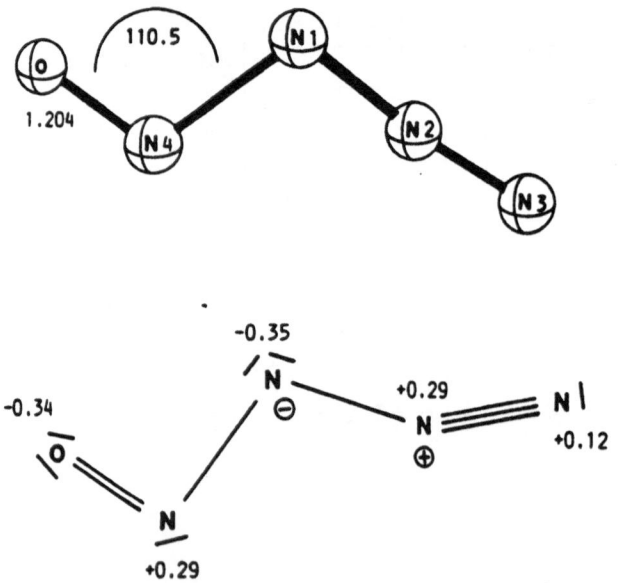

Figure 3. MP2 optimized geometry and Lewis representation with NBO charges for N_4O.

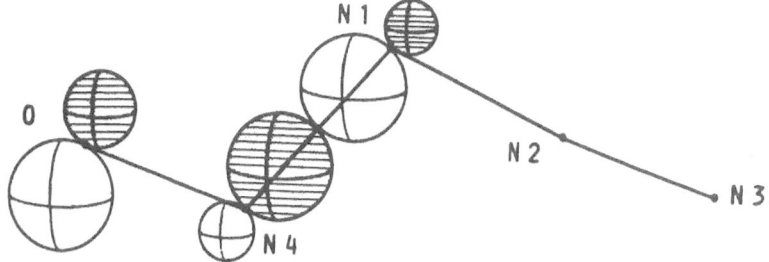

Figure 4. Negative hyperconjugation p-LP(O) $\rightarrow$ σ*(N4–N1) in N_4O.

determine experimentally (Otto *et al.*, 1992). The X–N bond strength was derived from the calculated dissociation enthalpies according to Eq. (7). The calculations predict that the X–N_3 bond strength decreases in the order H $\gg$ F > Cl > Br > I.

$$XN_3 \xrightarrow{D_0} N_3 + X \tag{7}$$

However, thermal fragmentation of XN_3 is not induced by breaking the X–N_3 bond but rather by dissociation into XN and N_2 (Eq. 8), although this process is spin-symmetry forbidden.

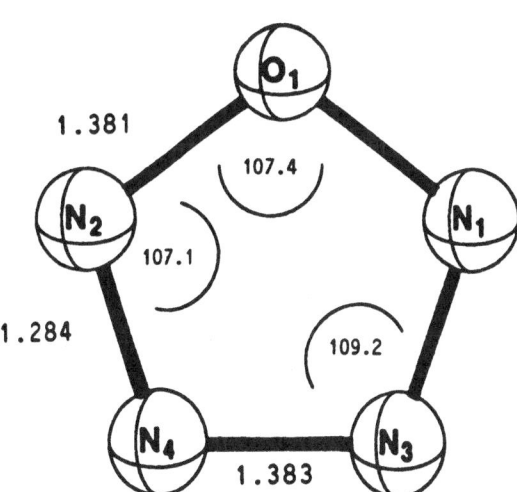

Figure 5. MP2 optimized geometry of the hitherto unknown cyclic isomer of N_4O (distances in angstroms, angles in degrees).

$$XN_3 \xrightarrow{D_0} N_2(^1\Sigma) + XN(^3\Sigma) \tag{8}$$

The reaction enthalpy for dissociation of XN_3 into X_2 and N_2 (Eq. 9) was also computed (Otto *et al.*, 1992).

$$2XN_3(g) \xrightarrow{\Delta H} 3N_2(g) + X_2(g) \tag{9}$$

The calculated reaction enthalpies for the reactions according to Eqs. (7)–(9) are summarized in Table II.

The dissociation according to Eq. (7) is strongly endothermic for all halogen azides. The calculated dissociation energies $D_0(7)$ show that the $X-N_3$ bond strength decreases in the order $F > Cl > Br > I$. As shown experimentally for HN_3, a dissociation corresponding to Eq. (8) can take place although it is spin-symmetry forbidden. These $XN-N_2$ dissociations are predicted to be exothermic for FN_3, ClN_3, and BrN_3 but are predicted to be slightly endothermic for IN_3. The heat of reaction for a decomposition according to Eq. (9) is strongly exothermic for all halogen azides and is consistent with the positive heat of formation of XN_3 from the elements in the gaseous state. The molar heat of formation ΔH°_f for XN_3 shows that IN_3 is thermodynamically the least stable species. Nevertheless, BrN_3 was experimentally shown to be (kinetically) less stable than IN_3 (Tornieporth-Oetting and Klapötke, 1993b). This observation cannot be explained by the reaction enthalpies according to Eqs. (7) and (9) because they are only insignificantly different. However, the reaction enthalpy of the dissociation of BrN_3 according to Eq. (8) is negative (exothermic process) contrary to the analogous reaction enthalpy value for the dissociation of IN_3 into IN and N_2, which is just slightly positive (endothermic). This explains why IN_3 is kinetically marginally more stable than BrN_3. The dissociation of (isolated, gas phase) N_4O into N_2O and N_2 (Eq. 10) was calculated as strongly exothermic (after temperature correction for H via C_p and inclusion of the work term): ΔH°_{298} [Eq. (10), MP2] $= -91$ kcal/mol (Table III; Schulz *et al.*, 1993). It is interesting to note that N_4O is one of the few examples of an intrinsically unstable molecule that has been isolated as an (essentially) pure compound.

$$N_4O(g) \rightarrow N_2O(g) + N_2(g) \tag{10}$$

Table II.   Calculated Reaction Enthalpies for Reactions (7)–(9)[a,b]

X	D_0 (7)	D_0 (8)	ΔH (9)	$0.5 \times \Delta H(9)$	$\Delta H^\circ_f(XN_3)$
F	+58	−25	−165	−82	+82
Cl	+49	−12	−186	−93	+93
Br	+47	−4	−194	−97	+102
I	+46	+7	−192	−96	+104

[a]From calculations at the MP2/6-31+G* level; Otto *et al.*, 1992.
[b]All values are in kcal/mol.

Table III. Absolute Energies E ($-$au) for
N_4O, N_2O, and N_2 at the HF/6-31+G* and
MP2/6-31+G* Levels[a]

	E [HF/6-31+G*]	E [MP2/6-31+G*]
N_4O	292.47954	293.32887
N_2O	183.68528	184.21442
N_2	108.94702	109.26193

[a]Schulz *et al.*, 1993.

4.2. Detonation Characteristics

In principle, the covalent halogen azides (and also nitrosyl azide) possess better detonation characteristics than lead azide (Table IV), and some research has been directed into this field (Tornieporth-Oetting and Klapötke, 1992). However, the binary halogen azides are less kinetically stabilized to far too great an extent to be used either as explosives, high-energetic formulations, or fuels.

On the other hand, whereas $[NMe_4]^+[I(N_3)_2]^-$ is explosive, one salt containing the unstable $I(N_3)_2^-$ anion was found to be kinetically stable: $[PPh_4]^+[I(N_3)_2]^-$ (Müller *et al.*, 1980). Moreover, the compound $[H_2N_3^+][SbF_6^-]$ is (kinetically) stable at room temperature whereas hydrazoic acid tends to explode violently. These results may very well encourage the preparation of further ionic azide salts which represent high-energetic, endothermic compounds but are kinetically stabilized. Especially, the combination of high-energetic cations and anions seems

Table IV.   Detonation Characteristics for IN_3, BrN_3, ONN_3, and $Pb(N_3)_2$

	IN_3(s)	BrN_3(l)	ONN_3(g)	$Pb(N_3)_2$(s)
Explosion heat[a]				
(kcal/mol)	-96.5	-98.2	-110	-106
(kcal/kg)	-571	-805	-1528	-367
Volume of detonation gas (liter/kg)	265	367	778	308
Estimated detonation temperature (K) (isochoric), T_{ex}	6000	6000	4300	6000
Moles of gas per kilogram, n	11.8	16.4	34.7	13.7
Specific energy (m·t/kg), $f = n \cdot R \cdot T_{ex}$[b]	60	83	177	50
Estimated lead test (bulge due to 10 g of explosive) (cm^{-3})	140	230	$\gg 600$	110
ΔH°_f (kcal/mol)	104 (s)	102 (l)	110 (g)	101 (s)
M (g/mol)	168.9	121.9	72.0	291.3

[a]Products: $0.5I_2(g) + 1.5N_2$; $0.5Br_2(g) + 1.5N_2(g)$; $0.5O_2(g) + 2N_2$; $Pb(g) + 3N_2(g)$.
[b]$R = 8.478 \times 10^{-4}$ m·t/K·kg; 1 m·t = 1000 kp = 2.3423 kcal.

to be interesting in terms of high-energetic formulations {e.g., $[H_2N_3]^+[I(N_3)_2]^-$ and $[I(N_3)_2]^+[I(N_3)_2]^-$}. So far, all of the prepared binary cationic N–I species $\{[I_2N_3]^+[SbF_6]^-$ (Tornieporth-Oetting et al., 1992), $[I(N_3)_2]^+[AsF_6]^-$ (Tornieporth-Oetting et al., 1993)\} do not show kinetic stabilization and are as explosive as the parent iodine azide (Tornieporth-Oetting, 1992).

5. CONCLUSIONS

Covalent inorganic azides (HN_3, halogen azides, nitrosyl azide) are well established entities. They are, however, thermodynamically unstable with respect to decomposition into the elements and are all explosive.

In principle, the partial charge is always slightly negative at N for N–X bonds (X = H, Cl, Br, I, NO) and positive for X = F.

In general, structures are as expected from simple valence theory (VSEPR) and the isoelectronic principle. However, some special structural principles should be mentioned:

- Contrary to ionic azides (e.g., NaN_3; N_3^-; D_∞), all covalent species of the XN_3 type (X = H, F, Cl, Br, I, NO) possess a trans-bent C_s configuration with a NNN bond angle of about $172 \pm 3°$ and two different N–N bond distances. The covalent bond order indicates that the XN–NN bond in XN_3 is intermediate between a single and a double bond and that the XNN–N bond has nearly triple bond character.

- Unexpectedly, IN_3 was found to exist in the solid state in a polymeric chain structure with linear, two-coordinated iodine. The solid-state structures of all other halogen azides are still unknown. The structural characterization of these species should be fascinating and may present many mysteries and should therefore be a challenging problem for the future.

Recently, computational work has intensified on XN_3 derivatives, having the advantage of no danger of explosion in the computer. However, only high-level quantum-mechanical *ab initio* computations on correlated levels give a satisfying agreement with the available experimental data. Moreover, for the heavy halogens, especially iodine, relativistic effects have to be taken into consideration, which is, for example, possible by the use of effective quasi-relativistic (core) potentials. In particular, the results of the last two years show the trend that computational chemistry is becoming more and more important in combination with modern inorganic preparative chemistry, especially for thermodynamically and kinetically unstable compounds.

ACKNOWLEDGMENTS. This research was supported by the Deutsche Forschungsgemeinschaft (Kl 636/2-2), the Fonds der Chemischen Industrie, the

German-Hungarian (TU Berlin, TU Budapest) partnership, and the North Atlantic Treaty Organization (CRG 920034).

REFERENCES

Barin, I. (ed.), 1993, *Thermochemical Data of Pure Substances*, VCH Verlagsgesellschaft, Weinheim.

Bauer, S. H., 1947, An electron diffraction investigation of the structure of difluorodiazine, *J. Am. Chem. Soc.* **69**:3104–3108.

Brauer, G., 1975, *Handbuch der präparativen anorganischen Chemie*, 3rd ed., F. Enke, Stuttgart, p. 463.

Buzek, P., Klapötke, T. M., Schleyer, P. v. R., Tornieporth-Oetting, I. C., and White, P. S., 1993, Iodine azide, *Angew. Chem. Int. Ed. Engl.* **32**:275–277.

Christe, K. O., Wilson, W. W., Dixon, D. A., Khan, S. I., Bau, R., Metzenthin, T., and Lu, J., 1993, The aminodiazonium cation $H_2N_3^+$, *J. Am. Chem. Soc.* **115**:1836–1842.

Christen, D., Mack, H. G., Schatte, G., and Willner, H., 1988, Structure of triazadienyl fluoride, FN_3, by microwave, infrared, and *ab initio* methods, *J. Am. Chem. Soc.* **110**:707–712.

Cook, R. L., and Gerry, M. C. L., 1970, Microwave spectrum and structure of chlorine azide, *J. Chem. Phys.* **53**:2525–2528.

Dehnicke, K., 1976, Isolation and infrared spectrum of iodine azide, *Angew. Chem. Int. Ed. Engl.* **15**:553–554.

Dehnicke, K., 1979, The chemistry of iodine azide, *Angew. Chem. Int. Edn. Engl.* **18**:507–514.

Dehnicke, K., 1983, The chemistry of the halogen azides, *Adv. Inorg. Radiochem.* **26**:169–200.

Doyle, M. P., Maciejko, J. J., and Busman, S. C., 1973, Reaction between azide and nitronium ions. Formation and decomposition of nitryl azide, *J. Am. Chem. Soc.* **95**:952–953.

Greenwood, N. N., and Earnshaw, A. 1984, *Chemistry of the Elements*, Pergamon, Oxford.

Hargittai, M., Tornieporth-Oetting, I. C., Klapötke, T. M., Kolonitz, M., and Hargittai, I., 1993, Bromine azide—determination of the molecular structure by electron diffraction in the gas phase, *Angew. Chem. Int. Ed. Engl.* **32**:759–761.

Hargittai, M., Molmar, J., Klapötke, T. M., Tornieporth-Oetting, I. C., Kolonits, M., and Hargittai, I., 1994, *J. Phys. Chem.* **98**:10095–10097.

Jones, K., 1973, Nitrogen, in *Comprehensive Inorganic Chemistry*, Vol. 2 (J. C. Bailar, H. J. Emeléus, R. Nyholm, and A. F. Trotman-Dickenson, eds.), Pergamon, Oxford.

Köhler, J., and Meyer, R., 1991, *Explosivstoffe*, VCH Verlagsgesellschaft, Weinheim, and references therein.

Krakow, B., Lord, R. C., and Neely, G. O., 1968, High resolution far infrared study of rotation in HN_3, HNCO, HNCS, and their deuterium derivatives, *J. Mol. Spectrosc.* **27**:148–176.

Müller, U., Dübgen, R., and Dehnicke, K., 1980, Diazidoiodat(I): Darstellung, IR-Spektrum und Kristallstruktur von $PPh_4[I(N_3)_2]$, *Z. Anorg. Allg. Chem.* **463**:7–13.

Otto, M., Lotz, S. D., and Frenking, G., 1992, Quantum mechanical *ab initio* studies of the structure of halogen azides XN_3 (X = F,Cl,Br,I), *Inorg. Chem.* **31**:3647–3655.

Schulz, A., Tornieporth-Oetting, I. C., and Klapötke, T. M., 1993, Nitrosyl azide, N_4O, an intrinsically unstable oxide of nitrogen, *Angew. Chem. Int. Ed. Engl.* **105**:1610–1612.

Tornieporth-Oetting, I. C., 1992, Neue Halogen—und Koordinationsverbindungen von Elementen der 15. Gruppe, Ph.D. Thesis, Technische Universität Berlin, Berlin.

Tornieporth-Oetting, I. C., and Klapötke, T. M., 1992, Work Report 102392, Technische Universität Berlin, Berlin.

Tornieporth-Oetting, I. C., and Klapötke, T. M., 1993a, Recent developments in the chemistry of binary nitrogen halogen species, *Comments Inorg. Chem.* **15**:137–169.

Tornieporth-Oetting, I. C., and Klapötke, T. M., 1993b, unpublished results. ·

Tornieporth-Oetting, I. C., Buzek, P., Schleyer, P. v. R., and Klapötke, T. M., 1992, Azidoiodo-
iodine(1+) hexafluoroantimonate: The first binary nitrogen–iodine cation, *Angew. Chem. Int. Ed.
Engl.* **31:**1338–1339.

Tornieporth-Oetting, I. C., Klapötke, T. M., Schulz, A., Buzek, P., and Schleyer, P. v. R., 1993, The
$I(N_3)_2^+$ cation—preparation, identification by Raman spectroscopy and *ab initio* quantum mechanical studies, *Inorg. Chem.* **32:**5640–5642.

Winnewisser, B. P., 1980, The substitution structure of hydrazoic acid, HNNN, *J. Mol. Spectrosc.*
82:220–223.

Chapter 4

Chemistry of a Burning Propellant Surface

Thomas B. Brill

1. INTRODUCTION

Combustion of energetic solids is the basis of rocket propulsion for space exploration and military technologies. Accurate models of combustion that contain chemical and fluid-mechanical details are greatly needed because atmospheric contamination and cost considerations limit ground-based testing. International disarmament treaties mandate disposal of stockpiled energetic materials. However, conventional disposal methods, such as open-pit burning and detonation, are increasingly restricted by environmental regulations. Description of the gaseous emission products frequently must be given before combustion is authorized. Manipulation of the combustion process may be necessary. Hence, combustion processes must be understood and predicted with ever greater accuracy.

Of the many chemistry issues that are important during the combustion of energetic solids, two are foremost in our laboratory. One program is focused on determining chemical details of the thin heterophase reaction zone on the surface of a burning energetic material. Identification of the early reaction steps and the first strongly exothermic reaction enables many characteristics of combustion to

THOMAS B. BRILL • Department of Chemistry, University of Delaware, Newark, Delaware 19716.

Combustion Efficiency and Air Quality, edited by István Hargittai and Tamás Vidóczy. Plenum Press, New York, 1995.

be understood and predicted. For instance, this knowledge makes it possible to couple the chemistry and transient characteristics of the near-surface flame zone to the condensed-phase processes. With the aggressive push toward "clean-burning" rocket propellant formulations (e.g., nonaluminized propellants and HCl-free exhaust), chemical details about the surface reaction zone must be incorporated into models that predict the performance and stability of the combustion process.

In a second program the relationship of the molecular structure and composition of the oxidizer or monopropellant to the gaseous products released upon fast thermolysis is being developed for a variety of energetic compounds. Such an understanding helps open the door to rational design of burn rate modifiers and new energetic molecules, whose combustion characteristics are predictable. Several advances in these two thrust areas are overviewed in this chapter.

2. EXPERIMENTAL APPROACHES

The chemical details of the surface reaction zone and near-surface, gas-phase reaction zone cannot be determined spectroscopically with the flame present because the most reactive species are consumed too rapidly. Moreover, the surface reaction zone is spatially very thin, has a steep temperature gradient, contains multiple phases, and may require three dimensions for an accurate description.

An alternate approach to gain the desired chemical description of the burning surface is to simulate the conditions, but in a manner that enables employment of spectroscopic diagnostics. For example, a small, thinly spread mass of sample in contact with a responsive heat source provides a snapshot view of the surface reaction zone. An outline of the chemical sequences is obtained from near-real-time spectroscopic monitoring of the gaseous products released from the surface and simultaneous measurement of the heat flow by a thermocouple or the resistance of the filament. Use of a cool, nonreactive atmosphere in the cell quenches the gaseous products as they form and enables them to be detected. The small amount of sample reduces the chance for ignition or explosion. The two methods of heating are ramp and temperature jump and are illustrated in Fig. 1.

Extensive description of the fast thermolysis/Fourier transform infrared (FTIR) experiments has been given by Oyumi and Brill (1985a), Cronin and Brill (1987), and Brill et al. (1992a). Figure 2 shows the design of the cell. The anti-reflection-coated 1.25-cm-thick $\times$ 2.5-cm diameter ZnSe windows are held in a 7.5-cm diameter aluminum cylinder by brass end caps. ZnSe was used because it is transparent in the mid-IR portion of the spectrum. It also has high tensile strength so that operation is possible at high pressure. The cell was designed to withstand a static internal pressure of 330 atm but is used primarily in the 0.1–60-atm range.

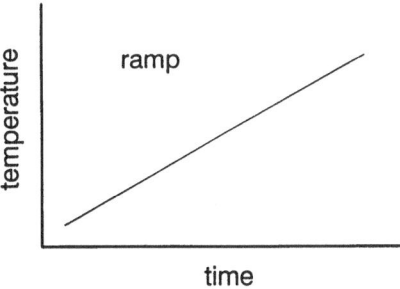

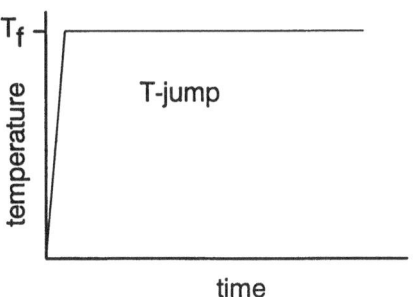

Figure 1. The two modes of fast thermolysis used for simulating surface pyrolysis conditions during combustion of a bulk material.

2.1. Fast Thermolysis by Ramp Heating

For the ramp heating mode in Fig. 1, the heater filament is a slightly creased nichrome IV ribbon (2.5 × 0.6 × 0.012 cm) supported on pressure-tight, electrical feed-through insulators. Although studies have not been conducted on a wide range of compounds, examination of compounds that are expected to be especially sensitive to catalysis by metals revealed little dependence of the thermolysis products on the filament material (Cronin and Brill, 1988). Typically, 1–2 mg of sample (solid, liquid, or a mixture) is heated on the filament by using a constant-voltage–variable-current controller. This method of heating has special value in the ramp-heating experiments whose results are described in Sections 3.2 and 4. In principle, any reasonable heating rate of the sample could be achieved, but 100–400°C/s is used because the spectral collection rate does not separate processes at higher heating rates. A static pressure of argon gas in the cell was set as desired in the 0.1–60-atm range.

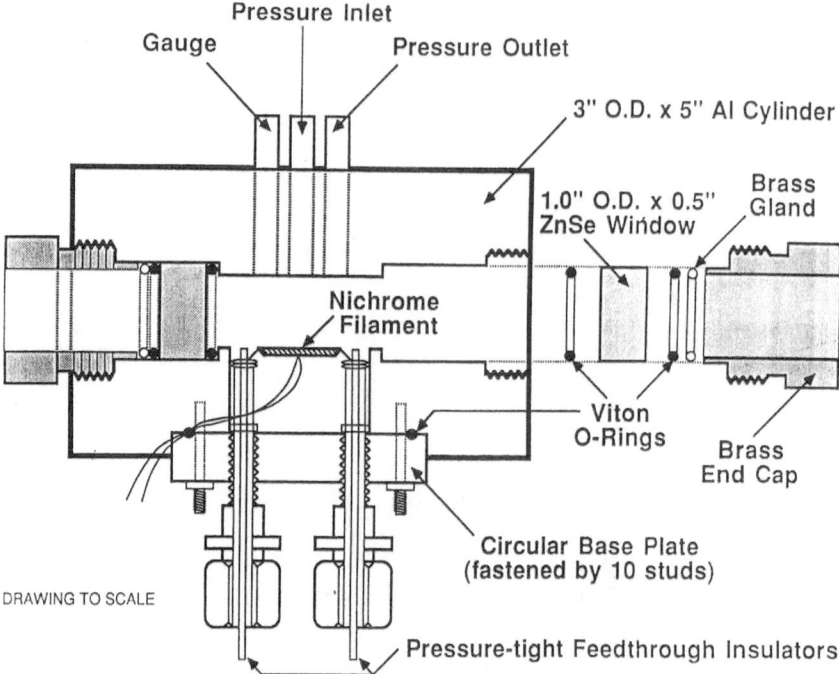

Figure 2. A cross-sectional drawing of the cell used for FTIR spectroscopy determination of the gaseous products from rapid thermolysis of solids and liquids.

2.2. Fast Thermolysis by Temperature Jump

The second mode of heating a sample is by temperature jumping the sample to a constant chosen temperature. T-jump/FTIR spectroscopy (Brill *et al.*, 1992a) permits isothermal decomposition studies to be performed on an approximately 200-μm-thick film of sample at temperatures up to 200°C above the onset of the conventionally accepted thermal decomposition temperature. This condition simulates the conditions that are present in the condensed phase in the heterophase reaction zone at the burning surface. Such studies require heating at a very high rate (e.g., 2000°C/s) to a high final filament temperature, T_f, and then holding at T_f for a period of time while monitoring the near-surface gas products and the heat flow. Since T_f can be set near the burning surface temperature, chemistry relevant to the surface zone during combustion is learned.

In experiments involving fast heating to a high temperature, the efficiency of heat transfer from the heat source to the sample is important, as is the ability of the heater circuit to respond to the endothermic and exothermic events taking place in the sample (Shepherd and Brill, 1993). The time of each thermal change is matched with the spectra to reveal the sequence of events that take place during decomposition at T_f. Because much of the slower "cooking" chemistry is by-passed by rapid heating, the products and their relative concentrations indicate how the various decomposition reaction branches of a bulk material depend on temperature.

The thinness of the filament on which about 200 μg of sample is spread helps optimize the heat transfer and the responsiveness of the filament to temperature changes (Brill et al., 1992a; Shepherd and Brill, 1993). A Pt filament replaces the nichrome filament shown in Fig. 2. A commercial pyroprobe controller (CDS Instruments) gives excellent heating control (Brill et al., 1992a). The heating rate dT/dt and T_f of the Pt filament can be controlled independently. Furthermore, the endothermic and exothermic events can be sensed by observing the control voltage of the heater circuit. A positive deflection in the control voltage represents an endotherm whereas a negative deflection represents an exotherm.

Because of the value of collecting IR spectra at high temporal resolution (50–100 ms), the rapid-scan mode of an FTIR spectrometer (e.g., Nicolet 800, 60SX, or 20SXB) was used for all studies. With the beam focused several millimeters above the filament surface, the IR-active gas products from the fast-heated sample are detected in near real time. No significant change of pressure in the cell occurred due to the evolved gases because of the small sample size. If smoke or an aerosol with particle diameters less than 50 μm forms during thermolysis, wavelength-dependent dispersion causes the baseline to slope downward from short to long wavelength. However, the gas product concentrations still can be calculated in most cases. In all experiments the relative concentrations of the gas products compared to CO_2 were obtained from the effective width factors and absolute intensities of noninterfering absorbances for each product (Brill, 1992).

3. "CLEAN-BURNING" ROCKET PROPELLANTS

Interest is growing in oxidizers that might replace ammonium perchlorate (AP) in solid propellants. Upon combustion, AP liberates HCl and H_2O, forming an environmentally undesirable plume of HCl(aq). Nucleation of H_2O into droplets by HCl contributes to a prominently visible signature. These detracting features have rekindled interest in ammonium nitrate (AN) as an oxidizer. Unfortunately, AN has a low surface temperature and a low burn rate. The decomposition chemistry of AN is largely responsible for the low energy release. For example, two major decomposition reactions of AN, represented by Eqs. (1) and (2), are endothermic and mildly exothermic, respectively.

$$NH_4NO_3 \rightarrow NH_3(g) + HNO_3(g) \tag{1}$$

$$NH_4NO_3 \rightarrow N_2O + 2H_2O \qquad (2)$$

The decomposition of AN can be compared with that of ammonium dinitramide (ADN), $NH_4[N(NO_2)_2]$ (Brill *et al.*, 1993a). Unlike AN, ADN decomposes very rapidly and is potentially a good propellant ingredient. Part of the additional energy release is attributable to the higher heat of formation of ADN [−35 kcal/mol (Swett, 1992)] compared to that of AN (−78 kcal/mol). Beyond this difference, the chemical reactions that cause ADN to decompose very exothermically are not obvious because the gas products from rapid thermolysis of ADN are similar to those of AN. Both compounds liberate HNO_3, NH_3, N_2O, NO_2, NO, H_2O, and N_2, although the mole fractions differ somewhat. As will be described below, the sequential chemistry leading up to the first large exothermic reaction of AN and ADN can be deduced by using T-jump/FTIR spectroscopy.

The fast decomposition sequence of HMX and RDX will also be presented below. HMX and RDX are well established as minimum-smoke, high-energy solid propellant ingredients. The sequential decomposition chemistry of these compounds and the first highly exothermic reaction in the surface zone are revealed by T-jump/FTIR spectroscopy.

3.1. The Chemistry of the Burning Surface Reaction Zone

The use of a sample thinly spread on the Pt foil filament described in Section 2 enables the burning surface to be simulated.

3.1.1. Ammonium Nitrate and Ammonium Dinitramide

Although pure AN will not burn at 1 atm, a sample can be driven by the T-jump method to a temperature that is at or above the measured surface temperature of AN burning at 25 atm (300–350°C) (Whittaker and Barham, 1964). Figure 3 shows the gas products and thermal response of a 200-μg mass film of AN heated at 2000°C/s to 383°C and held at 383°C. The concentration data in this plot are based on the scaled growth of the IR absorbance values for each product. The superposition of several stoichiometric reactions is indicated. Of course, many elementary steps are embedded in each of these stoichiometric reactions, but they are not determinable by T-jump/FTIR spectroscopy.

The first event is rapid endothermic melting of AN, as indicated by the upward deflection in the difference control voltage trace. The control voltage decreases upon completion of melting during the initial 0.5–1 s. The process turns markedly endothermic again at about 1 s. This second endothermic event corresponds to the appearance and growth of AN aerosol. AN aerosol forms from the endothermic dissociation of AN and desorption to $HNO_3(g) + NH_3(g)$, followed by recombination of NH_3 and HNO_3 in the gas phase (Table I, reaction A). Only

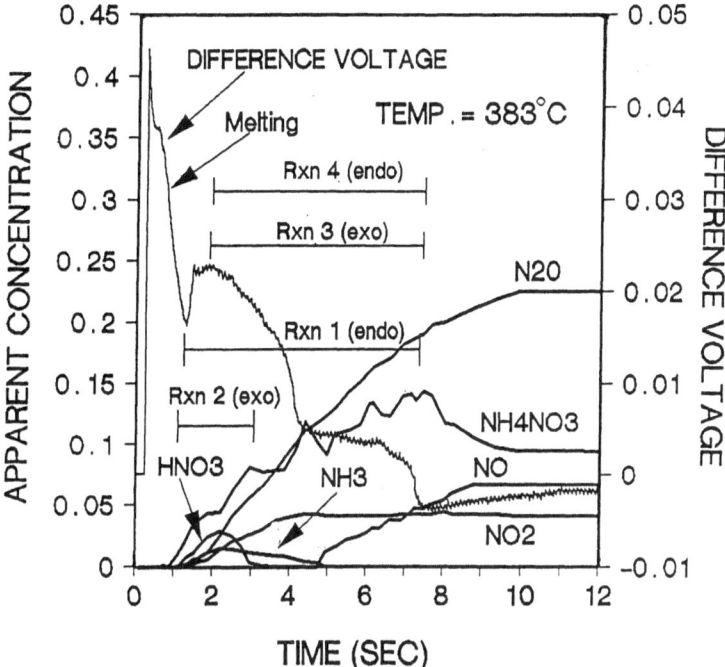

Figure 3. Gaseous products and heat changes of NH_4NO_3 (AN) at 383°C under 1 atm of Ar. The reactions are given in Scheme 1.

the endothermic first step of reaction A is included in the value of ΔH given for reaction A because the second step occurs in the cooler region of the cell away from the filament. Hence, the exothermic second step is not sensed by the filament. However, a white smoke of AN aerosol is visually observed. Despite the continuation of reaction A throughout the decomposition process as evidenced by the growth of the AN aerosol concentration, the decomposition process becomes less endothermic again at about 2 s. H_2O (not quantified) and excess HNO_3 form at this time, which is consistent with the occurrence of reaction B (Table I). This reaction is known, and its enthalpy has been deduced (Federoff, 1960). It is exothermic and would reduce the overall endothermicity of the decomposition process, as is found.

The process becomes still less endothermic from 2 to 4 s as the amount of HNO_3 diminishes. However, N_2O grows rapidly in concentration through this time, suggesting that the exothermic reaction C plays an increasingly important role. However, there is evidence of yet another reaction that occurs in parallel as indicated by the appearance of NO_2 and the eventual decrease in exothermicity again between 4 and 6 s. Also, NO, whose IR absorbance is very small, probably

Table I.   Proposed Reactions That Account for the Products of High-Temperature Decomposition of NH_4NO_3 (AN) in Fig. 3

		Approx ΔH (kcal)
A.	$4[NH_4NO_3(l) \rightarrow HNO_3(g)+NH_3(g) \rightarrow NH_4NO_3(\text{solid aerosol})]$	4 (44)[a]
B.	$3[5NH_4NO_3(1) \rightarrow 2HNO_3 + 4N_2 + 9H_2O]$	3 (−35)
C.	$5[NH_4NO_3(1) \rightarrow N_2O + 2H_2O]$	5 (−13)
D.	$4NH_4NO_3(1) \rightarrow 2NH_3 + 3NO_2 + NO + N_2 + 5H_2O$	81
A–D.[b]	$28NH_4NNO_3(1) \rightarrow 6HNO_3 + 3NO_2 + NO + 2NH_3 + 5N_2O + 13N_2 +$ $42H_2O + 4NH_4NO_3(\text{aerosol})$	87

[a]ΔH for the desorption step only (see text).
[b]Gives the approximate IR-active gas product ratios at 2 s for AN at 383°C (Fig. 3).

forms earlier than is indicated in Fig. 3. Reaction D (Kaiser, 1935) accounts for these observations. Its endothermicity is superimposed on the exothermicity of reaction C and results in a leveling of the control voltage trace (heat flow is balanced) at 4–7 s. Reaction D is also a source of NH_3, which appears as a product for a much longer time than does HNO_3.

The multiplicative factors of the reactions in Table I were determined by the need to match the approximate relative concentrations of the gas products at a time when all of the reactions contribute. The concentrations at 2 s were chosen. The stoichiometry of the net reaction in Table I approximates that found at 2 s in Fig. 3. The enthalpy of the net reaction is slightly endothermic as written.

The formation of NH_3 and NO_2 by reaction D raises the possibility that the process could become exothermic when confined by pressure. The reaction of NH_3 and NO_2 becomes rapid and exothermic in the 330–530°C range (Rosser and Wise, 1956; Bedford and Thomas, 1972). However, significant generation of heat requires confinement to enhance the concentration of NH_3 and NO_2 in the hot zone around the condensed phase.

Figure 4 shows the decomposition process of a 200-μg film of AN heated at 2000°C/s to 415°C under 33 atm of Ar. The concentrations are shown in relative percentages throughout so that the behavior early in the decomposition process can be clearly seen. The melting endotherm initially dominates. The heats of reactions A–D leading to the formation of AN aerosol, N_2O, HNO_3, NH_3, and NO_2 are overall endothermic until 1.5 s. At this time the concentrations of NH_3 and NO_2 formed by reaction D drop markedly, and this drop is accompanied by an exotherm which suggests that the following reaction occurs:

$$2NH_3 + 2NO_2 \rightarrow N_2O + N_2 + 3H_2O \qquad (3)$$

ΔH is about −148 kcal for this reaction as written. Under pressure, this nominally gas-phase reaction could occur in the heterogeneous gas-condensed phase (e.g.,

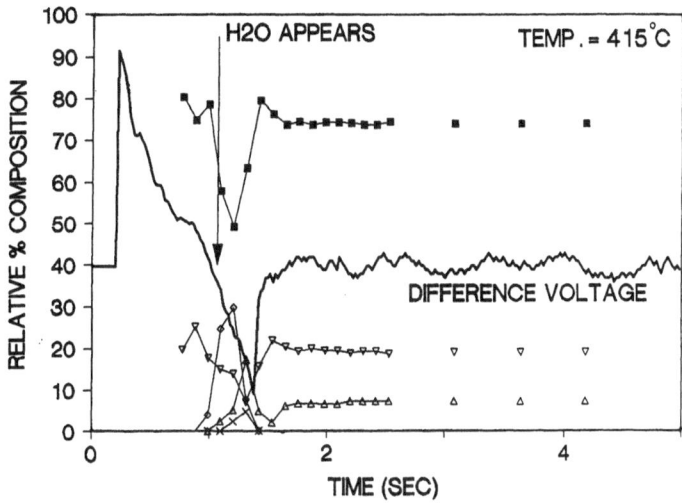

Figure 4. Gaseous products and heat changes of NH$_4$NO$_3$ (AN) at 415°C under 33 atm of Ar. $\triangle$, NO$_2$; $\lozenge$, HNO$_3$; $\times$, NH$_3$; $\blacksquare$, N$_2$O; $\triangledown$, NH$_4$NO$_3$.

bubbles and voids) and contribute to the condensed-phase heat balance under combustion conditions.

The thermal decomposition behavior of bulk ADN is very different from that of AN despite the fact that similar gas products are formed upon rapid decomposition. Figure 5 shows T-jump/FTIR data for a 200-µg film of ADN heated at 2000°C/s to 260°C. This temperature compares with a preliminary surface temperature measurement of burning ADN of about 300°C (Fetherolf and Litzinger, 1992), which is surprisingly similar to that of AN. At the onset of decomposition, gas products form and, in contrast to the findings for AN, sharp exothermicity occurs instantly. The first detected products are mostly HNO$_3$, NH$_3$, and N$_2$O in roughly similar amounts. Minor quantities of NO$_2$, AN, and H$_2$O are also present in the initial spectrum.

The formation of HNO$_3$, NH$_3$, and N$_2$O in comparable amounts at the beginning suggests the presence of branch A in Table II. This mildly endothermic reaction may have a role during slow decomposition at lower temperatures. It appears to be a minor branch during rapid heating, especially because it does not account for the major heat release that is experimentally observed.

Branch B of Table II is proposed to dominate under rapid thermolysis conditions. Reaction a of branch B is dissociation of ADN to produce NH$_3$ and HN(NO$_2$)$_2$. HN(NO$_2$)$_2$ is not detected and probably homolyzes in the condensed phase at high temperature by reaction b to NO$_2$ and HNNO$_2$. Reactions a and b are endothermic. Because relatively large quantities of NH$_3$ and NO$_2$ occur early in branch B, much heat can be generated by reaction k in the gas phase near the

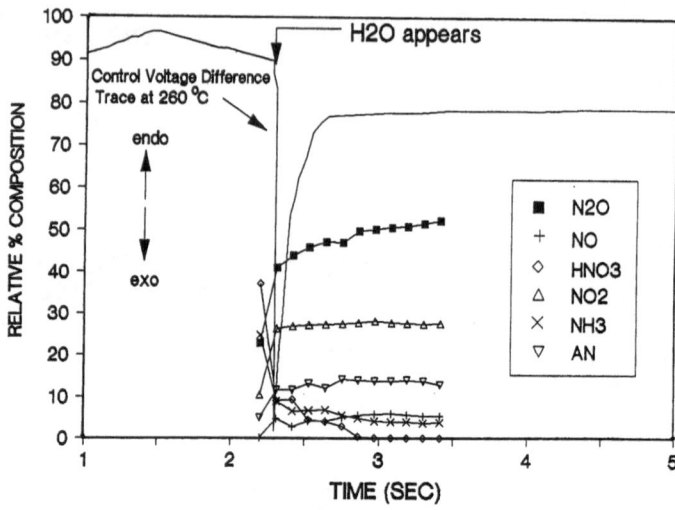

Figure 5. Gaseous products and heat changes of $NH_4[N(NO_2)_2]$ (ADN) at 260°C under 1 atm of Ar.

surface or even as part of the heterogeneous gas–liquid zone at the surface. The high exothermicity is evident in the large control voltage deflection at 2.3 s and provides the energy to complete the decomposition process very rapidly. Some NH_3 and NO_2 remain unreacted because they escape to the cooler atmosphere. Reactions c–h are plausible subsequent steps for decomposition of $HNNO_2$, but they are not determined by T-jump/FTIR spectroscopy. They are simply proposed as reasonable sources of stable products in the quantities detected. The net reaction i of branch B is mildly exothermic. Combining branches A and B yields the exothermic reaction j. Adding some gas-phase recombination of NH_3 and HNO_3 (reaction m) to account for the observed AN solid aerosol yields reaction n, whose stoichiometry approximates the experimentally observed gas product ratios observed at 2.5 s in Fig. 5. Reaction n is strongly exothermic largely because of reaction k, which is the reaction of NH_3 with NO_2.

For both AN and ADN, the exothermic NH_3 + NO_2 reaction appears to dominate the heat release stage. In the case of AN, the exotherm occurs only under a large applied pressure and is accompanied by a drop in the amounts of NH_3 and NO_2 that appear in the gas phase. Although the reaction of NH_3 with NO_2 appears to be responsible for this exotherm, the amount of NH_3 and NO_2 is smaller for AN than for ADN, and, therefore, much less heat is generated.

The rapid decomposition process of ADN is strongly exothermic early in the reaction scheme. This behavior is consistent with the ease of formation of a large amount of NH_3 and NO_2 in the early decomposition steps. Because the reaction of

Table II. Proposed Reactions Responsible for the Gases Released by $NH_4[N(NO_2)_2]$ (ADN) during High-Rate Pyrolysis

	Approx ΔH (kcal)
Branch A[a]	
$3\{NH_4[N(NO_2)_2] \rightarrow NH_3 + HNO_3 + N_2O\}$	$3(+11.5)$
Branch B	
a. $9\{NH_4[N(NO_2)_2] \rightarrow NH_3 + HN(NO_2)_2\}$	
b. $9[HN(NO_2)_2 \rightarrow NO_2 + HNNO_2]$	
c. $6[HNNO_2 \rightarrow N_2O + OH]$	
d. $2[HNNO_2 + OH \rightarrow 2NO + H_2O]$	
e. $HNNO_2 + NO \rightarrow NO_2 + HNNO$	
f. $HNNO + OH \rightarrow N_2O + H_2O$	
g. $3[NH_3 + OH \rightarrow H_2O + NH_2]$	
h. $3[NH_2 + NO \rightarrow N_2 + H_2O]$	
i.[b] $9NH_4[N(NO_2)_2] \rightarrow 6NH_3 + 7N_2O + 10NO_2 + 9H_2O + 3N_2$	-49
j.[c] $12NH_4[N(NO_2)_2] \rightarrow 9NH_3 + 10N_2O + 10NO_2 + 9H_2O + 3N_2 + 3HNO_3$	-14
k. $4NH_3 + 4NO_2 \rightarrow 3N_2 + 2NO + 6H_2O$	-309
l.[d] $12NH_4[N(NO_2)_2] \rightarrow 5NH_3 + 10N_2O + 6NO_2 + 15H_2O + 2NO + 6N_2 +$ $3HNO_3$	-323
m.[e] $2NH_3 + 2HNO_3 \rightarrow 2NH_4NO_3$(aerosol)	
n.[f] $12NH_4[N(NO_2)_2] \rightarrow 3NH_3 + 10N_2O + 6NO_2 + 15H_2O + 2NO + 6N_2 +$ $HNO_3 + 2NH_4NO_3$	-323

[a]Assumes ΔH_f(ADN) $= -35$ kcal/mol, and $\Delta H[HNO_3(g)] = -33$ kcal/mol.
[b]Sum of reactions a–h.
[c]Sum of branches A and B.
[d]Sum of reactions j and k.
[e]Occurs in the gas phase away from surface so reaction m is not included in ΔH.
[f]Sum of l and m gives the approximate gas-phase stoichiometry at the end of the exotherm (Fig. 5).

NH_3 and NO_2 can dominate early and produces a large amount of heat, the overall decomposition and gasification process occurs at a much high rate for ADN than AN. Therefore, for both AN and ADN the reaction of NH_3 with NO_2 is implicated as the main source of heat when the pure material is decomposed at high temperature. For AN, this reaction only becomes important under confinement, such as by the application of pressure.

3.1.2. Nitramines

Nitramines, such as HMX and RDX (Scheme 1), are especially important as minimum-smoke, high-energy explosive and propellant ingredients. The decom-

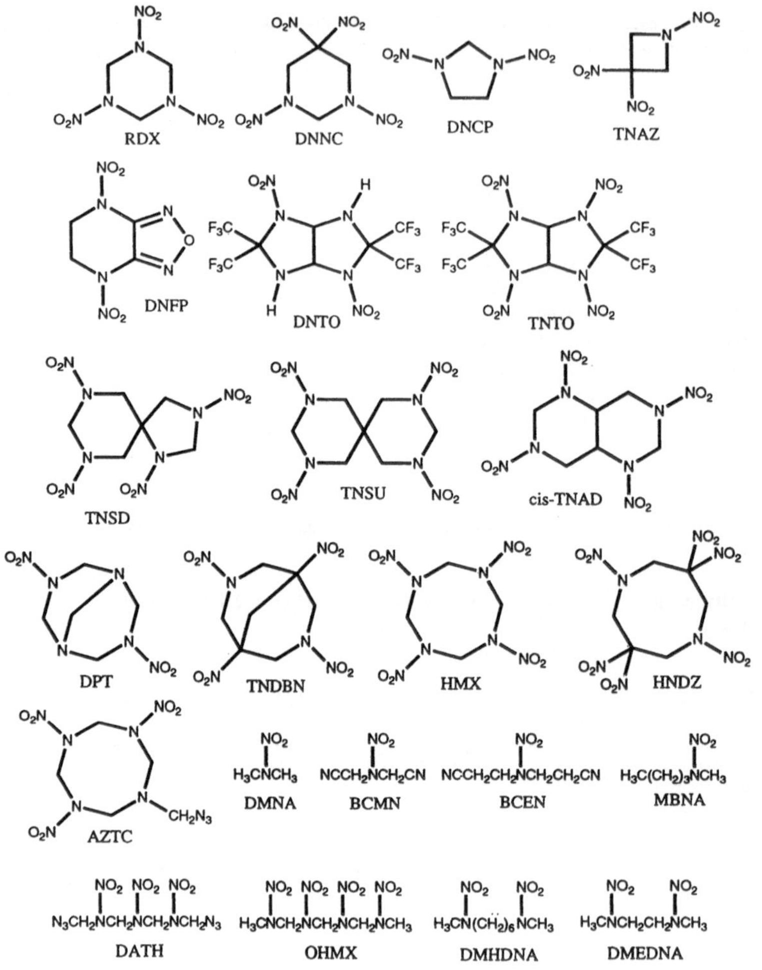

Scheme 1

position processes of HMX (Brill and Brush, 1992) and RDX (Brill *et al.*, 1992b) are similar, and so only HMX will be described.

Figure 6 shows T-jump/FTIR data for a thin film of HMX heated at 2000°C/s to 298°C and then held isothermally while IR spectra of the near-surface gas products are recorded. The global decomposition branches represented by the following reactions occur for bulk RDX and HMX:

$$HMX \rightarrow 4(NO_2 + HCN + H) \tag{4}$$

$$HMX \rightarrow 4(N_2O + CH_2O) \tag{5}$$

Reactions (4) and (5) imply that N_2O and NO_2 should form simultaneously with CH_2O and HCN. This is not found at any temperature studied. Rather, N_2O and

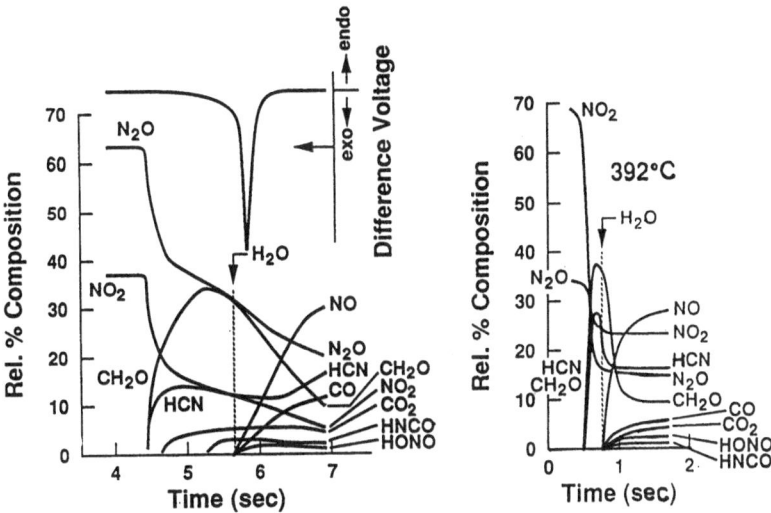

Figure 6. Gaseous products and heat changes of HMX at 298°C (left) and products at 392°C (right) under 2.5 atm of Ar.

NO_2 appear before CH_2O and HCN, which form from the residue left by elimination of N_2O and NO_2. This residue is a mixture of products like hydroxy-methylformamide and acetamide (Cosgrove and Owen, 1974; Kimura and Kubota, 1980; Karpowicz and Brill, 1984; Behrens, 1990), which decompose and delay the release of CH_2O, HCN, and HNCO (Behrens, 1990; Palopoli and Brill, 1991).

The growth of the total IR absorbance of the products accelerates between 4.0 and 5.5 s despite the constant heat flow from the filament, which implies that autocatalysis occurs in this stage of decomposition of HMX and RDX. Moreover, the control voltage trace in Fig. 6 reveals only mild exothermicity between 4 and 5.5 s when reactions (4) and (5) dominate. Thus, these reactions release little energy in the condensed phase.

A sharp exotherm develops at 5.5 s. The following secondary reaction appears to be responsible because CH_2O and NO_2 are consumed as NO, CO, and H_2O appear:

$$5CH_2O + 7NO_2 \rightarrow 7NO + 3CO + 2CO_2 + 5H_2O \qquad (6)$$

Figure 6 shows that more NO than CO forms, in accordance with reaction (6). Reaction (6) is highly exothermic as written ($\Delta H = -320$ kcal), and, by the large exotherm in Fig. 6, it is the main source of heat in the heterogeneous condensed-phase and near-surface gas-phase reaction zone. These conclusions also apply at 392°C (Fig. 6) except that the time scale is compressed. Hence, this description of the decomposition of HMX at 298°C applies as well at the surface reaction zone temperature (350–400°C) during combustion. However, the branching ratio of reactions (4) and (5) depends on temperature and favors reaction (4) at higher temperature (Brill and Brush, 1992).

3.2. Structure/Decomposition Relationships among Nitramines

As noted in Section 3.1.2, nitramines are an important class of compounds which produce high energy and minimum smoke on combustion. The decomposition products NO_2, NO, HONO, and N_2O can function as oxidizers in the heterophase surface zone and the primary and secondary flame zone, while CH_2O and HCN are fuels in the surface and the flame zone. However, the relative amounts of these fuel and oxidizer molecules depend on the structure and composition of the parent molecule. This relationship has been developed from fast thermolysis/FTIR spectroscopy in which ramp heating is employed. The relative concentrations of the initially detected gas products are useful to achieve this description.

3.2.1. Formation of NO_2

The formation of NO_2 by reaction (4) is important for nitramine combustion because NO_2 is an oxidizer in the primary (near surface) flame zone. It seems plausible that the length of the N–N bond in secondary nitramines might be an important factor in the tendency of the N–N bond to homolyze and liberate NO_2. It is necessary to draw these conclusions from rapidly heated samples and to detect the species in rapidly recorded FTIR spectra (e.g., at 50–100-ms intervals). Otherwise, secondary reactions of NO_2 obliterate the relationship of the NO_2 concentration to the structure of the parent molecule. However, even simple homolysis of the N–N bond in the condensed phase is, overall, a complicated process. Intermolecular activity possibly occurs. At the very least, some of the NO_2 must diffuse through and desorb from the heterogeneous environment before it is detected. The most severe complication would be if N–N bond homolysis followed or competed with another decomposition route, such that the initial N–N bond distance is no longer the controlling factor. If this were the case, then structure–decomposition relationships involving NO_2 formation would be disguised.

The crystal structures of nitramines are needed for this comparison. Fortunately, many are available. In Fig. 7 (Brill and Oyumi, 1986a; Oyumi et al., 1987a) the average N–N bond distance for a series of secondary nitramines, R_2NNO_2, is plotted versus the average asymmetric $-NO_2$ stretching frequency from the IR spectrum. The compound identities are given in Scheme 1. A reasonably good relationship exists between these two parameters, suggesting that

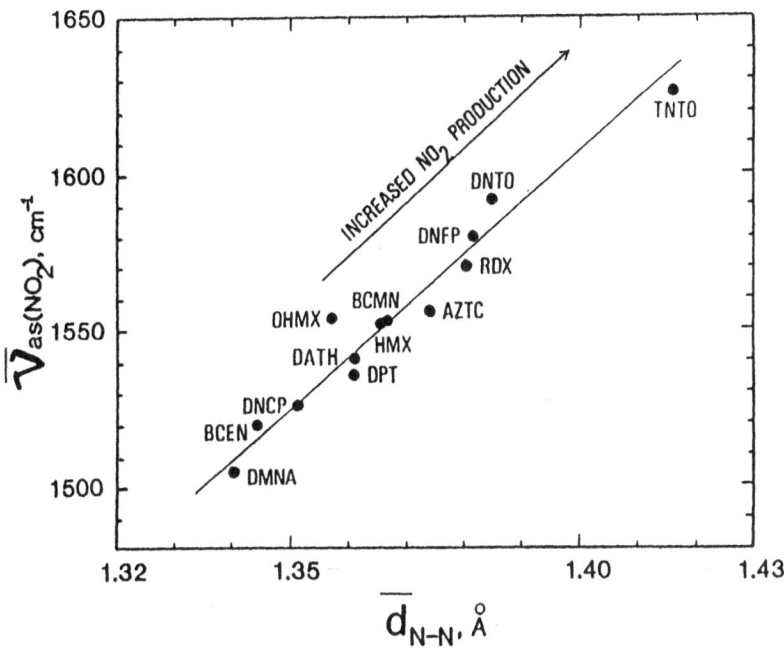

Figure 7. The average asymmetric $-NO_2$ stretching frequency plotted against the average $N-NO_2$ bond distance for selected secondary nitramines. The compounds on the right release a large amount of $NO_2(g)$ upon fast thermolysis.

the force constant of $-NO_2$ stretching greatly depends on the amount of electron density that the $-NO_2$ group shares with the adjacent N–N bond. From fast thermolysis of these compounds, it is found that $NO_2(g)$ is the dominant, initially detected product for compounds toward the right side of this plot. NO_2 was either not detected or was detected as a less abundant product from compounds on the left side. Thus, long N–N bonds favor N–N scission during fast thermolysis. With reasonable confidence, one can predict the amount of NO_2 that is likely to be generated by fast thermolysis of a secondary nitramine from its IR spectrum or its crystal structure.

Although several exceptions exist to the structure/reactivity relationship for $NO_2(g)$ generation (Oyumi and Brill, 1988), it is encouraging that most secondary nitramines thermolyze in the condensed phase in a systematically predictable way. Thermal reactions in the neat condensed phase are complex, but the correlation above gives hope for uncovering and refining patterns that can be applied in practice to propellant ignition and explosives initiation.

3.2.2. Formation of HONO(g)

Closely related to NO_2 formation is the formation of HONO. At least one additional process, that of H· transfer, is required before HONO(g) is detected from nitramine decomposition. Kinetic modeling indicates that HONO formation plays a role in the N–N bond fission process (Melius and Binkley, 1986). The formation of HONO has been invoked in many previous studies of nitramines to rationalize the formation of other products, but HONO itself was very rarely detected prior to fast thermolysis/FTIR spectroscopy studies (Brill and Oyumi, 1986b). HONO is a reactive and, thus, transient molecule that is not observed without rapid heating and near-real-time product detection.

After examination of a large number of nitramines heated at 145–180°C/s under 1 atm of Ar, it was discovered that the initial relative percent composition of HONO(g) depends strongly on the composition of the parent secondary nitramine (Brill and Oyumi, 1986b). In one instance, HONO represented nearly 60% of the initially detected gaseous products. A qualitative relationship was discovered between the ratio of the number of H atoms to NO_2 groups in the parent molecule and the initial percentage of HONO in the gas products. The relationship is shown in Fig. 8. The leveling off at the higher H/NO_2 values is attributable to dilution of the HONO by other gases. Qualitative insight into the processes by which

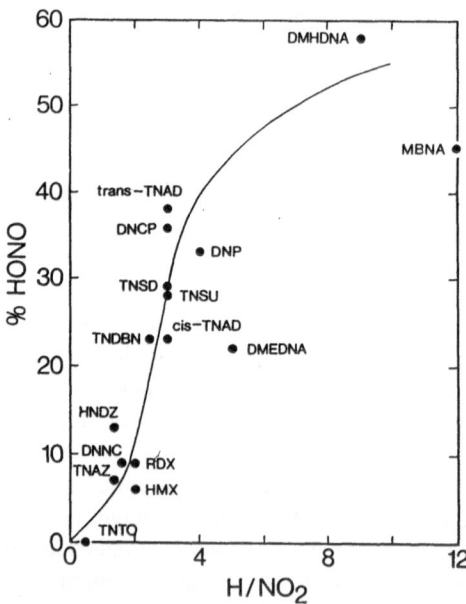

Figure 8. The percentage of HONO in the quantified gaseous products after thermolysis of secondary nitramines compared to the H/NO_2 ratio in the parent molecule.

HONO is formed from the condensed phase is contained in the pattern of Fig. 8. The concentration of HONO must depend on the adventitious encounter of H· and NO_2· (or an NO_2· source) in the condensed phase. Statistically increasing this chance for contact (the H/NO_2 ratio) enhances the HONO concentration. Determination of the molecularity, the nature of the transition state, and any other aspects of the reaction mechanism is beyond extension of these condensed-phase results. Many species could liberate H·, but it is interesting to note that C–H bond fission has been proposed to be rate determining in nitramine decomposition (Shackelford et al., 1985; Bulusu et al., 1986) and combustion (Shackelford, 1987). H_2O assists in HONO formation in the condensed phase (Melius et al., 1991), but the pattern in Fig. 8 would still need to be accommodated if this were the case. Compounds producing the largest amount of HONO are not the ones expected to produce the largest amount of H_2O because of their low oxygen content. Also, the four- and five-center concerted reactions, whose connectivities are shown in Scheme 2 and which could be important in the gas phase (Shaw and

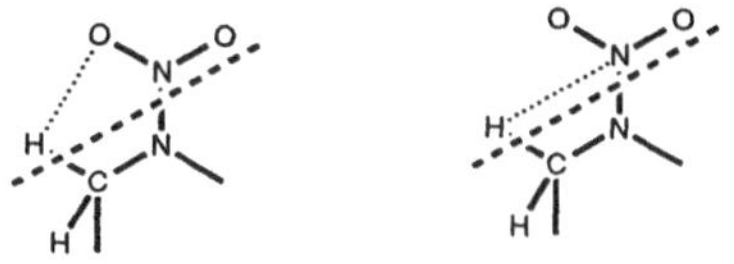

5 - center   4 - center

Scheme 2

Walker, 1977), do not appear to dominate in the condensed phase. If they did, only a $-CH_2-$ fragment adjacent to the nitramine would be required to produce HONO. All nitramines having this linkage should produce the same amount of HONO, which they do not.

The most plausible explanation for HONO formation in the condensed phase remains the chance contact between H· and NO_2·, possibly with H_2O as a catalyst. In keeping with this notion, steric crowding of H and NO_2 groups in a molecule enhances the initial HONO concentration (Brill and Oyumi, 1986b). It is interesting to note that if the H/NO_2 ratio is calculated using only the H atoms from secondary carbon atoms ($-CH_2-$), the fit is slightly more linear, as shown in Fig. 9. This suggests that $-CH_2-$ groups in these molecules may be somewhat more efficient H· sources than $-CH_3$ groups. Nitramines having only tertiary carbon atoms ($\equiv CH$) do not produce HONO (Brill and Oyumi, 1986a; Oyumi et al., 1987a).

3.2.3. Formation of $N_2O(g)$

As noted in Section 3.1.2, the relative role of reactions (4) and (5) is temperature dependent for a given compound. A few secondary nitramines are found to liberate a large quantity of N_2O compared to NO_2 or HONO (Brill et al.,

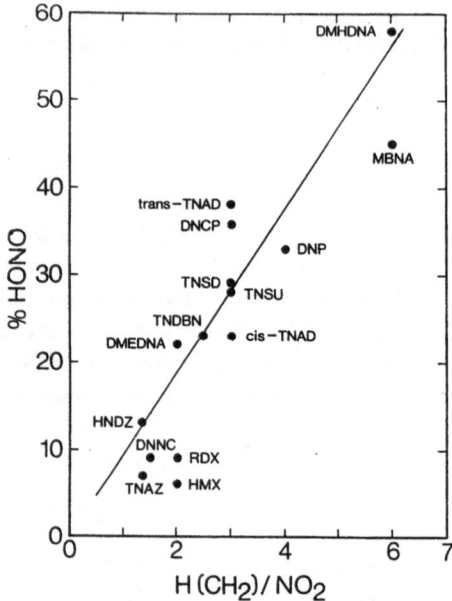

Figure 9. The plot equivalent to Fig. 8 but using only H from the $-CH_2-$ groups in the H/NO$_2$ ratio.

1984; Oyumi and Brill, 1985a; Oyumi *et al.*, 1986a, 1987b,c). Conversely, other nitramines generate N_2O as only a few percent of the total gaseous products (Brill and Oyumi, 1986b; Oyumi and Brill, 1985b; Oyumi *et al.*, 1985, 1986b), while still others produce no N_2O at all (Oyumi *et al.*, 1987a; Oyumi and Brill, 1988).

All of the compounds liberating N_2O have in common a $-CH_2-$ group straddled by two nitrogen atoms, i.e., $-(NO_2)NCH_2N(NO_2)-$. This feature appears to be the only structural requirement for N_2O formation by a secondary nitramine.

Three secondary nitramines that liberate virtually all of their nitramine nitrogen atoms as N_2O are AZTC (Brill *et al.*, 1984), DPT (Oyumi *et al.*, 1986a), and DATH (Oyumi *et al.*, 1987b). The N–N bond distances in these compounds (1.355–1.375 Å) suggest that the abundance of gaseous NO_2 and N_2O should be competitive based on the observations in Section 3.2.1. NO_2 and N_2O appear together at temperatures slightly above the decomposition point of DATH, but N_2O dominates around the decomposition point. The observation that these molecules produce little or no NO_2 at the decomposition temperature emphasizes that the balance of the multiple decomposition pathways is easily shifted in nitramines by the temperature and structure. Both DATH and AZTC produce HN_3 (and no doubt N_2) initially. Thus, the azide group begins the thermolysis reaction. By either radical or electron-pair migration, the backbone tends toward depolymerization in preference to N–N bond homolysis. During depolymerization, oxygen must transfer, but the data do not indicate how. Possibly the same type of reaction in DPT is initiated by the homolysis of a C–N bond in the $-CH_2-$ group bridging the non-nitrated amine groups.

3.2.4. Formation of $CH_2O(g)$

Formaldehyde is an important fuel in the primary flame of nitramines. As we have just noted that a $-CH_2-$ group straddled by two nitramine groups appears to be important, if not required, to produce N_2O from a pure secondary nitramine compound, a logical implication is that transfer of an oxygen atom from the NNO_2 group, perhaps to the adjacent $-CH_2-$ group, liberates CH_2O along with N_2O. $H\cdot$, H^+, $OH\cdot$, or H_2O could assist in lowering the barrier of this oxygen transfer reaction. While DPT, AZTC, and DATH do indeed have this unit and liberate a large amount of CH_2O along with N_2O, several compounds do not follow this simple pattern with respect to CH_2O formation. DNCP and most bicyclonitramines (Brill and Oyumi, 1986b) have the $-(NO_2)NCH_2N(NO_2)-$ unit but liberate no CH_2O. Reasonable explanations exist. DNCP has an unusually thermally stable five-membered ring (Oyumi and Brill, 1988), so that $N-NO_2$ homolysis is the preferred decomposition route. The bicyclonitramines are strong HONO generators (Brill and Oyumi, 1986b). Loss of NO_2 and the removal of H atoms to form HONO could interfere with the tendency to form CH_2O.

The notion that CH_2O and N_2O form simultaneously during the condensed-phase decomposition of nitramines should be discarded. As illustrated in Section 3.1.2, in experiments using T-jump/FTIR spectroscopy and simultaneous thermogravimetry coupled with modulated beam mass spectrometry (Behrens, 1987), N_2O [and NO_2 (Brill and Brush, 1992)] was found to reach the gas phase distinctly ahead of CH_2O and HCN. CH_2O is retained in the condensed phase as part of a nonvolatile residue that decomposes more slowly than the parent nitramines decompose.

4. MODIFICATION OF BURNING RATES

The difficulty of accelerating the burning rate of RDX and HMX by the use of additives and catalysts has stimulated much research and discussion (Schwartz *et al.*, 1984). In principle, potential burn rate modifiers could be intelligently identified with knowledge of (1) the rate-determining decomposition reactions in the condensed and the gas phase and (2) the dominant heat-generating reactions in the near-surface gas phase and surface regions. From the results described in Section 3.1.2, it is easy to reconcile why the burn rates of nitramine-containing propellants are difficult to modify. A dominating early heat generation reaction of many nitramines is experimentally revealed to be reaction (6) (Brill and Brush, 1992; Brill *et al.*, 1992b, 1993b). Note, however, that the reactants CH_2O and NO_2 of reaction (6) are themselves products of the two separate decomposition branches of RDX and HMX, reactions (4) and (5). If either reaction (4) or (5) were accelerated at the expense of the other, then the burn rate of the nitramine would be unlikely to increase significantly. Both reactions must be about equally prevalent to provide the reactants for reaction (6). Therefore, strategies for greatly accelerating the burning rate of nitramines must either completely alter the decomposition mechanism or be based on additives that accelerate the rate of both reactions (4) and (5) without altering the branching ratio. For these reasons, it is not surprising that acceleration of the burning rate of a nitramine propellant is difficult.

Burn rate modification research in our laboratory has emphasized *suppression* of fast-burning propellants, such as ammonium perchlorate-based composite propellants. In principle, the approach is applicable to any solid rocket propellant. The concept is to form a thin layer of a relatively thermally stable material on the surface of the propellant during combustion. This notion evolved from studies of the mechanism (Stoner and Brill, 1991) by which DAF and DAG (Scheme 3)

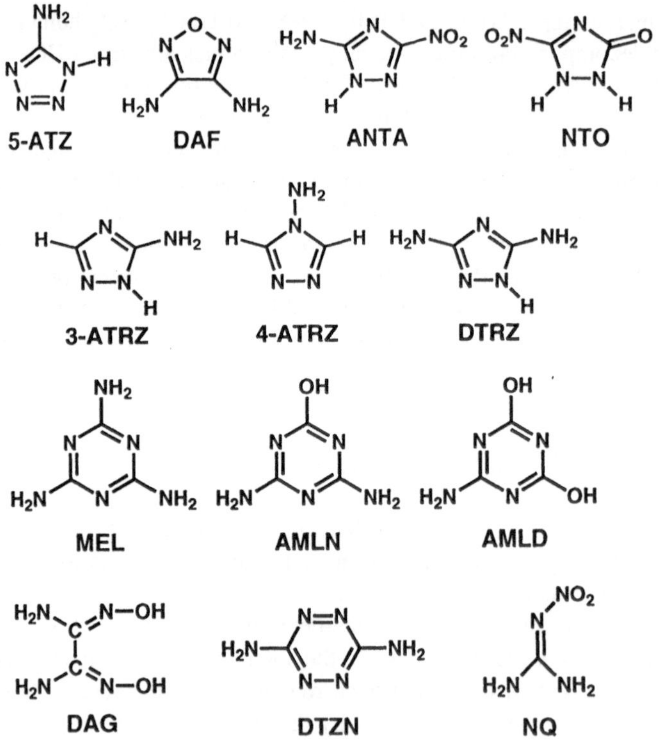

Scheme 3

suppress the burn rate of ammonium perchlorate-based composite propellants (Willer *et al.*, 1991). Upon thermal decomposition, DAG and DAF form high-molecular-weight, thermally stable, cyclic azines in addition to low-molecular-weight gaseous products (Stoner and Brill, 1991). The cyclic azines melem and melon (Scheme 4) appear to form from DAG and DAF via the ammonium

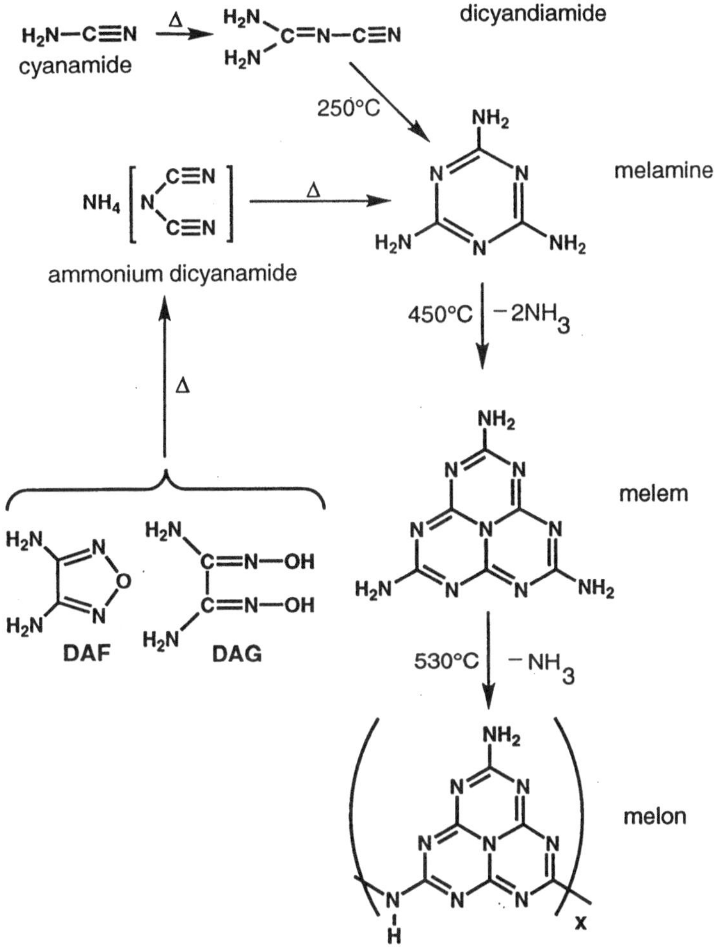

Scheme 4

dicyanamide intermediate. Melon is stable up to at least 650°C. Melon could transiently accumulate on the burning surface and suppress the regression rate by retarding the mass transfer from the condensed phase to the gas phase, and retarding the heat transfer from the flame zone to the reacting heterogeneous condensed phase (Stoner and Brill, 1991).

It has long been known that cyanamide dimerizes to dicyandiamide (Scheme 4) and, upon heating, cyclizes to melamine, melem, and melon (Bann and Miller, 1958). Nitroguanidine (NQ) is known to form melamine upon heating (Stals and Pitt, 1975; Volk, 1985; Lee and Back, 1988). The observation that DAG and DAF react similarly (Stoner and Brill, 1991) suggested that the presence of a NH_2–C–N linkage in the parent molecule might lead to thermally stable cyclic azines. Subsequently, 5-ATZ, which possesses this linkage, was found to produce $NH_2CN(g)$ and a solid residue upon heating (Gao et al., 1991). The residue has an IR spectrum characteristic of melon. In fact, as shown in Fig. 10, all of the compounds in Scheme 3 form a melon-like solid residue when heated above about 400°C (Williams et al., 1994). The NH_2–C–N linkage is present in each of these molecules except for two. NTO only possesses the NH–C–N linkage, implying that H atom migration occurs to produce the NH_2–C–N linkage in the decomposition scheme. 4-ATRZ lacks the NH_n–C–N linkage altogether, which implies

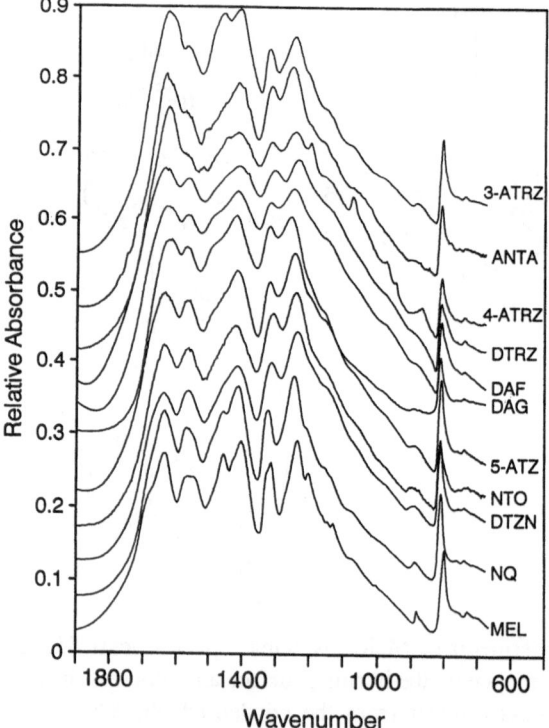

Figure 10. The IR spectra in the ring mode region of compounds (Scheme 5) that upon heating to 600–650°C form residues which are melon-like cyclic azines (Scheme 6).

that relocation of the $-NH_2$ group onto a carbon atom must occur in the decomposition scheme.

The compounds in Scheme 3 are probably "the tip of the iceberg" as potential suppressants of the burn rate of solid propellants by the mechanism of formation of cyclic azines. The presence of the NH_2-C-N unit enhances the chance of formation of cyclic azines but is not a necessary condition.

Information about rapid pyrolysis processes at the surface of energetic and nonenergetic materials provides considerable insight into how solids burn and how burning can be controlled. In turn, these details stimulate design criteria and models based on experimental fact, as opposed to empirical supposition, that improve combustion performance and control.

ACKNOWLEDGMENTS.   The students referenced in this chapter have performed the experimental work that is summarized: Peter Brush, Dilip Patil, Aiming Gao, Yoshio Oyumi, Charles Stoner, and Graylon Williams. I am grateful for financial support from the Air Force Office of Scientific Research, Aerospace Sciences Directorate, and Thiokol Corporation.

REFERENCES

Bann, B., and Miller, S. A., 1958, Melamine and derivatives of melamine, *Chem. Rev.* **58**:131–172.

Bedford, G., and Thomas, J. H., 1972, Reaction between ammonia and nitrogen dioxide, *J. Chem. Soc., Faraday Trans. 1*, **1972**:2163–2170.

Behrens, R., 1987, Simultaneous thermogravimetric modulated beam mass spectroscopy and time-of-flight velocity spectra measurements: Thermal decomposition mechanisms of RDX and HMX, in *Chemical Propulsion Information Agency Publication*, Vol. 476, Part I, pp. 333–342.

Behrens, R., 1990, Thermal decomposition of energetic materials: Temporal behaviors of the rates of formation of the gaseous pyrolysis products of the condensed-phase decomposition of HMX, *J. Phys. Chem.* **94**:6706–6718.

Brill, T. B., 1992, Connecting the chemical composition of a material to its combustion characteristics, *Prog. Energ. Combust. Sci.* **18**:91–116.

Brill, T. B., and Brush, P. J., 1992, Condensed phase chemistry of explosives and propellants at high temperature: HMX, RDX and BAMO, *Philos. Trans. R. Soc. London Ser. A* **339**:377–385.

Brill, T. B., and Oyumi, Y., 1986a, Thermal decomposition of energetic materials 10. A relationships of molecular structure and vibrations to decomposition: Polynitro-3,3,7,7-tetrakis(trifluoromethyl)-2,4,6,8-tetraazabicyclo[3.3.0]octanes, *J. Phys. Chem.* **90**:2679–2682.

Brill, T. B., and Oyumi, Y., 1986b, Thermal decomposition of energetic materials 17. A relationship of molecular composition to HONO formation: Bicyclo and spiro tetranitramines, *J. Phys. Chem.* **90**:6848–6853.

Brill, T. B., Karpowicz, R. J., Haller, T. M., and Rheingold, A. L., 1984, A structural and Fourier transform infrared spectroscopy characterization of the thermal decomposition of 1-(azidomethyl)-3,5,7-tetrazacyclooctane, *J. Phys. Chem.* **88**:4138–4143.

Brill, T. G., Brush, P. J., James, K. J., Shepherd, J. E., and Pfeiffer, K. J., 1992a, T-Jump/FTIR spectroscopy: A new entry into the rapid, isothermal pyrolysis chemistry of solids and liquids, *Appl. Spectrosc.* **46**:900–911.

Brill, T. B., Brush, P. J., Patil, D. G., and Chen, J. K., 1992b, Chemical pathways at a burning surface, in *Twenty-Fourth Symposium (International) on Combustion*, pp. 1907–1914, The Combustion Institute, Pittsburgh.

Brill, T. B., Brush, P. J., and Patil, D. G., 1993a, Thermal decomposition of energetic materials 58. Chemistry of ammonium nitrate and ammonium dinitramide near the burning surface temperature, *Combust. Flame* **92:**178–186.

Brill, T. B., Patil, D. G., Lengellé, G., and Duterque, J. R., 1993b, Thermal decomposition of energetic materials 63. Surface reaction zone chemistry of simulated 1,3,5,5-tetranitrohexahydropyrimidine (DNNC or TNDA) compared to RDX, *Combust. Flame* **95:**183–190.

Bulusu, S., Weinstein, D. I., Autera, J. R., and Velicky, R. W., 1986, Deuterium kinetic isotope effect in the thermal decomposition of 1,3,5-trinitro-1,3,5-triazacyclohexane and 1,3,5,7-tetranitro-1,3,5,7-tetrazacyclooctane: Its use as an experimental probe for their shock-induced chemistry, *J. Phys. Chem.* **90:**4121–4126.

Cosgrove, J. D., and Owen, A. J., 1974, The thermal decomposition of 1,3,5-trinitro hexahydro-1,3,5-triazine (RDX)—part II: Effects of the products, *Combust. Flame* **22:**19–22.

Cronin, J. T., and Brill, T. B., 1987, Thermal decomposition of energetic materials 26. Simultaneous temperature measurements of the condensed phase and rapid-scan FTIR spectroscopy of the gas phase at high heating rates, *Appl. Spectrosc.* **41:**1147–1151.

Cronin, J. T., and Brill, T. B., 1988, Thermal decomposition of energetic materials 29. The fast thermal decomposition characteristics of a multicomponent material: Liquid gun propellant 1845, *Combust. Flame* **74:**81–89.

Federoff, B. T. (ed.), 1960, *Encyclopedia of Explosives and Related Items*, Vol. 1, Picatinny Arsenal, Dover, New Jersey, p. A3111.

Fetherolf, B. L., and Litzinger, T. A., 1992, Penn State University, personal communication.

Gao, A., Oyumi, Y., and Brill, T. B., 1991, Thermal decomposition of energetic materials 49. Thermolysis routes of mono- and diaminotetrazoles, *Combust. Flame* **83:**345–352.

Kaiser, R., 1935, The explosiveness of ammonium nitrate, *Angew. Chem.* **48:**149–150.

Karpowicz, R. J., and Brill, T. B., 1984, *In situ* characterization of the melt phase of RDX and HMX by rapid-scan FTIR spectroscopy, *Combust. Flame* **56:**317–325.

Kimura, J., and Kubota, N., 1980, Thermal decomposition of HMX, *Prop. Explos.* **5:**1–8.

Lee, P. R., and Back, M. H., 1988, Kinetic studies of the thermal decomposition of nitroguanidine using accelerating rate calorimetry, *Thermochim. Acta* **127:**89–100.

Melius, C. F., and Binkley, J. S., 1986, Thermochemistry of decomposition of nitramines in the gas phase, in *Twenty-First Symposium (International) on Combustion*, pp. 1953–1963, The Combustion Institute, Pittsburgh.

Melius, C. F., Bergan, N. E., and Shepherd, J. E., 1991, Effects of water on combustion kinetics at high pressure, in *Twenty-Third Symposium (International) on Combustion*, pp. 217–223, The Combustion Institute, Pittsburgh.

Oyumi, Y., and Brill, T. B., 1985a, Thermal decomposition of energetic materials 3. A high-rate, *in situ*, FTIR study of the thermolysis of HMX and RDX with pressure and heating rate as variable, *Combust. Flame* **62:**213–224.

Oyumi, Y., and Brill, T. B., 1985b, Thermal decomposition of energetic materials 4. High-rate, *in situ*, thermolysis of four, six, and eight membered, oxygen-rich, *gem*-dinitroalkyl cyclic nitramines, TNAZ, DNNC and HNDZ, *Combust. Flame* **62:**225–231.

Oyumi, Y., and Brill, T. B., 1988, Thermal decomposition of energetic materials 28. Predictions and results for nitramines of Bis-imidazolidinedione: DINGU, TNGU and TDCD, *Prop. Explos. Pyrotech.* **13:**69–73.

Oyumi, Y., Brill, T. B., and Rheingold, A. L., 1985, Thermal decomposition of energetic materials 7. High-rate FTIR studies and the structure of 1,1,1,3,6,8,8,8-Octanitro-3,6-diazaoctane, *J. Phys. Chem.* **89:**4824–4828.

Oyumi, Y., Brill, T. B., and Rheingold, A. L., 1986a, Thermal decomposition of energetic materials 9. Polymorphism, crystal structures and thermal decomposition of polynitroazabicyclo[3.3.1]nonanes, *J. Phys. Chem.* **90**:2526–2533.

Oyumi, Y., Rheingold, A. L., and Brill, T. B., 1986b, Thermal decomposition of energetic materials 16. Solid-phase structural analysis and the thermolysis of 1,4-dinitrofurazano[3,4-*b*]piperazine, *J. Phys. Chem.* **90**:4686–4690.

Oyumi, Y., Rheingold, A. L., and Brill, T. B., 1987a, Thermal decomposition of energetic materials 18. Bis(cyanomethyl)nitramine and bis(cyanoethyl)nitramine, *Prop. Explos. Pyrotech.* **12**:1–7.

Oyumi, Y., Rheingold, A. L., and Brill, T. B., 1987b, Thermal decomposition of energetic materials 19. Unusual condensed phase and thermolysis properties of a mixed azidomethyl nitramine: 1,7-Diazido-2,4,6-trinitro-2,4,6-triazaheptane, *J. Phys. Chem.* **91**:920–925.

Oyumi, Y., Brill, T. B., and Rheingold, A. L., 1987c, Thermal decomposition of energetic materials 20. A comparison of the structure properties and thermal reactivity of an acyclic and cyclic tetramethylenetetranitramine pair, *Thermochim. Acta* **114**:209–225.

Palopoli, S. F., and Brill, T. B., 1991, Thermal decomposition of energetic materials 52. On the foam zone and surface chemistry of rapidly decomposing HMX, *Combust. Flame* **87**:45–60.

Rosser, W. A., and Wise, H., 1956, Gas phase oxidation of ammonia by nitrogen dioxide, *J. Chem. Phys.* **25**:1078–1079.

Schwartz, W. W., Askins, R. E., and Flanigan, D. A., 1984, Nitramine Combustion, Report AFRPL TR-84-012, Air Force Rocket Propulsion Laboratory, Edwards AFB, California, April.

Shackelford, S. A., 1987, *In situ* determination of exothermic transient phenomena: Isotopic labelling studies, *J. Phys.* **48**(C4):193–207.

Shackelford, S. A., Coolidge, M. B., Goshgarian, B. B., Loving, B. A., Rogers, R. N., Janney, J. L., and Ebinger, M. H., 1985, Deuterium isotope effects in condensed-phase thermochemical decomposition reactions of octahydro-1,3,5,7-tetranitro-1,3,5,7-tetrazocine, *J. Phys. Chem.* **89**:3118–3126.

Shaw, R., and Walker, F. E., 1977, Estimated kinetics and thermochemistry of some initial unimolecular reactions in the thermal decomposition of 1,3,5,7-tetrazacyclooctane in the gas phase, *J. Phys. Chem.* **81**:2572–2576.

Shepherd, J. E., and Brill, T. B., 1993, Interpretation of time-to-explosion tests, in *Tenth International Symposium on Detonation*, Office of Naval Research, in press.

Stals, J., and Pitt, M. J., 1975, Investigations of the thermal stability of nitroguanidine below its melting point, *Aust. J. Chem.* **28**:2629–2640.

Stoner, C. F., and Brill, T. B., 1991, Thermal decomposition of energetic materials 46. The formation of melamine-like cyclic azines as a mechanism for ballistic modification of composite propellants by DCD, DAG and DAF, *Combust. Flame* **83**:302–308.

Swett, M., 1992, Naval Weapons Center, China Lake, California, personal communication.

Volk, F., 1985, Determination of gaseous and solid decomposition products of nitroguanidine, *Prop. Explos. Pyrotech.* **10**:139–146.

Whittaker, A. G., and Barham, D. C., 1964, Surface temperature measurements on burning solids, *J. Phys. Chem.* **68**:196–199.

Willer, R. L., Chi, M. S., Gleeson, B., and Hill, J. C., 1991, DAG and DAF in propellants based on ammonium perchlorate, U.S. Patent 5,071,495.

Williams, G. F., Palopoli, S. F., and Brill, T. B., 1994, Thermal decomposition of energetic materials 66. Thermal conversion of insensitive explosives and related compounds to polymeric cyclic azine flame retardants, *Combust. Flame* **98**:197–204.

Chapter 5

Competitive Reactions of Methyl Radicals in Partial Oxidation of Methane

Ernő Kulcsár, Paul Benedek, and András Németh

1. INTRODUCTION

The partial oxidation of methane known as the Sachsee process, later modified by others several times, is a procedure for the production of acetylene from methane. In the course of the process a mixture of methane and pure oxygen in a ratio of 2:1, preheated to 973 K, is fed into a flame reactor, where in an autotherm reaction the combustion takes place within 0.02–0.03 s, producing acetylene (Fig. 1). The product gas is quenched to protect acetylene from polymerization. Acetylene and oxidation products are separated from the cooled product gas by suitable methods (Németh *et al.*, 1987).

The conversion of methane into acetylene is relatively small, since the greater part of the methane is needed to heat the reaction mixture to the desired temperature by its oxidation (1573 K) and to supply reaction heat. The competition between dimerization and oxidation is characterized by the ratio $[C_2]/([CO] + [CO_2])$ in the product gas. Also, the methane:oxygen ratio of the feed cannot be

ERNŐ KULCSÁR AND PAUL BENEDEK • Department of Chemistry, Eötvös University, H-1518 Budapest, Hungary.   ANDRÁS NÉMETH • Central Research Institute for Chemistry, H-1525 Budapest, Hungary.

Combustion Efficiency and Air Quality, edited by István Hargittai and Tamás Vidóczy. Plenum Press, New York, 1995.

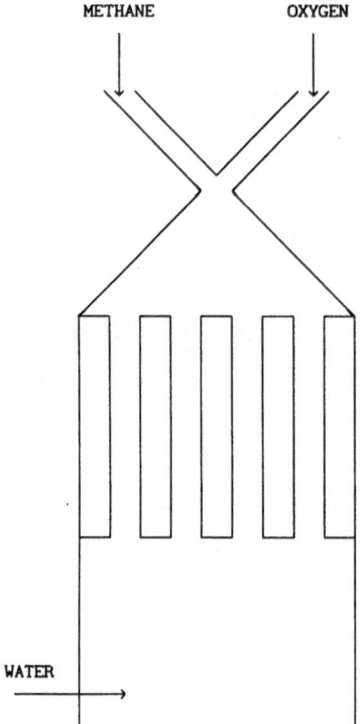

Figure 1. The scheme of the flame reactor.

further increased, due to the fact that the 2:1 feed composition corresponds to the lower inflammability limit.

The improvement of the process has been attempted by modifying the reactor design and by chemical means. The modification of the reactor design has involved the partitioning of the preheater and reactor sections (Fig. 2). The first chamber is designated to burn one part of the methane with a stoichiometric amount of oxygen. In this chamber, methane is principally combusted to carbon dioxide and water, the temperature of the gas is sufficiently high, and the composition of the product gas corresponds to the shift gas equilibrium. The gas leaving the first chamber does not contain acetylene. It enters the second chamber, where the pyrolysis of methane takes place, yielding acetylene as well as a small amount of soot. The residence time in both chambers is long compared to that in the one-chamber flame reactor. This technological solution has not found wide industrial application so far.

Attempts to improve the partial oxidation process in a flame reactor by chemical means have involved the introduction of suitable chloro compounds into

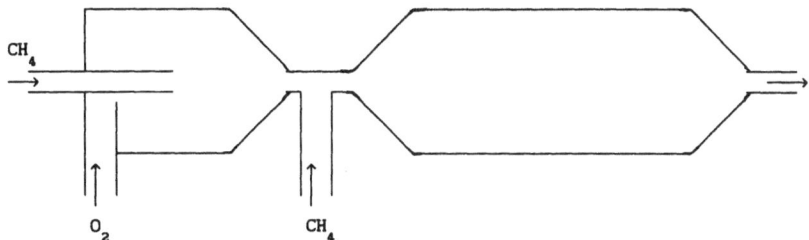

Figure 2.   The scheme of the modified reactor.

the reaction mixture. This led, however, to unpleasant separation problems. Another possibility, the use of a catalyst, does not present such difficulties, and thus it has been investigated by several authors (see Wolf, 1992).

What is expected from a catalyst? First of all, it accelerates the reaction and thus allows the process to be conducted at lower temperature. At lower temperature, a product distribution can be expected which better meets the market demand (at present, ethylene is the favored component). Furthermore, soot formation is decreased, and the feed ratio is not restricted by the lower inflammability limit.

2. SIMULATION AND EVALUATION

We assumed that catalysts affect only a few elementary reactions of the reaction mechanism accepted for the homogeneous gas-phase process. This implies that a suitably selected catalyst will modify only the frequencies of the individual reactions, while the original net reaction remains valid. Starting from this working hypothesis, for the time being, we investigated the influence of temperature and initial concentration of the reactants as independent parameters on the frequencies of the reactions included in the reaction mechanism. The experimental design consisted of three temperatures and three feed compositions, the values chosen in such a way as to conform with published data of other authors (Benedek *et al.*, 1976; Pereira *et al.*, 1990; Zanthoff and Baerns, 1990). The initial concentrations were set by the methane:oxygen ratio and the relative amounts of inert components. The temperature range was from 873 to 1200 K (Table I). The resulting nine experimental points span a wide range of experimental conditions for the investigations.

For the model of the chemical reaction, the set of Zanthoff and Baerns (1990) consisting of 164 elementary reactions was taken, since these authors also published computer simulations.

Figures 3a and 3b display the consumption and accumulation of oxygen, C_2, and $CO + CO_2$ as a function of time at 873 and 1044 K, respectively, for the feed

Table I.   Operating Conditions for the Oxidation of Methane

Temperature (k)	Ratio of inlet concentrations				Reference
	CH_4	O_2	H_2O	Inert compounds	
1200	2	1	—	—	Benedek *et al.*, 1976
873	3	1	6.5	—	Pereira *et al.*, 1990
1044	10	1	—	1.5	Zanthoff and Baerns, 1990

composition examined by Zanthoff and Baerns (1990). It is striking that the overall reaction exhibits three distinct periods: the induction period (up to 1% oxygen conversion), the main period (the oxygen conversion increases up to 99%), and the post period. The periods are rather arbitrarily defined since the transitions between them are not pronounced.

Comparison of the Figs. 3a and 3b shows the effect of temperature on the course of the reaction at the same feed composition. When the temperature is increased from 873 to 1044 K, the reaction rate increases by two orders of magnitude. Also, the amount of C_2 increases somewhat. The length of the induction period decreases by a factor of 100, while that of the main period decreases by a factor of 10.

This is also demonstrated in Table II, where the computed time intervals of the main periods are summarized for the nine points included in the experimental design. They show that—in accordance with preliminary expectations—the investigated experimental parameters influence essentially the oxidation. It is an important observation that when the oxygen has been consumed, i.e., in the post period, the reaction ceases, and, although methane is abundantly present, the amount of C_2 does not increase. This means that the formation of C_2 hydrocarbons by pyrolysis of methane is out of the question (cf. the two-chamber reactor design mentioned above).

Next, we proceed to the details. From the set of elementary reactions, we constructed four reaction bundles (Table III):

1. formation of methyl radicals from methane
2. formation of methane from methyl radicals
3. conversion of methyl radicals into C_2 hydrocarbons
4. oxidation of methyl radicals

These reaction bundles show that

(a) the same oxygen-containing radicals participate both in the formation and in the recombination of methyl radicals, and
(b) there is competition between conversion into C_2 hydrocarbons and oxidation of methyl radicals.

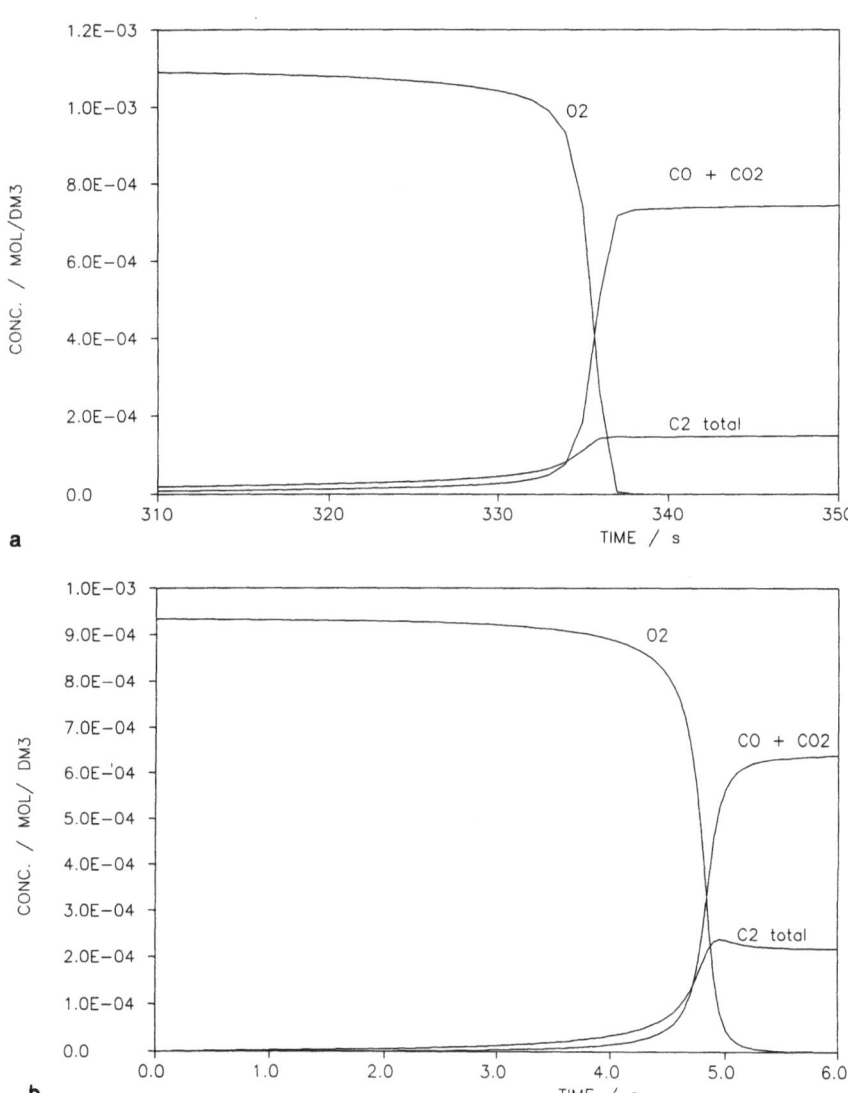

Figure 3. Consumption of oxygen and accumulation of C_2 and $CO + CO_2$ as a function of time. Inlet ratio of compounds, $CH_4:O_2:N_2 = 10:1:1.5$. (a) $T = 873$ K; (b) $T = 1044$ K.

Table II. Time Intervals for the Main Period of the Reaction in the Gas Phase and under Atmospheric Pressure Obtained from the Simulations for Various $CH_4:O_2$ Ratios and Temperatures

Inlet ratio, $CH_4:O_2$	Time interval		
	873 K	1044 K	1200 K
10:1	333–337 s	4.4–5.1 s	0.16–0.36 s
3:1	193–200 s	2.6–3.2 s	0.1–0.4 s
2:1	65–71 s	0.7–0.9 s	0.04–0.1 s
1:1	31–43 s	0.3–0.5 s	0.02–0.04 s

Table III. Reaction Bundles Based on Instantaneous Rate Data

1. Formation of $\cdot CH_3$

28. $CH_4 + O_2 \;\;\; \rightarrow \cdot CH_3 + HO_2\cdot$
29. $CH_4 + HO_2\cdot \;\; \rightarrow \cdot CH_3 + H_2O_2$
30. $CH_4 + \cdot OH \;\; \rightarrow \cdot CH_3 + H_2O$
31. $CH_4 + H\cdot \;\;\;\; \rightarrow \cdot CH_3 + H_2$
32. $CH_4 + \cdot C_2H_5 \rightarrow \cdot CH_3 + C_2H_6$
33. $CH_4 + \cdot C_2H_3 \rightarrow \cdot CH_3 + C_2H_4$
34. $CH_4 \;\;\;\;\;\;\;\;\;\; \rightarrow \cdot CH_3 + H\cdot$
35. $CH_4 + \cdot O\cdot \;\;\; \rightarrow \cdot CH_3 + \cdot OH$

2. Formation of CH_4 from $\cdot CH_3$

39. $\cdot CH_3 + HO_2\cdot \rightarrow CH_4 + O_2$
43. $\cdot CH_3 + \cdot OH \rightarrow CH_4 + \cdot O\cdot$
44. $\cdot CH_3 + H_2O_2 \rightarrow CH_4 + HO_2\cdot$
45. $\cdot CH_3 + H_2O \rightarrow CH_4 + \cdot OH$
46. $\cdot CH_3 + H_2 \;\; \rightarrow CH_4 + H\cdot$
47. $\cdot CH_3 + H\cdot \;\; \rightarrow CH_4$

3. Conversion of $\cdot CH_3$ into C_2 hydrocarbons

49. $\cdot CH_3 + \cdot CH_3 \rightarrow C_2H_6$
50. $\cdot CH_3 + \cdot CH_3 \rightarrow C_2H_4 + H_2$
51. $\cdot CH_3 + \cdot CH_3 \rightarrow \cdot C_2H_5 + H\cdot$

4. Oxidation of $\cdot CH_3$

36. $\cdot CH_3 + O_2 \;\;\; \rightarrow CH_3O\cdot + \cdot O\cdot$
37. $\cdot CH_3 + O_2 \;\;\; \rightarrow CH_2O + \cdot OH$
38. $\cdot CH_3 + \cdot O\cdot \;\;\; \rightarrow CH_2O + H\cdot$
40. $\cdot CH_3 + HO_2\cdot \rightarrow CH_3O + \cdot OH$
41. $\cdot CH_3 + \cdot OH \rightarrow CH_3O\cdot + H\cdot$

Figure 4 displays the reaction rates for the four reaction bundles as a function of time corresponding to the points of the experimental design. The plots demonstrate that the methyl radicals formed are consumed by reaction bundles 2, 3, and 4, and a steady-state concentration cannot be observed. Subsequent to the induction period, not shown in the figure, the rates reach a maximum, then decrease considerably, and converge to zero. This is more clearly illustrated in Fig. 5, where the rates are plotted against oxygen conversion. The rates of formation of methyl radicals and their conversion into C_2 hydrocarbons as well as the rate of oxidation of methyl radicals converge to zero.

A plot of the sum of the concentrations of radicals and the sum of the concentrations of carbon-containing compounds against oxygen conversion is given in Fig. 6. Both curves start from zero, then reach a maximum, and finally converge to zero. The similarity of the curves in Figs. 5 and 6 can be explained by the fact that the rate of the elementary reactions is governed by the concentrations of the radicals.

We computed the material balance of the net reaction for various branching points determined by the production of C_2 (Table IV). Whereas the selection of reactions in Table III is based on the instantaneous rate data, here we dealt with integral accumulation and consumption of the methyl radicals, respectively, in the complete main period, i.e., up to 99% conversion of oxygen.

The data demonstrate that, for the experimental parameters employed in the

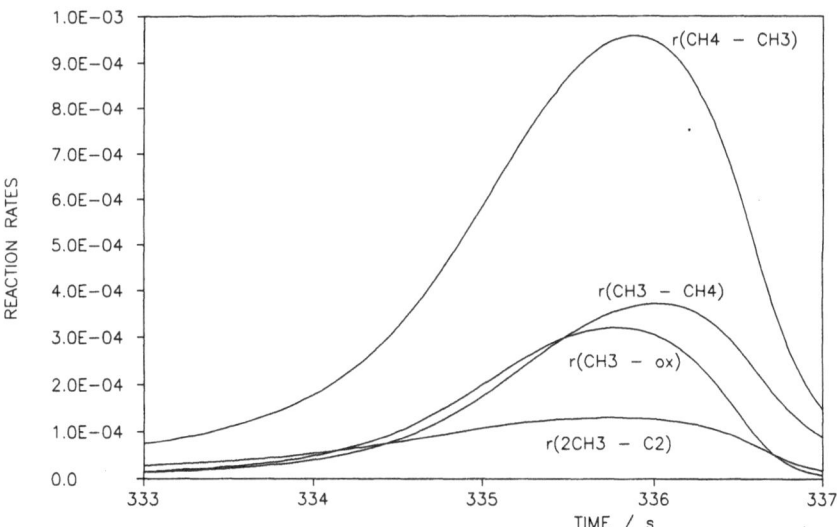

Figure 4. Reaction rates of the four reaction bundles as a function of time. Inlet ratio of compounds, $CH_4:O_2:N_2 = 10:1:1.5$. $T = 873$ K.

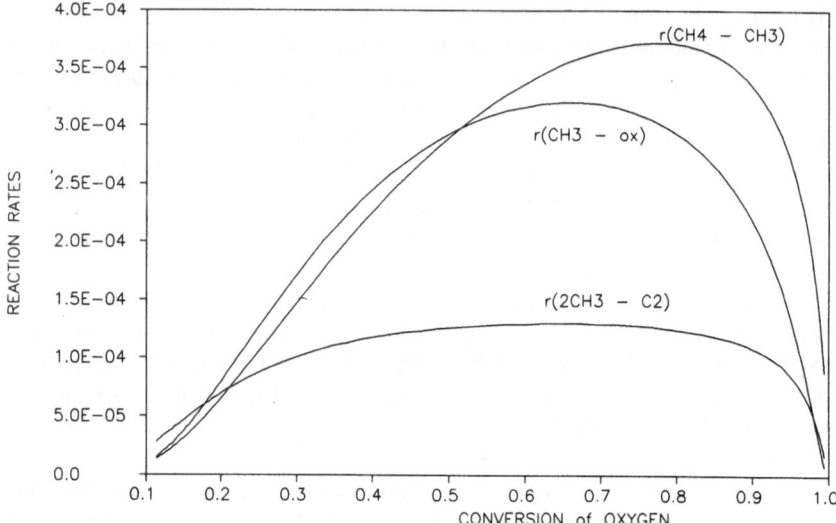

Figure 5.   Reaction rates of the four reaction bundles as a function of oxygen conversion. Inlet ratio of compounds, $CH_4:O_2:N_2 = 10:1:1.5$. $T = 873$ K.

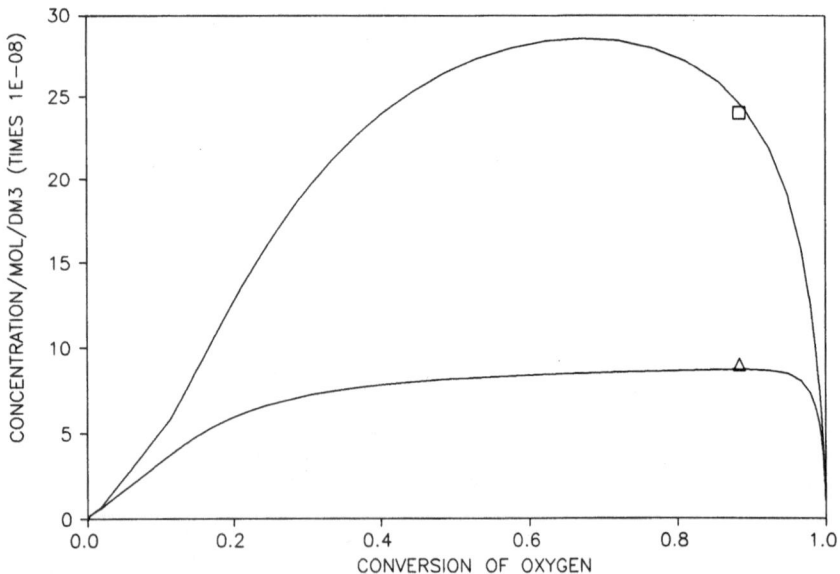

Figure 6.   Total concentration of radicals (□) and of carbon-containing compounds (△) as a function of oxygen conversion. Inlet ratio of compounds, $CH_4:O_2:N_2 = 10:1:1.5$. $T = 873$ K.

Table IV. The $\cdot CH_3$ Material Balance of the Net Reaction for the Integral Main Period

Reaction	Contribution[a]
Generation of $\cdot CH_3$ radicals (reaction bundle 1)	
30. $CH_4 + \cdot OH \rightarrow \cdot CH_3 + H_2O$	10.43
29. $CH_4 + HO_2\cdot \rightarrow \cdot CH_3 + H_2O_2$	6.764
31. $CH_4 + H\cdot \rightarrow \cdot CH_3 + H_2$	4.557
35. $CH_4 + \cdot O\cdot \rightarrow \cdot CH_3 + \cdot OH$	0.076
32. $CH_4 + \cdot C_2H_5 \rightarrow \cdot CH_3 + C_2H_6$	0.010
28. $CH_4 + O_2 \rightarrow \cdot CH_3 + HO_2\cdot$	0.005
Total $\cdot CH_3$ radical generation	21.848
Recombination of $\cdot CH_3$ radicals to methane (reaction bundle 2)	
58. $\cdot CH_3 + CH_2O \rightarrow CH_4 + \cdot CHO$	2.727
44. $\cdot CH_3 + H_2O_2 \rightarrow CH_4 + HO_2\cdot$	2.669
46. $\cdot CH_3 + H_2 \rightarrow CH_4 + H\cdot$	2.272
52. $\cdot CH_3 + C_2H_6 \rightarrow CH_4 + \cdot C_2H_5$	1.706
54. $\cdot CH_3 + C_2H_4 \rightarrow CH_4 + \cdot C_2H_3$	0.985
39. $\cdot CH_3 + HO_2\cdot \rightarrow CH_4 + O_2$	0.140
45. $\cdot CH_3 + H_2O \rightarrow CH_4 + \cdot OH$	0.127
47. $\cdot CH_3 + H\cdot \rightarrow CH_4$	0.102
145. $\cdot CH_3 + C_3H_6 \rightarrow CH_4 + \cdot C_3H_5$	0.005
144. $\cdot CH_3 + C_3H_8 \rightarrow CH_4 + \cdot C_3H_7$	0.004
142. $\cdot CH_3 + \cdot C_2H_5 \rightarrow CH_4 + C_2H_4$	0.002
59. $\cdot CH_3 + \cdot CHO \rightarrow CH_4 + CO$	0.002
Total $\cdot CH_3$ recombination	10.742
Net $\cdot CH_3$ generation	11.106
Consumption of $\cdot CH_3$ radicals into C_2 hydrocarbons (reaction bundle 3)	
49. $\cdot CH_3 + \cdot CH_3 \rightarrow C_2H_6$	7.706
53. $\cdot CH_3 + \cdot C_2H_5 \rightarrow CH_4 + C_2H_4$	
50. $\cdot CH_3 + \cdot CH_3 \rightarrow C_2H_4 + H_2$	
51. $\cdot CH_3 + \cdot CH_3 \rightarrow \cdot C_2H_5 + H\cdot$	
Total	7.706
Consumption of $\cdot CH_3$ radicals into oxidized carbon compounds (reaction bundle 4)	
40. $\cdot CH_3 + HO_2\cdot \rightarrow CH_3O\cdot + \cdot OH$	3.322
36. $\cdot CH_3 + O_2 \rightarrow CH_3O\cdot + \cdot O\cdot$	0.061
37. $\cdot CH_3 + O_2 \rightarrow CH_2O + \cdot OH$	0.005
38. $\cdot CH_3 + \cdot O\cdot \rightarrow CH_2O + H\cdot$	
Total	3.388
Total consumption of $\cdot CH_3$ radicals	11.094

[a]Contribution in relative units.

simulation, a large part of the methyl radicals formed by the first reaction bundle are rechanneled into methane via reaction bundle 2. The greater part of the remaining methyl radicals are converted into C_2 hydrocarbons via reaction bundle 4, while the radicals that are oxidized via reaction bundle 3 are lost for the C_2 production.

Let us define the following ratio for the consumption of $\cdot CH_3$ radicals:

$$DO = \frac{\text{Rate of conversion into } C_2 \text{ hydrocarbons}}{\text{Rate of oxidation}} \qquad (1)$$

This ratio is an indication of the number of $\cdot CH_3$ radicals converted into C_2 hydrocarbons relative to the number converted into oxidized compounds under the conditions specified by the initial parameters and at the given oxygen conversion. It is to our interest to have a large DO ratio. If we introduce the corresponding rate expressions into both the numerator and the denominator, we arrive at the following expression:

$$DO = \frac{(k_{49} + k_{50} + k_{51})[\cdot CH_3]}{(k_{36} + k_{37})[O_2] + k_{38}[\cdot O\cdot] + k_{40}[HO_2^{\cdot}] + k_{41}[\cdot OH]} \qquad (2)$$

Equation (2) indicates several interesting features:

1. Due to the great excess of methane, the numerator varies only slightly with increasing oxygen conversion.
2. Two of the four terms in the denominator are smaller by two orders of magnitude than the others; consequently, O_2 and $HO_2^{\cdot}$ play dominant roles in the oxidation.
3. With increasing oxygen conversion, the $HO_2^{\cdot}$ concentration increases initially, passes through a maximum, and decreases.

Figure 7 shows the plot of DO ratio against oxygen conversion. At the beginning, DO is greater than 1; that is, the rate of conversion into C_2 hydrocarbons is greater than the rate of oxidation. This can be explained by the fact that the conversion into C_2 hydrocarbons depends only on the concentration of methyl radicals, whereas the oxidation needs also $\cdot O\cdot$, $\cdot OH$, and $HO_2^{\cdot}$, which accumulate over a rather long induction period. Subsequently, the rate of re-forming methane from methyl radicals and that of the oxidation of methyl radicals increase, and, as a consequence, the DO ratio reaches unity at 20% oxygen conversion. The value of DO then decreases further, reaching a minimum at 70% oxygen conversion. Owing to the consumption of $HO_2^{\cdot}$ and to the continuing conversion of methyl radicals into C_2 hydrocarbons, DO then increases to values greater than 1.

The above reasoning leads also to the conclusion that the initial methane:oxygen ratio has a large influence on the value of DO, as seen in Figs. 8a and

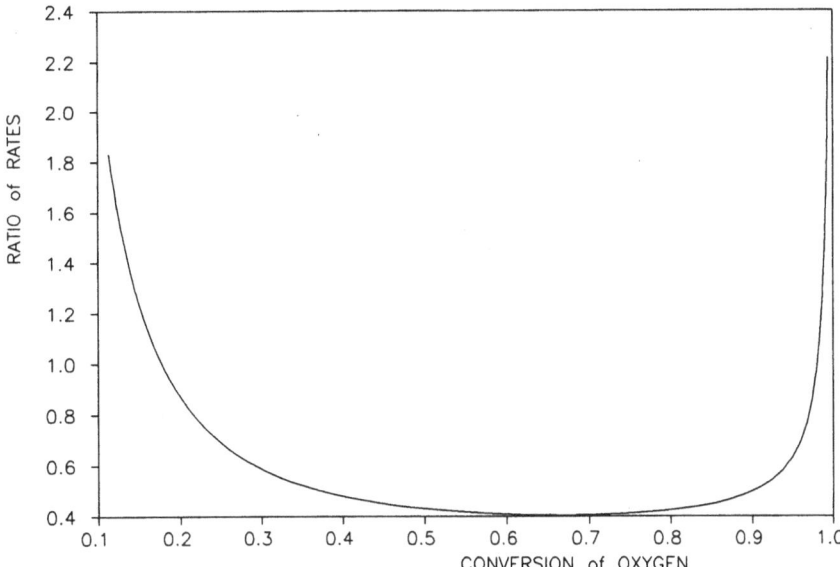

Figure 7. Plot of the ratio of the rate of dimerization to that of oxidation as a function of oxygen conversion. Inlet ratio of compounds, $CH_4:O_2:N_2 = 10:1:1.5$. $T = 873$ K.

8b, where we show plots of DO for two different methane:oxygen ratios at three temperatures. The increase in the DO values with increasing temperature can be explained by the fact that the average activation energy of reaction bundle 1 increases faster with increasing temperature than that of the oxidation reaction bundle.

3. CONCLUSIONS

The main reaction pathway for the methyl radicals produced by the partial oxidation of methane leads to their oxidation. Under the conditions investigated, a plot of conversion into C_2 hydrocarbons/oxidation of $\cdot CH_3$ radicals against oxygen conversion exhibits a minimum. Increasing the methane:oxygen ratio is advantageous for the production of C_2 species. Increasing the temperature also favors conversion into C_2 hydrocarbons compared to oxidation. This is in agreement with the experimental findings that methanol production is favored at lower temperature while C_2 production is favored at higher temperature. The simulation calls attention to the fact that the homogeneous gas-phase reaction is not highly efficient for the industrial production of C_2 compounds.

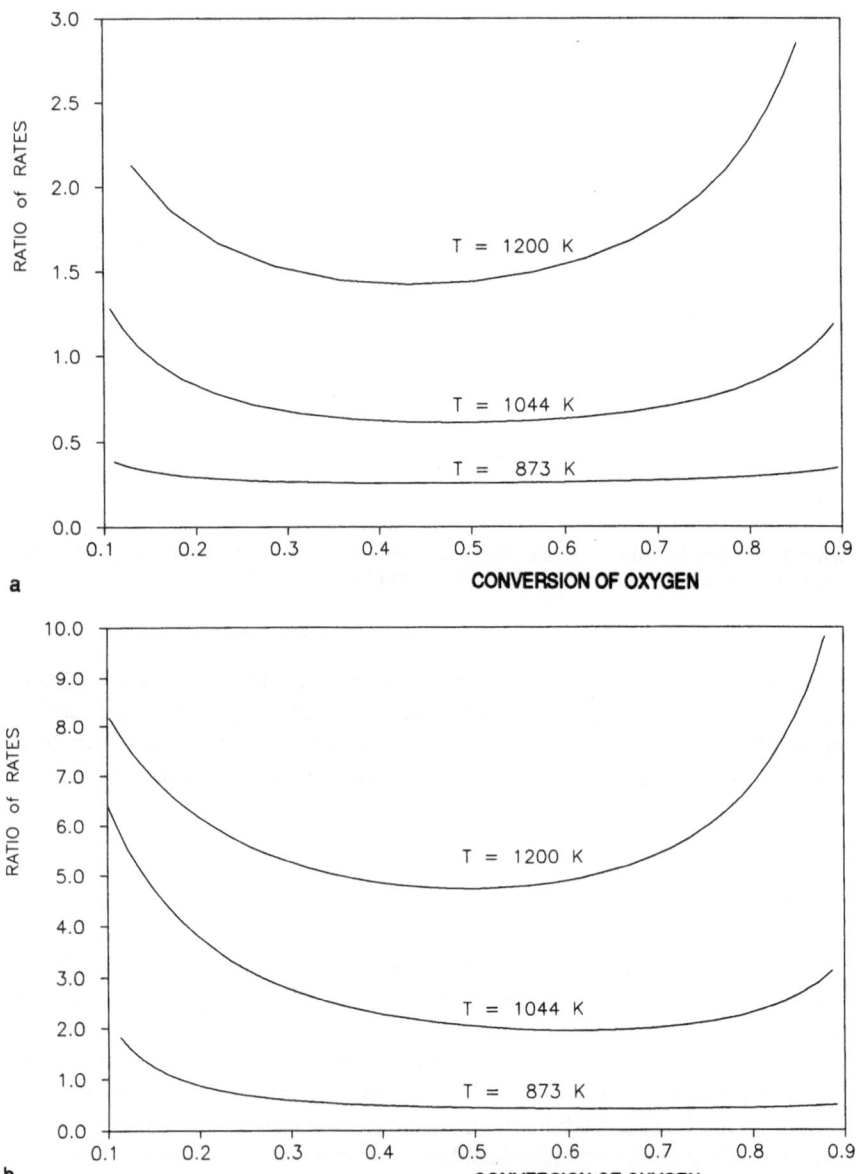

Figure 8. The influence of temperature on the ratio of the rate of dimerization to that of oxidation. Inlet ratio of compounds: (a) $CH_4:O_2:H_2O = 3:1:6.5$; (b) $CH_4:O_2:N_2 = 10:1:1.5$.

REFERENCES

Benedek, P., László, A., Németh, A., and Váczi, P., 1976, Mathematical model of the partial oxidation flame-reaction of methane, *Hung. J. Ind. Chem.* **4**:77–92.

Németh, A., Benedek, P., and Váczi, P., A combined sensitivity and contribution analysis for construction of the reaction mechanism in modelling chemical reactors, The Use of Computers in Chemical Engineering, Chemical Engineering Fundamentals XVIII, Congress, Giardini Naxos, Italy, 1987.

Pereira, P., Lee, S. H., and Somorjai, G. A., 1990, The conversion of methane to ethylene and ethane with near total selectivity by low temperature (<610°C) oxydehydrogenation over a calcium–nickel–potassium oxide catalyst, *Catal. Lett.* **6**:255–262.

Wolf, E. E. (ed.), 1992, *Methane Conversion by Oxidative Processes: Fundamental and Engineering Aspects*, Van Nostrand Reinhold, New York.

Zanthoff, H., and Baerns, M., 1990, Oxidative coupling of methane in the gas phase: kinetic simulation and experimental verification, *Ind. Eng. Chem. Res.* **29**:2–10.

REFERENCES

Chapter 6

Theoretical Quantum-Mechanical Study on the Soot Formation Process

Jerzy Leszczyński

1. INTRODUCTION

Recent developments in computational chemistry and advances in computer technology allow for a successful application of quantum-mechanical methods to broad areas of chemistry where experimental methods are difficult to apply. For species that are awkward to handle experimentally, computational modeling is an attractive source of information. Currently, *ab initio* methods based on quantum-mechanical postulates prove to provide accurate data on structures and properties of small- and medium-size molecules. These data could supplement experimental results or serve as a valuable source of information when experimental data are scarce. Computerized modeling is especially attractive for systems that are dangerous to study experimentally and for short-lived reactive species as well as transient ions. A detailed knowledge of structures and properties of these agents would help in the understanding of their chemistry and might be of importance in technology.

Knowledge of reaction mechanisms of processes leading to soot formation is

---

JERZY LESZCZYŃSKI • Department of Chemistry, Jackson State University, Jackson, Mississippi 39217.

Combustion Efficiency and Air Quality, edited by István Hargittai and Tamás Vidóczy. Plenum Press, New York, 1995.

an important scientific goal. Creation of soot particles takes place in technological processes as well as in many areas of everyday life. Such various processes as operation of a diesel engine, burning wood in a fireplace in a winter afternoon or a candle in an outdoor hurricane lamp in a summer evening, and work of industrial furnaces all are linked together through formation of microscopic, black soot particles. A satisfactory solution to improvement of fuel economy and reduction of air pollution during combustion reactions may rest on our ability to understand, predict, and manage soot formation processes.

A review of the chemical kinetics of soot formation was given by Harris and Weiner (1985). Several steps have been proposed for this process (Calcote, 1981; Homann, 1985; Glassman, 1989). During the first step of particle inception (or nucleation), a series of homogeneous reactions between hydrocarbon "species" take place. Initially, hydrocarbons with few carbon atoms form large species with molecular weights up to 1000. This process is very fast and usually takes no more than a few milliseconds. In atmospheric flames the particle nucleation rate is on the order of 10^{16} cm^{-3} s^{-1}.

The second step leading to soot formation is called growth. Species formed during the first step of the process collide and coalesce to create large spherical species. Simultaneously, hydrocarbons in the gas phase are involved in hetero- geneous reactions on the soot surface, increasing the mass of soot by surface growth processes. The second step is an order of magnitude faster than the first one. The final, chain formation step of the process includes formation of the chainlike particles composed of hundreds of thousands of carbon atoms that make up soot.

The nature of the "species" reacting in the first step of the soot formation process is not currently known. If the predominant forms of species in this reaction are ions, an ionic mechanism must be governing the process. However, if free- radical species are the major forms, a radical mechanism is the essential one.

Experimental studies have not led to a definitive explanation of the soot formation process. Both ionic (Calcote, 1981; Keil et al., 1985; Calcote et al., 1988) and radical species (Homann and Wagner, 1967; Homann, 1985) have been observed in soot flames. However, under normal flame conditions, the relative concentration of the neutral forms seems to be higher than that of ions. This observation might suggest a radical mechanism for the process (Calcote, 1981; Homann, 1985; Frenklach et al., 1985; Harris et al., 1988; Westmoreland et al., 1989; Miller and Melius, 1992). In spite of this suggestion, the chemistry of soot formation in flames is still an enigma.

An understanding of the chemistry of processes resulting in soot formation requires a detailed knowledge of the first few phases. Since $C_3H_3^+$ is the most intense ion observed in sooting flames (Tanner et al., 1981; Calcote, 1981), the first step in computational studies involved a theoretical characterization of its possible isomers. This study recovered possible forms of the cations that could be identified in sooting flames. Acetylene can react with $C_3H_3^+$ to form $C_5H_5^+$, which creates a challenge for a theoretical description. The final step involves modeling of chemical reactions between $C_8H_7^+$ and C_2H_2. Products of these reactions can further react with neutral molecules, forming higher hydrocarbons.

2. COMPUTATIONAL TECHNIQUES

Since this chapter is intended to promote the role of quantum-mechanical, computational methods among experimentalists, traditionally involved in the investigation of combustion processes, we shall give a short review of the terminology used in discussing results of quantum-mechanical studies and refer readers interested in more details to recent books (Hehre *et al.*, 1986; Hirst, 1990) and review articles (Lipkowitz and Boyd, 1990, 1991, 1992, 1993, 1994).

2.1. General Assumptions

The solution of the Schrödinger equation for the studied species is the aim of any quantum-mechanical computational method. Unfortunately, the accurate analytical solution of this equation is not feasible even for the smallest molecules. Therefore, simplifications of the Schrödinger equation are introduced. The orbital (one-electron) approximation and separation of the motions of electrons and nuclei (e.g., the Born–Oppenheimer approximation) are among the most important assumptions. In many applications, molecular orbitals are approximated by linear combinations of atomic orbitals (LCAO-MO). The coefficients of such combinations are calculated by a variational method in which the best one-electron functions (MOs) are obtained by minimizing the electronic energy of the molecular system (Hartree–Fock–Roothaan, HFR, HF, or SCF method). Such calculations are carried out for fixed positions of the nuclei (molecular geometry). This is not a trivial task, and larger molecular systems require calculations of a vast number of integrals. Each step of the molecular geometry optimization (setting up nuclei at new fixed positions) involves new calculations of the LCAO coefficients.

Two approaches—*ab initio* and semiempirical—are used in quantum-mechanical calculations. In *ab initio* methods all the integrals are calculated without any further approximations or the introduction of experimental parameters. The semiempirical approach significantly reduces the number of calculated integrals by neglecting some and/or replacing some by experimental atomic data and introducing additional "nonphysical" parameters to compensate for the approximations employed.

2.2. *Ab Initio* Methods

The term *ab initio* refers to different levels of calculations ranging from very crude to extremely accurate. The two parameters characterizing the level of *ab initio* calculations are the size of the one-particle basis set and the level of electron correlation. The smallest, minimal basis set commonly used in contemporary *ab initio* calculations is an STO-3G basis set. Somewhat better is the split-valence (double-zeta) 3-21G basis set, followed by basis sets with polarization functions 6-31G* (*d*-polarization functions included on heavy elements) and 6-31G** (augmentation of the 6-31G* basis set by *p*-polarization functions on hydrogens). In more accurate calculations, larger (e.g., valence triple-zeta 6-311) basis sets, augmented in many cases by a larger number of polarization functions, are used. The size of the basis set is an important parameter characterizing the quality of the calculated data and the computer resources required.

2.3. Electron Correlation Contributions

The consideration of electron correlation in *ab initio* predictions ranges from none [Hartree–Fock (HF) formalism] to full in infinite-order perturbation theory and in full configuration interaction calculations. The HF method ignores correlation of electronic motions by assuming that electrons can move independently. Consequently, the calculated results are less accurate; however, in many cases, though the predicted energies may not be quite reliable, the molecular equilibrium geometries obtained are relatively dependable. In order to correct the deficiencies of the Hartree–Fock approach, the electron correlation energy can be introduced into calculations by many different methods. Second-order many-body perturbation theory, often designated MP2 (second order of Møller–Pleseet theory), is a very popular and relatively computer-nonintensive approximation which accounts for a large fraction of the correlation energy. More correlation energy can be recovered by the more accurate, though much more costly, fourth-order (MP4) approximation.

The combination of the size of the basis set and the level of electron correlation applied in calculations determines the quality of the results. High-level *ab initio* calculations produce results of experimental (or in many cases higher) accuracy. However, an increase of the basis set and more accurate treatment of the electron correlation escalate cost in terms of the computer resources used in the calculations. Also, larger molecular systems cannot generally be treated as accurately as small model molecules.

2.4. Semiempirical Methods

One way to overcome the intense computer demands of accurate *ab initio* calculations for large molecules is to develop theories based on quantum-mechanical principles but introducing input from experimental atomic data instead of calculating all integrals, as is necessary in nonempirical calculations. In addition, a large number of very small integrals can be neglected in this approach.

If the results of such approximations are not accurate, a number of "fitted" parameters can be included. These "fitted" parameters are usually chosen to mimic an assumed molecular property, e.g., molecular geometry, heat of formation, or the electronic spectrum. As a result, such a chosen property is often well reproduced by a semiempirical scheme parameterized on this property, but other molecular parameters may be poorly mimicked by such a method. Generally, semiempirical methods are much faster than *ab initio* calculations; as a consequence, semiempirical predictions are possible for much larger molecular systems.

The two semiempirical schemes which will be mentioned in the chapter are MNDO and its newer version, the AM1 method, derived by Dewar's group, and the ZINDO scheme proposed by Zerner and co-workers.

The MNDO method was developed by Dewar and Thiel (1977) as an improved version of the MINDO approximation, introduced by Dewar's group in the mid-seventies [for a review, see Stewart (1990)]. This popular method is based on the intermediate neglect of the differential overlap (INDO) approximation and is parameterized to reproduce experimental heats of formation, dipole moments, ionization potentials, and molecular geometries. However, this method was not able to reproduce correctly hydrogen bonding in biomolecules, and ten years after it was introduced, Dewar *et al.* (1985) proposed an improved scheme, called AM1, which corrected this problem.

The ZINDO program package was developed by Zerner *et al.* (1980), Bacon and Zerner (1979), Anderson *et al.* (1986), and Zerner (1991). It includes a number of semiempirical programs. The most unique part of the package is a semiempirical method based on the INDO approximation designed to predict UV spectra. This model was calibrated at the configuration interaction singles (CIS) level and has had a lot of success in its application to predictions of UV–Vis spectra when calculations are carried out using accurate (experimental or *ab initio*) geometries.

2.5. Calculation of Energy Derivatives

Energy derivatives with respect to molecular parameters provide much important information about molecular systems. One of the most important goals of theoretical calculations is characterization of the potential energy surface (PES) of the studied species. A stationary point on the PES is one for which there are no forces acting on any of the molecular nuclei. This can be tested by calculations of the energy gradients, which possess zero values for stationary points. Using geometry optimization procedures, one can predict for a polyatomic molecule a number of stationary points on the PES which differ in molecular geometries.

In the subsequent calculations, all optimized structures are characterized via their predicted harmonic vibrational frequencies. These calculations provide a few important findings. Foremost, they are necessary to determine the nature of the predicted structures. Though a zero value of the gradients is a necessary condition for the existence of stationary points, their full characterization is also based on the second derivatives of energy with respect to atomic coordinates. Calculated vibrational frequencies (or, equivalently, eigenvalues of the matrix of these second derivatives) characterize stationary points on the potential energy surfaces. Minimum-energy species possess only real (positive) frequencies, while saddle points on the PES are characterized by one (first-order transition species), two (second-order transition species), or more (higher order transition species) imaginary (negative) vibrational frequencies.

Theoretically predicted harmonic vibrational frequencies are also sources of identification of short-lived or reactive species. Comparison of the experimentally detected IR bands and accurate *ab initio* frequencies and intensities could guide an assessment of the molecular structures of these species.

Best calculated molecular geometries (those obtained by the highest level of applied theory) are used as reference molecular geometries for prediction of molecular properties. Electronic energies obtained by *ab initio* calculations might be further improved through single-point calculations. In such calculations, the best obtained geometries are used for calculations of energy and/or molecular properties with larger basis sets and with inclusion of electron correlation contributions.

Widely applied abbreviations used to denote levels of *ab initio* calculations furnish information about the basis set and electron correlation contributions. HF/3-21G is the abbreviation for calculations at the Hartree–Fock level with a split-valence basis set, while MP2/6-31G** stands for calculations at the level of second-order perturbation theory carried out with a split-valence basis set with d- and p-polarization functions on heavy elements and hydrogens, respectively. To abbreviate information on single-point calculations, details about the geometry optimization level as well as the approximations used for the best energy calculations are necessary. MP4/6-31G**//HF/6-31G* denotes single-point calculations at the fourth order of perturbation theory using a split-valence basis set with d- and p-polarization functions, performed at (//) molecular geometry optimized using Hartree–Fock theory with the split-valence basis set augmented by d-polarization functions.

3. QUANTUM-MECHANICAL PREDICTIONS

3.1. Structure and Properties of $C_3H_3^+$ Forms

The fact that the $C_3H_3^+$ ion was observed as the most intense ionic species in sooting flames (Tanner *et al.*, 1981; Calcote, 1981; Gaydon and Wolfhard, 1979; Miller, 1973; Hayhurst and Kittelson, 1978; Harris and Weiner, 1985) suggests this molecule as a starting point of any computational investigation of the soot formation process. Interest in the properties of different isomeric forms of $C_3H_3^+$ has been boosted by discoveries of its potential role in interstellar chemistry (Herbst *et al.*, 1984; Lepp *et al.*, 1988; Turner, 1989; Smith and Adams, 1978; Schiff and Bohme, 1979). Because of its small size, this ion also remains a fascinating subject and a reference for a number of theoretical investigations aiming toward improvement of accuracy of predicted data. Among them are the *ab initio* investigations of the energetics and molecular parameters of different possible conformations by Radom *et al.* (1976), Raghavachari *et al.* (1981), and Cameron *et al.* (1989).

Even such a small species as $C_3H_3^+$ could display a number of forms. Proper characterization of all these forms is of importance for further experimental studies and identification of these species in sooting flames. A detailed study by Cameron *et al.* (1989) provides information on the five most stable structures. The reported calculations consisted of few steps. Both *ab initio* and semiempirical approaches were combined, reducing significantly the required computer resources and supplying reasonable estimates of spectroscopic parameters for the most important forms.

Initially, a number of possible conformers was optimized using the ZINDO method (Zerner, 1991). These calculations were not restricted by any symmetry constraints and delivered initial geometries for more rigorous, *ab initio* optimizations. Several levels of *ab initio* theory were applied in the study. A first phase of *ab initio* calculations was performed to characterize the potential energy surface of the $C_3H_3^+$ ion. Predicted stationary points on the PES differ in the molecular parameters. The best calculated molecular geometries (those obtained by the highest level of applied theory) were used as reference molecular geometries for prediction of molecular properties.

Minimal, STO-3G basis set, Hartree–Fock level calculations followed by consecutively more accurate geometry optimizations with a split-valence double-zeta 3-21G basis set and finally a 6-31G* basis set augmented by *d*-polarization functions on carbon atoms were performed on all structures selected by ZINDO calculations. Also, during *ab initio* level optimizations no symmetry restrictions were imposed on initial structures. When an optimized conformer exhibits molecular symmetry, this was the result of a geometry search. All geometry optimizations were carried out using gradient-driven optimization procedures.

In the subsequent step of *ab initio* calculations, all optimized structures were characterized via harmonic vibrational frequency calculations. Electronic energies obtained by HF/6-31G* calculations were further improved through single-point MP4 level calculations.

Five structures (Figs. 1–5) were obtained at the *ab initio* 3-21G level of geometry optimization. As expected, D_{3h} symmetry cyclopropenyl cation **I** (Fig. 1) was calculated to be the global minimum structure. This highly symmetrical ion

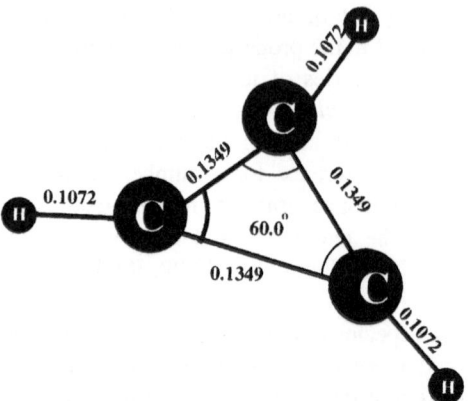

Figure 1. Optimized structure (HF/6-31G*) of cyclopropenyl cation **I** (c-$C_3H_3^+$). The bond lengths of this structure and all others is given in nm.

has the lowest energy at all levels of predictions. It is an interesting model system, the first member of the $4n + 2$ ($n = 0$) aromatic series. The aromatic character of the cyclopropenyl cation was confirmed by the predicted molecular parameters. The calculated (HF/6-31G* level) bond distance of 0.1349 nm between carbon atoms agrees well with its postulated aromaticity.

The next predicted structure, propargylium cation **II** (Fig. 2), possesses C_{2v} symmetry. This conformer is an important, reactive form of $C_3H_3^+$. It is the second lowest energy structure. Two resonance structures can be derived for this form:

$$\begin{array}{c} H \\ \diagdown \\ \diagup \\ H \end{array} C=C=C^{\pm}H \qquad\qquad \begin{array}{c} H \\ \diagdown \\ \diagup \\ H \end{array} C^{\pm}C\equiv C-H$$

The knowledge of the electronic structures of these forms is important in predicting reaction paths for the reactions of propargylium cation with electron-rich species present in sooting flames.

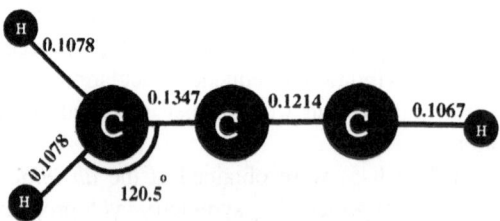

Figure 2. Optimized structure (HF/6-31G*) of propargylium cation **II** (l-$C_3H_3^+$).

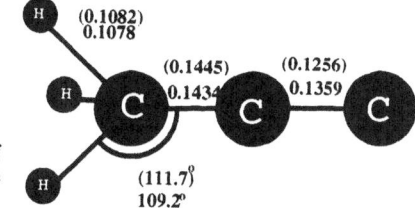

Figure 3. Optimized structure (HF/6-31G*) of 1-propargyl cation **III**. The numbers in brackets are for the triplet state.

The 1-propynyl cation **III** (Fig. 3) has high (C_{3v}) symmetry. The highest occupied orbitals in this ion are a degenerate π pair. Only two electrons share this degenerate pair of orbitals. As a consequence, forms with different multiplicities can exist. The calculations yielded both singlet and triplet state structures. As depicted in Fig. 3, their geometries are significantly different.

Prop-2-en-1-yl-3-ylidiene cation **IV** (Fig. 4) and cycloprop-1-yl-2-ylidine cation **V** (Fig. 5) represent low (C_s) symmetry, high-energy forms. Though semiempirical calculations, as well as low-level (STO-3G, 3-21G) *ab initio* calculations, predict the cycloprop-1-yl-2-ylidine cation to be a stable form, higher level HF/6-31G* predictions furnish an interesting observation. Addition of *d*-polarization functions to the carbon basis set (6-31G* vs. 3-21G basis set) significantly influences the predicted molecular properties of the high-energy forms *IV* and **V**. The C–C–C angle of form **IV** is notably affected by the inclusion of the polarization functions, and it decreases from 118° (HF/3-21G level geometry optimization) to 108° (HF/6-31G* level geometry optimization). The effect of the basis set quality is even more pronounced for form **V**, which at the HF/6-31G* level is not stable, and during geometry optimization the C–C–C angle opens to yield finally structure **IV**.

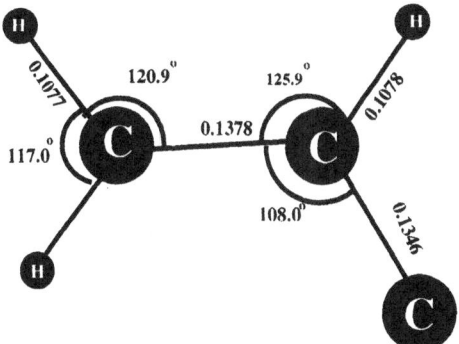

Figure 4. Optimized structure (HF/6-31G*) of pro-2-en-1-yl-3-ylidine cation **IV**.

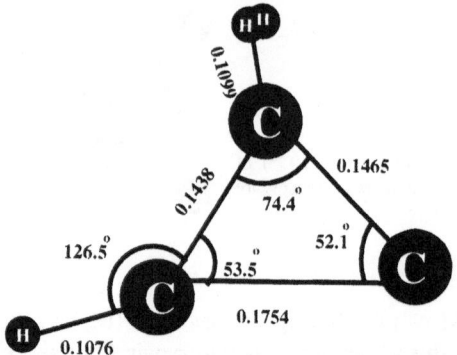

Figure 5. Optimized structure (HF/3-21G) of cycloprop-1-yl-2-ylidine cation **V**.

Figure 6 displays another example of the dependence of predicted properties upon the level of theory used. Relative energies of all studied forms are displayed versus the basis set and level of electron correlation used in obtaining the predictions. The cyclopropenyl cation (**I**) is the global minimum species at all applied levels, followed by the propargylium cation (**II**). Forms **III**, **IV**, and **V** lie more than 70 kcal/mol (MP4/6-31G** level) above the reference compound **I**. For

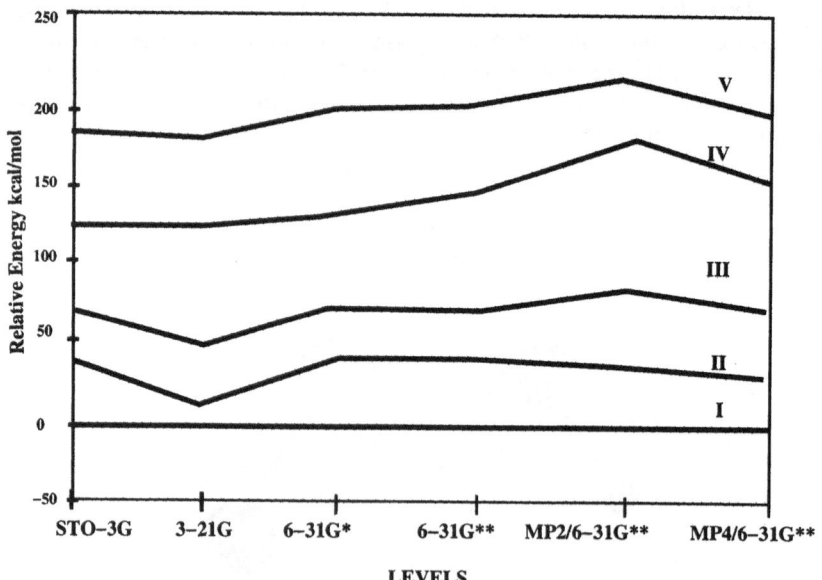

Figure 6. Plot of relative energies of $C_3H_3^+$ forms versus theoretical levels. All energies are relative to **I**.

all forms, the relative stabilities change by more than 20 kcal/mol as a function of basis set and electron correlation level. For experimental applications, the most important prediction is the energy gap between the two lowest energy structures. This calculated energy difference ranges from 12.6 (3-21G//3-21G) to 35.4 kcal/mol (6-31G**//6-31G*). The most accurate value of 28.9 kcal/mol predicted at the correlated MP4/6-31G**//HF/6-31G* level can be further improved by adding corrections for the zero-point energy (ZPE) terms. The relative energy of **II** derived in this manner amounts to 27.7 kcal/mol, which agrees well with the experimental data of Lossing (1972) of 25 ± 4 kcal/mol and more recent estimations by Lias *et al.* (1984) of 24.3 kcal/mol.

More advanced *ab initio* calculations of the energetics of cyclopropenyl and propargylium cations have improved agreement between experiment and theory. Wong and Radom (1989) optimized molecular structures of **I** and **II** at the electron-correlated MP2 level with the 6-31G* basis set functions (MP2/6-31G* level). Using such optimized reference geometries, they performed single-point calculations at the MP4(SDQ) approximation level applying the 6-311G($2d$, $2p$) basis set. They estimated the relative energy of **II** to be equal to 24.1 kcal/mol.

Recently, results of higher-level calculations on the relative stability of **II** were published by Maclagan (1992). He also used MP2/6-31G* reference geometries of **I** and **II** and calculated single-point energies in the QCISD(T) approximation with the 6-311G($2df$, p) basis set including ZPE corrections. This high level of theory increases the previously estimated energy difference to 25.7 kcal/mol.

Experimental IR spectra could be used for characterization of species present in flames and in model flames. To guide identifications of cyclopropenyl and propargylium ions, their theoretically predicted harmonic vibrational frequencies and intensities are provided in Table I (Cameron *et al.*, 1989). Again, there are noticeable differences between data predicted under the HF/3-21G and HF/6-31G* approximations; usually, frequencies predicted at a higher level of theory are smaller. The most pronounced discrepancy is observed for the lowest frequencies. When predicted frequencies are compared with experimental ones, it is recommended (Hehre *et al.*, 1986) to scale them by multiplying all of the predicted data by the same, constant factor. This factor includes corrections for anharmonicity of the experimental data, as well as for the deficiencies of the basis sets used and the limited accounting for electron correlation contributions. Generally, predicted frequencies are too high, and less accurate calculations require smaller scaling factors, e.g., 0.85 and 0.90 for the HF/3-21G and HF/6-31G* levels, respectively. However, when the basis set used in the calculations is adequate and vibrational frequencies are calculated at the correlated level, the scaling factors between predicted and experimental "harmonized" frequencies approach 1.0. Careful studies by Kwiatkowski and Leszczyński (1992, 1993) and Leszczyński and Goodman (1994) on the basis set and electron correlation effects in *ab initio* calculations of the vibrational spectra of formaldehyde clearly demonstrate such trends. A valence triple-zeta basis set augmented by three sets of *d-* and a

Table I. Calculated Vibrational Frequencies (cm^{-1}) of the Two Most Stable Forms of $C_3H_3^+$ Using 3-21G Optimized Structures and Wave Functions and 6-31G* Optimized Structures and Wave Functions, Cameron *et al.*, 1989

	Form I					Form II			
	3-21G	6-31G*	I(IR)	I(R)[a]		3-21G	6-31G*	I(IR)	I(R)
a_2''	907.9	842.5	83	00	b_1	351.7	280.3	45	00
e'	1041.8	1029.2	27	6	b_2	396.4	364.2	25	03
					b_2	842.0	741.6	74	01
e''	1136.2	1099.6	00	00	b_1	1089.1	1023.9	02	01
					b_2	1182.9	1131.5	01	00
a_2'	1209.2	1164.0	00	00	a_1	1222.8	1216.6	77	57
e'	1373.6	1418.3	74	04	b_1	1281.5	1265.1	06	00
					a_1	1613.0	1611.1	03	02
a_1'	1702.3	1795.2	00	53	a_1	2216.5	2194.9	816	28
e'	3466.0	3467.5	86	28	a_1	3275.2	3331.1	16	101
					b_2	3380.2	3443.3	45	58
a_1'	3519.1	3518.5	00	91	a_1	3535.8	2569.5	75	47
ZPE	0.048	0.0486				0.0464	0.0460		
E^{MP4}		-115.4081^b					-115.3621^b		

[a]The IR intensities are in units of km/mol and the Raman intensity (R) is in A^4/amu.
[b]MP4/6-31G**//HF/6-31G* energy (in a.u.).

set of f-polarization functions on carbon and oxygen and two sets of p-polarization functions on hydrogens [6-31G($3df$, $2p$) basis set] was applied. Geometry optimization was carried out at the correlated levels, and the same approximations were used for prediction of harmonic vibrational frequencies. At the MP2 level, a scaling factor of 0.988 was suggested (Kwiatkowski and Leszczyński, 1992, 1993), while at higher MP4(SDTQ) and couple-cluster singlets, doublets plus perturbative triplets [CCSD(T)] levels no scaling factor (multiplication by 1) was necessary to match experimental and predicted frequencies (Leszczyński and Goodman, 1994).

UV–Vis spectroscopy is another effective experimental technique. The calculated electronic spectra of cyclopropenyl and propargylium cations are presented in Table II. A semiempirical ZINDO/S-CI spectroscopically parameterized method (Ridley and Zerner, 1973; Zerner *et al.*, 1980) was applied for both cations by Cameron *et al.* (1989), while for the propargylium ion more accurate calculations by Eyler *et al.* (1984) using good basis set and polarization-propagator methodology have been reported. There is fair correspondence between the ZINDO/S-CI results and those of Eyler *et al.* The agreement is particularly notable for the two lowest excitations, but also predicted oscillator strengths and ordering of the remaining states are in good accord.

Table II. Calculated Electronic Spectra of the Two Most Stable Forms of $C_3H_3^+$ (1000 cm^{-1}) Cameron *et al.*, 1989

Form I			Form II				
ZINDO			ZINDO			b	
$^1A_1''$	56.1		$^1A_2$	17.9		18.8	
$^1E''$	60.6		$^1A_1$	41.0	(0.092)	42.2	(0.154)
	60.6						
$^1E'$	64.4	(0.166)[a]	$^1A_2$	49.4		60.1	
	64.4	(0.166)	$^1A_2$	52.4		66.2	(0.000)
$^1A_2''$	62.2	(0.003)	$^1B_1$	59.6	(0.003)	73.0	(0.009)
			$^1A_1$	59.8	(0.807)	73.8	(0.787)
			$^3A_2$	14.8			

[a]The numbers in parentheses are oscillator strengths.
[b]Eyler *et al.*, 1984.

3.2. Structure and Properties of $C_5H_5^+$ Species

The next series of species considered in the soot formation process are $C_5H_5^+$ cations. Because of the larger number of atoms, these species create more challenges in terms of obtaining theoretical predictions than the smaller $C_3H_3^+$ ions. First of all, the number of structures lying on the potential surface of $C_5H_5^+$ is larger than in the case of $C_3H_3^+$. Second, the computational time necessary for calculations is roughly proportional to the fourth (HF level) or sixth (MP2 level) power of the number of basis functions. Cox and Williams (1981) found that the average relative central processor unit (CPU) time used for *ab initio* calculations with STO-3G, 6-31G (similar to 3-21G), and 6-31G** basis sets was approximately 1:3:8. On the other hand, for calculations carried out with the same basis sets for the $C_3H_3^+$ and $C_5H_5^+$ systems, the CPU time for $C_5H_5^+$ increases by a factor of 7.8 and 21.6 at the HF and MP2 levels, respectively. Also, the requirements for computer memory and disk storage increase rapidly with an increase in the number of basis functions used. In addition, the number of molecular parameters which are optimized is larger for the ten-atomic than for the six-atomic species.

Despite all these constraints, the properties and structures of $C_5H_5^+$ ions have been the subject of both semiempirical and *ab initio* level predictions (Kollmar *et al.*, 1973; Hehre and Schleyer, 1973; Stohrer and Hoffman, 1972; Dewar and Haddon, 1973; Kohler and Lischka, 1979; Feng *et al.*, 1989).

The most thorough study on the potential energy surface of $C_5H_5^+$ was published by Feng *et al.* (1989). The computational methodology was analogous to that applied in a previous study of the $C_3H_3^+$ system (Cameron *et al.*, 1989). As in the latter study, investigations started with geometry optimizations using the semiempirical ZINDO method. This approach led to 21 stable structures. All these

Table III. Calculated Vibrational Frequencies (cm^{-1}) of $C_5H_5^+$ Forms Using 3-21G Optimized Structures and Wave Functions, Feng et al., 1989

Form:	I	II	III	IV	V	VI[b]	VII	VIII	IX
	167.1	247.4	150.7	505.3	255.6	254	167.8	154.1	155.9
	251.6		358.9		360.2	360	198.4	195.4	274.7
	412.1	553.9	425.1	794.4	469.0	445.9	365.4	401.7	380.9
	579.5	858.9	774.2	940.9	807.2	811.5	395.3	453.5	389.7
	768.1		793.2		895.4	825.8	554.1	622.0	574.0
	804.0		815.9	977.4	900.1	931.9	633.0	664.1	591.5
	994.1	1002.4	954.1		935.5	933.7	833.1	898.0	672.2
	1089.6	1003.5	1021.6	1006.1	1007.7	1005.3	846.5	1036.6	1015.8
	1120.2		1070.6		1107.3	1070.6	1104.1	1038.8	1050.4
	1129.8	1059.5	1104.9	1092.6	1129.8	1127.9	1181.3	1106.8	1075.5
	1131.9	1152.9	1190.2		1158.6	1159.0	1212.8	1138.9	1106.8
	1235.7	1155.3	1204.8	1290.8	1176.2	1178.2	1238.1	1211.1	1112.5
	1245.9		1224.4	1297.7	1179.4	1213.5	1267.6	1271.0	1115.6
	1335.9	1246.9	1257.8		1267.2	1268.6	1430.5	1359.2	1173.1
	1466.4	1259.6	1306.0	1549.5	1355.0	1356.3	1469.9	1434.1	1490.6
	1479.9	1295.4	1425.0		1429.4	1439.1	1582.0	1500.6	1522.0
	1589.6	1316.7	1540.5	1566.7	1500.2	1475.0	1660.2	1642.6	1539.3
	1777.3	1445.6	1610.3	1580.6	1581.4	1691.1	1721.8	1726.2	2041.0
	1807.6		1911.6		1800.0	1715.9	2309.5	2308.8	2112.9
	3331.7	3480.7	3331.9	3474.2	3404.0	3389.2	3289.6	3316.9	3274.4
	3394.8	3589.1	3408.0		3404.7	3430.7	3332.6	3324.4	3279.1
	3425.5		3410.0	3480.8	3447.6	3435.9	3380.7	3400.3	3345.3
	3474.3	3509.2	3419.1		3465.7	3465.9	3395.9	3425.7	3360.6
	3512.8	3585.9	3495.4	3688.8	3474.0	3472.0	3431.6	3594.0	3360.9
	0.0903[a]	0.0886[a]	0.0848[a]	0.0897[a]	0.0855[a]	0.0855[a]	0.0843[a]	0.0899[a]	0.0873[a]
	−192.5468[c]	−192.5263[c]	−192.5124[c]	−192.511[c]	−192.5322[c]	−191.9105[c]	−191.8891[c]	−191.8919[c]	−191.8844[c]

[a]ZPE (in a.u.)

[b]The symmetric structure of Fig. 12 has an imaginary frequency of 274i cm^{-1}, leading to the asymmetric structure of that figure.

[c]MP2/6-31G**//HF/6-31G* energy (in a.u.).

Table IV. Calculated Electronic Spectra $(1000 \text{ cm}^{-1})^a$ of $C_5H_5^+$ Forms by Using the ZINDO Spectroscopic Method for 6-31G* Optimized Geometries, Feng et al., 1989

Form:	I	II	III	IV	V	VI	VII	VIII	IX
	singlets	singlets	singlets	triplets	singlets	singlets	singlets	singlets	singlets
	43.1	61.7	18.5	28.6	10.0	7.6	15.7	22.6	23.0
	(0.582)	(0.177)	(0.024)	(0.000)	(0.005)	(0.009)	(0.000)	(0.000)	(0.000)
	50.3	61.7	38.4	28.6	25.3	22.1	34.6	33.1	27.9
	(0.008)	(0.177)	(0.006)	(0.000)	(0.019)	(0.004)	(0.762)	(0.662)	(0.000)
	50.4	71.1	44.3	33.7	40.5	41.1	42.9	4.6	42.3
	(0.001)	(0.005)	(0.667)	(0.001)	(0.004)	(0.004)	(0.000)	(0.000)	(1.002)
	55.4	71.1	48.8	33.7	45.9	43.7	46.4	46.6	48.6
	(0.003)	(0.005)	(0.004)	(0.001)	(0.124)	(0.137)	(0.001)	(0.363)	(0.001)
	57.2		49.6	49.0	49.0	48.2	46.5	47.8	53.6
	(0.001)		(0.000)	(0.510)	(0.031)	(0.070)	(0.476)	(0.000)	(0.009)
	58.3			49.0	51.9	51.3	54.0	53.3	53.7
	(0.000)			(0.510)	(0.000)	(0.000)	(0.010)	(0.002)	(0.361)
	59.6			49.7	52.3	51.7	54.1		
	(0.000)			(0.004)	(0.000)	(0.000)	(0.002)		
				51.6	53.0	51.9			
				(0.000)	(0.000)	(0.007)			
				51.6					
				(0.000)					

^{a}The numbers in parentheses are oscillator strengths.

structures were subsequently examined by *ab initio* calculations using STO-3G, 3-21G, and, finally, 6-31G* basis sets. Harmonic vibrational frequencies and intensities were calculated at the HF/3-21G level, and electronic spectra were predicted using the ZINDO/S-CI method (Ridley and Zerner, 1973; Zerner et al., 1980).

Out of the initial group of 21 species, 9 stable structures were retrieved by *ab initio* optimizations (Figs. 7–15). The global minimum, the vinylocyclopropenium cation (**I**) has only C_s symmetry, as do forms **VII** and **VIII**. Four of the structures obtained, **III**, **V**, **VI**, and **IX**, are characterized by C_{2v} symmetry, while structure **II** and the triplet state of form **IV** have higher symmetry, C_{5v} and D_{5h}, respectively. Pyramidal form **II**, predicted by early calculations of Stohrer and Hoffman (1972) to lie lowest on the potential energy surface, has been detected in solution in the experiments of Masamune et al. (1972a,b) and Saunders et al. (1973).

The 1,3-dimethyleneallyl cation (**IX**) furnishes an example of the failure of the semiempirical method to predict correctly molecular geometry. The form obtained by an optimization from the ZINDO calculations is characterized by a closed, three-membered ring consisting of central carbons. During *ab initio* optimization, the ring opens, resulting in structure **IX**.

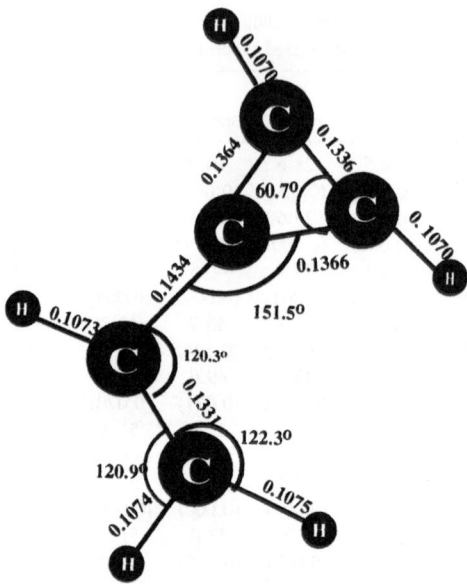

Figure 7. Optimized structure (HF/6-31G*) of vinylcyclopropenium cation **I**.

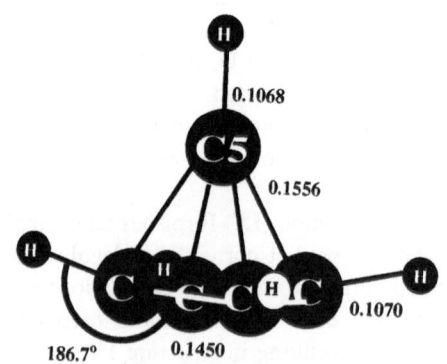

Figure 8. Optimized structure (HF/6-31G*) of form **II**.

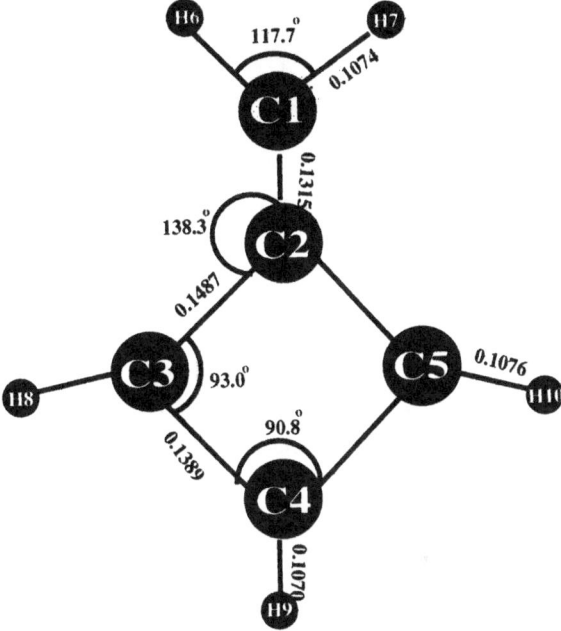

Figure 9.   Optimized structure (HF/6-31G*) of methylenecyclobutenium cation **III**.

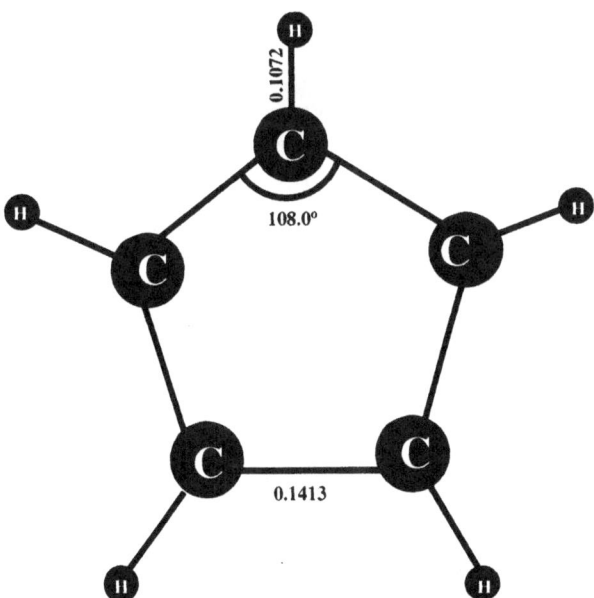

Figure 10.   Optimized structure (HF/6-31G*) of triplet state D_{5h} symmetry form **IV**.

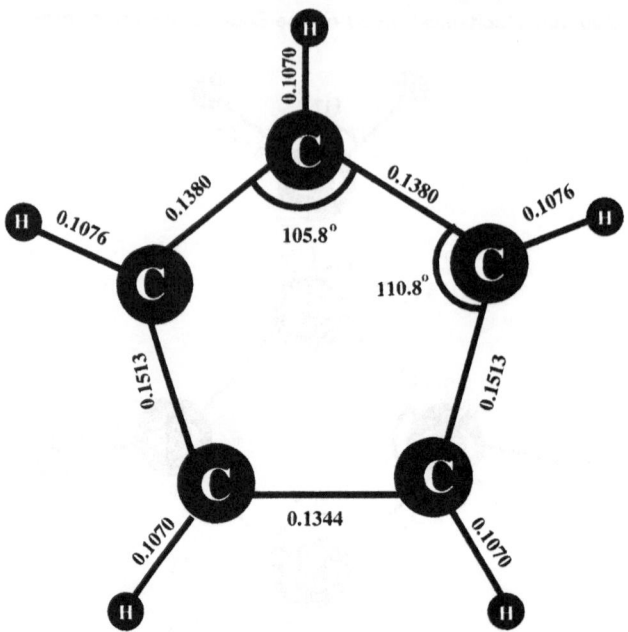

Figure 11. Optimized structure (HF/6-31G*) of singlet state C_{2v} symmetry cyclopentadienium cation, **V**.

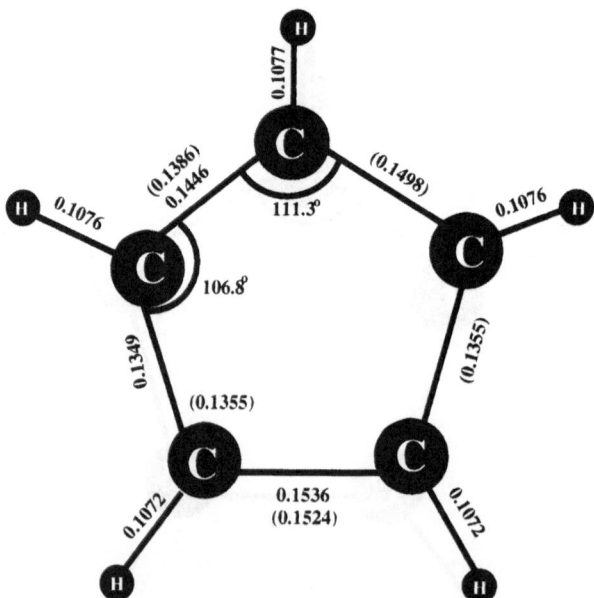

Figure 12. Optimized structure (HF/6-31G*) of transition form **VI**. The numbers given in parenthesis give rise to a minimum structure that is lower in energy by 0.0002 a.u.

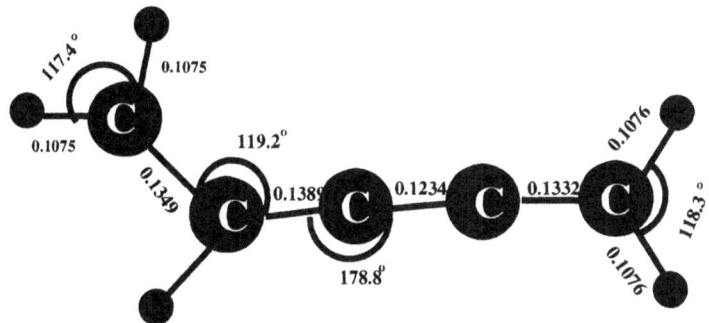

Figure 13. Optimized structure (HF/6-31G*) of vinyl-substituted propagylium cation **VII**.

Figure 16 displays the predicted relative energies of nine forms of $C_5H_5^+$ as a function of the theory level. Also for these systems, a notable variation in relative stabilities is observed. Even the order of the most stable species is reversed, on going from calculations at the STO-3G (3-21G) levels to the MP2/6-31G**//HF/6-31G* approximation. The relative stabilization for form **II** is the most sensitive to the level of calculations. Since the relative energy of even the most unstable form **IX** is smaller than 30 kcal/mol, this structure can be accessed in high temperatures during soot formation reactions. Accordingly, Tables III and IV show calculated harmonic vibrational frequencies and electronic spectra for all the structures. Because of the size of the systems, vibrational frequencies were predicted at the HF/3-21G level. The spectroscopically parameterized ZINDO/ S-CI method developed by Zerner and co-workers (Ridley and Zerner, 1973; Zerner *et al.*, 1980) was applied for prediction of the UV spectrum. These

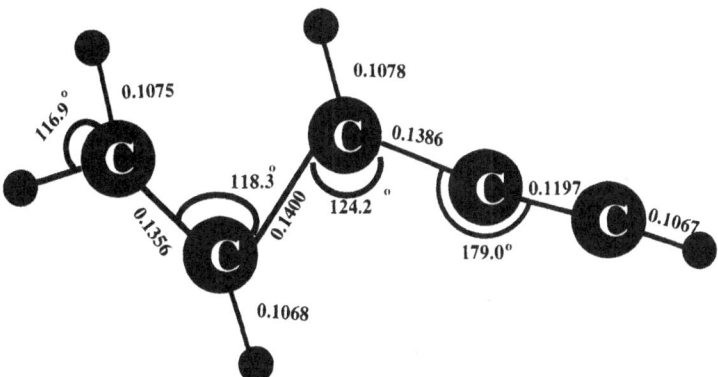

Figure 14. Optimized structure (HF/6-31G*) of ethynyl-substituted allyl cation **VIII**.

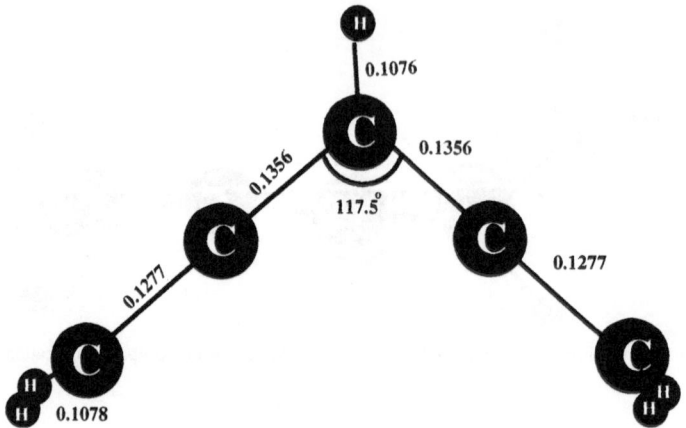

Figure 15. Optimized structure (HF/6-31G*) 1,3-dimethyleneallyl cation **IX**.

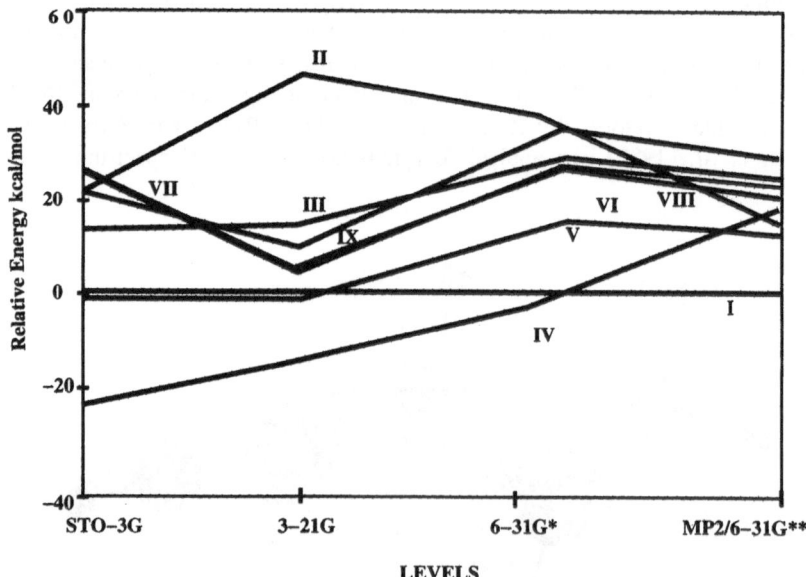

Figure 16. Plot of relative energies of $C_5H_5^+$ forms versus theoretical levels. All energies are relative to **I**.

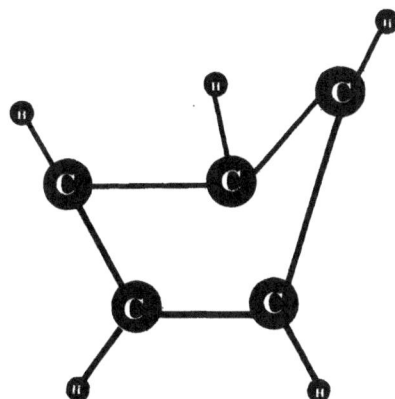

Figure 17. Suggested by Dewar and Haddon (1973) bent cyclic structure of $C_5H_5^+$ form.

calculations were carried out on previously optimized HF/6-31G* geometries of the cations.

Another serious failure of simple, semiempirical methods to predict properties of the $C_5H_5^+$ species should be noted. Based on the results of MNDO/3 calculations, Dewar and Haddon (1973) predicted that the bent cyclic form shown in Fig. 17 might be the most favored structure. However, nonempirical calculations do not recover this form (Feng *et al.*, 1989). A few attempted *ab initio* geometry optimizations at different levels starting from the bent geometry of Dewar and Haddon led to the planar forms **V** or **VI**.

3.3. Reaction of $C_3H_3^+$ with Acetylene

Reactions of $C_3H_3^+$ with acetylene and different products of its polymerization were proposed by Calcote (Calcote, 1981; Olson and Calcote, 1981) in his ion/molecular soot formation mechanism. Such fast reactions lead to the formation of large polycyclic aromatic hydrocarbon ions reaching molecular weights on the order of 10^3. Many reactions might occur between positive ions and neutral molecules. The initial reactions between $C_3H_3^+$ and acetylene can be described by the following scheme (Leszczyński *et al.*, 1988):

$$C_3H_3^+ + C_2H_2 \rightarrow [C_5H_5^+]^* \rightarrow C_5H_5{}^+$$
$$\downarrow$$
$$C_5H_3^+ + H_2$$
$$\downarrow$$
$$[C_3H_3^+]^* + H_2$$
$$\downarrow$$
$$C_3H_3^+$$

This chain of reactions yields then $C_5H_5^+$, $C_5H_3^+$, and perhaps different isomeric forms of the reactant $C_3H_3^+$.

Comprehensive theoretical studies on reactions of different forms of $C_3H_3^+$ with acetylene were published by Leszczyński et al. (1988) and Feng et al. (1989). An *ab initio* method at the Hartree–Fock level was applied for geometry optimizations of the possible reaction products. Basis sets used in the calculations ranged from the minimal, STO-3G basis set to a split-valence, double-zeta 6-31G* basis set with *d*-polarization functions on carbons and a 6-31G** basis set with polarization functions on both carbon and hydrogen atoms. The 6-31G** basis set was applied at the single-point, MP2 level calculations of the total energies of the predicted structures.

Initially, the reactions of the global minimum form, the cyclopropenyl cation, were examined. Three different approaches of acetylene to the cyclopropenyl cation were investigated, one out-of-plane and two in-plane, as shown in Figs. 18 and 19. The only symmetry restriction assumed for the starting structures was a plane of symmetry of the cyclopropenyl ion. The C_2H_2 molecule was placed sufficiently far from the target cation to account for the possible rearrangements of the acetylene molecule during reaction. During geometry optimizations, the acetylene molecule was repelled by the cyclopropenyl cation; the final geometries of the products were very similar to the initial geometries of the reactants, but the optimized geometries were characterized by larger distances from the cyclopropenyl ion to acetylene than those assumed for the input data. All structures displayed a very flat potential toward the motion of C_2H_2. The cyclic isomer formed encounter complexes $C_3H_3^+ \cdots C_2H_2$ of no specific structures. No further process occurred without a profound reaction barrier. The stabilization energies

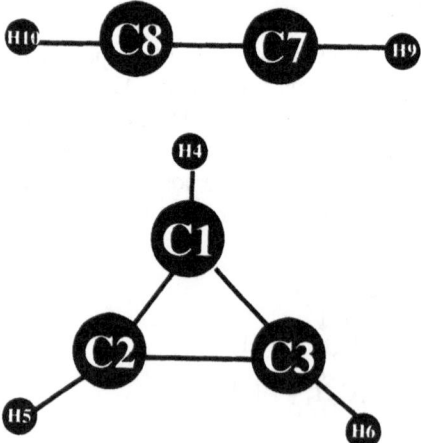

Figure 18. Initial and optimized structures (HF/STO-3G) of weekly bonded complex between c-$C_3H_3^+$ and acetylene.

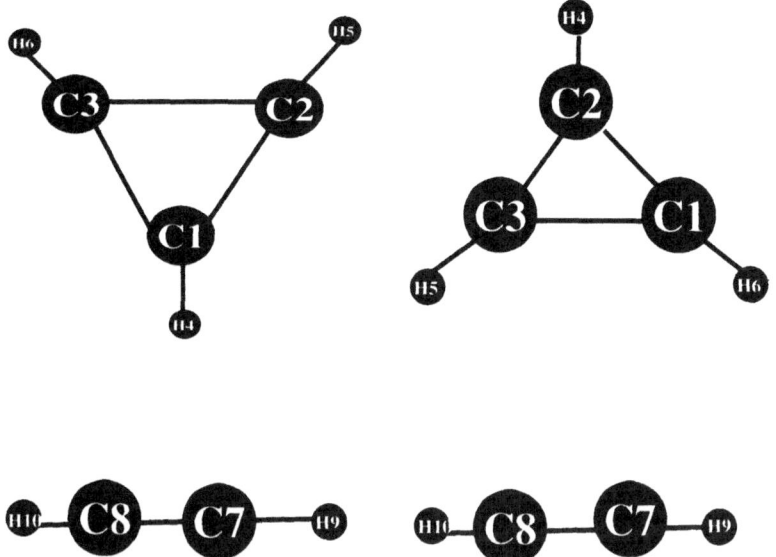

Figure 19. Initial and optimized structures (HF/STO-3G) of the two encounter complexes between acetylene and c-$C_3H_3^+$.

calculated at the MP2/6-31G** level amount to about 6 kcal/mol for all three complexes studied.

More fruitful results were obtained by the same authors (Leszczyński *et al.*, 1988; Feng *et al.*, 1989) when the starting material was the propargylium cation, the second most stable $C_3H_3^+$ form. They examined various reaction paths between acetylene and the linear propargylium ion. Figures 20 and 21 display initial in-plane approaches of C_2H_2 to $C_3H_3^+$ together with the optimized structures of the reaction products, while out-of-plane initial reaction conditions were assumed for the reactions in Figs. 22 and 23. Figure 24 shows the starting geometry, the intermediate geometry, and two views of the final geometry of the third, out-of-plane reaction. Initially, acetylene was directed over the central atom of the propargylium ion but slightly twisted from the perpendicular orientation. During geometry optimization, the acetylene drifted to the most positive carbon atom at the end of the ion. Finally, it attached to this carbon and formed a three-carbon ring. Only for the reaction depicted in Fig. 20 was a plane of symmetry imposed; no symmetry restrictions apply to the other initial geometries. All reactions were optimized at different levels, up to the HF/6-31G* theory level.

The product of the reaction displayed in Fig. 23 has been characterized by Feng *et al.* (1989) as one of the stable forms (**III**) lying on the potential energy

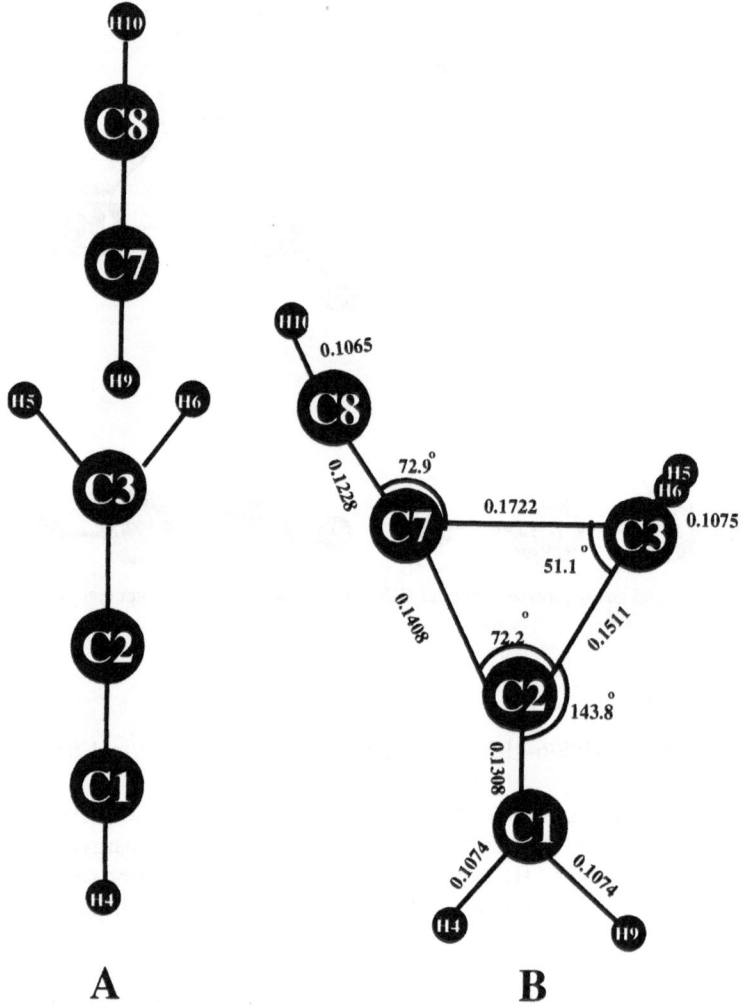

Figure 20. Initial (A) and optimized (B) structures (HF/6-31G*) of one of the reactions between 1-$C_3H_3^+$ and acetylene.

surface of $C_5H_5^+$. The calculated MP2/6-31G**//HF/6-31G* energies and HF/3-21G harmonic vibrational frequencies for the products shown in Figs. 20, 21, 23, and 24 are reported in Table V.

Products of the studied reactions are stable relative to dissociation back into acetylene and the global-minimum cyclic $C_3H_3^+$ ion if their energies are smaller

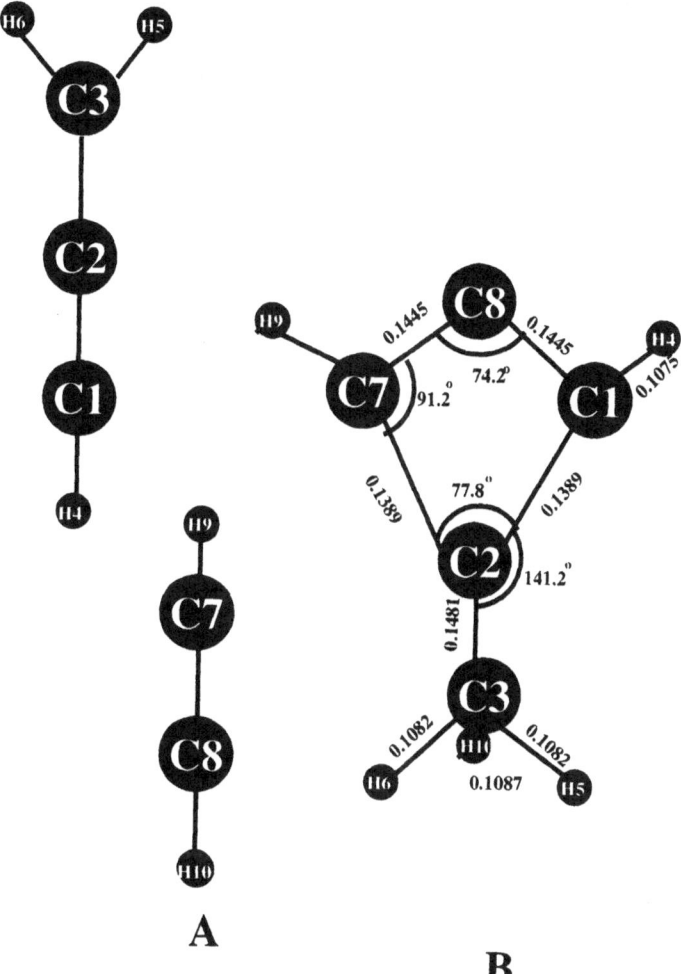

Figure 21. Initial (A) and optimized (B) structures (HF/6-31G*) of the second reaction between 1-$C_3H_3^+$ and acetylene.

than -115.3694 (MP2/6-31G**//HF/6-31G* energy of the propargylium cation) and -77.0795 hartrees (MP2/6-31G**//HF/6-31G* energy of C_2H_2). After corrections for the scaled zero-point vibrational energy (ZPE), the predicted value amounts to -192.3758 hartrees. This estimation can be further improved by including the entropy effects. All these contributions are incorporated in the equation yielding the free energy of the reaction (Leszczyński *et al.*, 1988):

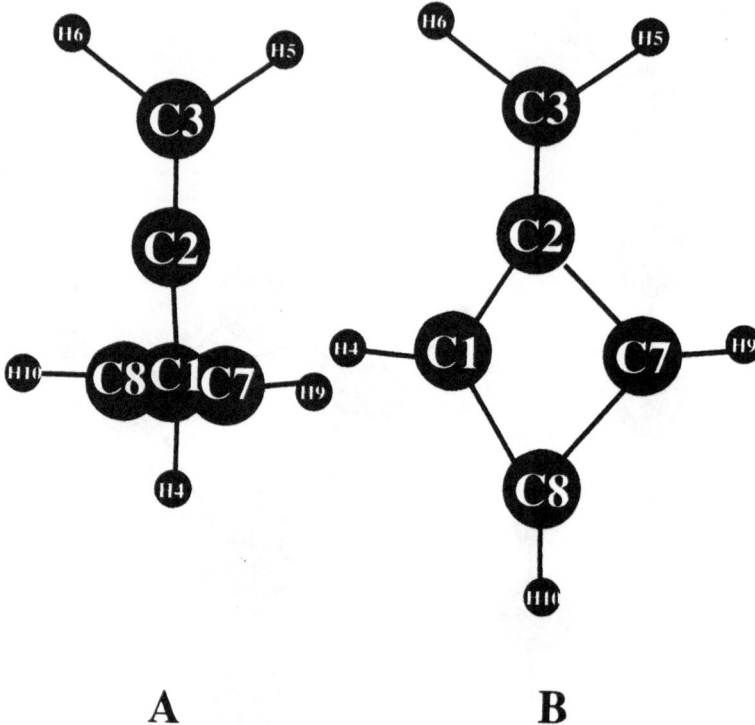

A   **B**

Figure 22. Initial (A) and optimized (B) structures (HF/6-31G*) of the third reaction between 1-$C_3H_3^+$ and acetylene.

$$\Delta G = E^{MP2} + 0.9ZPE - 192.3758 + 5.0 \times 10^{-5}T$$

where E^{MP2} is the MP2/6-31G**//HF/6-31G* energy of the product, and T is the temperature in degrees Kelvin.

An examination of the energies of the reaction products indicates that at 0 K all species, except that of Fig. 23, are stable toward dissociation. However, at 1000 K, which corresponds to the temperatures in flames, the entropy effects become larger, and only the reaction shown in Fig. 22 leads directly to a thermodynamically stable product. Nevertheless, because of the excess energy available at such temperatures, the high-energy products predicted for the other reactions could easily isomerize to the most stable structure for the $C_5H_5^+$ ion. Further collisions might lead to consecutive reactions with acetylene and products of higher carbon content.

The results obtained are in agreement with the experimental data obtained from Fourier transform ion cyclotron resonance (FT-ICR) studies on model flames.

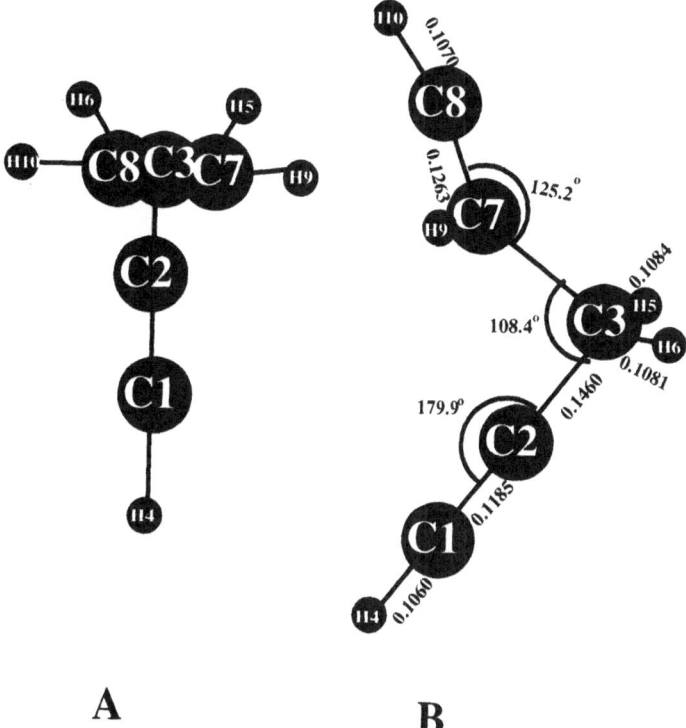

A   **B**

Figure 23. Initial (A) and optimized (B) structures (HF/6-31G*) of the fourth reaction between 1-$C_3H_3^+$ and acetylene.

Investigations by Ozturk *et al.* (1987) indicated that $C_3H_3^+$ does react with acetylene; however, under the conditions of their experiments, no stable products were formed. The same study confirmed the existence of two forms of $C_3H_3^+$, only one of which was reactive.

3.4. Larger Ions

Though the initial reactions during formation of small $C_nH_n^+$ ions were studied by rigorous *ab initio* calculations, the formation of larger species ($n \geqslant 7$) has not been investigated by nonempirical calculations. For these systems, semiempirical predictions could elucidate further steps in the ionic mechanism of the soot formation process. Nevertheless, results of such calculations may be misleading, as discussed in previous sections, and their conclusions should be used with caution.

A number of minimum and first-order transition forms of $C_7H_7^+$ and $C_7H_8^+$

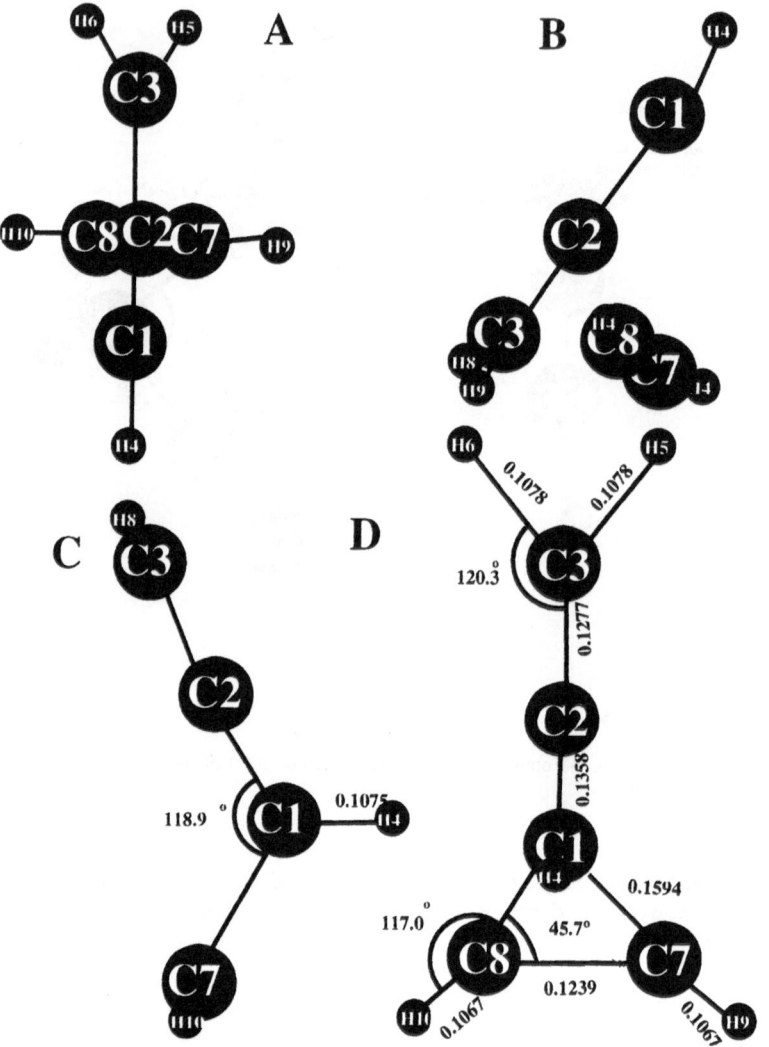

Figure 24. Initial (A), intermediate (B) and two views of optimized (C and D) structures (HF/6-31G*) of fifth reaction of 1-$C_3H_3^+$ and acetylene.

were studied by Dewar's group (Dewar and Landman, 1977; Cone *et al.*, 1977) using the MNDO/3 semiempirical method. The calculations yielded the next member of the aromatic series, the tropylium ion. This symmetrical (D_{7h} symmetry) cation was calculated to possess C–C bond distances of 0.1405 nm and a heat of formation of 1033 kJ/mol. Based on the calculated data, Cone *et al.* (1977) suggested a mechanism for the arrangement of the benzyl cation to the tropylium ion.

Table V. Calculated Vibrational Frequencies (cm^{-1}) of Products of Reactions Between 1-C$_3$H$_3^+$ and Acetylene by Using 3-21G Optimized Structure and Wave Functions, Feng et al., 1989

Fig. 20B	Fig. 21B	Fig. 23B	Fig. 24D
169.4	73.2	104.5	135.0
291.2	293.5	237.8	231.3
305.7	364.8	270.7	335.9
467.6	519.9	419.3	517.1
475.5	665.3	542.0	532.5
765.5	699.8	747.6	560.7
818.2	860.1	759.6	733.0
834.2	1006.4	888.0	869.9
908.3	1010.5	959.3	922.4
986.7	1151.1	977.6	985.9
1023.7	1165.2	996.6	1043.1
1148.3	1204.3	1047.3	1089.2
1206.6	1243.7	1125.0	1115.4
1243.3	1322.0	1240.3	1125.3
1256.3	1390.5	1367.9	1159.7
1579.0	1566.0	1457.5	1427.1
1616.0	1619.8	1616.3	1534.3
1914.8	1624.1	1836.0	1947.1
2249.6	1651.6	2415.6	2106.3
3297.6	3187.9	3243.7	3285.0
3345.7	3261.8	3271.6	3368.4
3425.1	3309.4	3309.3	3409.3
3440.3	3440.0	3482.6	3500.9
3579.6	3448.2	3638.6	3586.2
-192.4844^a	-192.4588^a	-192.4323^a	-192.4787^a

^{a}MP2/6-31G**//HF/6-31G* energy (in a.u.).

Very recently, results of semiempirical studies on the reaction between the phenylvinyleum cation C$_8$H$_7^+$ and acetylene were published by Wang et al. (1993). Using the AM1 method, they identified 7 major forms for the product of the reaction

$$C_8H_7^+ + C_2H_2 \rightarrow C_{10}H_9^+$$

Also, 11 possible reaction channels were predicted for the reaction. All species were characterized through their equilibrium geometries, vibrational frequencies, and enthalpies of formation. Protonated naphthalenes were found to be the most stable forms of the C$_{10}$H$_9^+$ cations. These results are in agreement with the experimental work of Stein (1983).

Calculated properties were subsequently applied by the authors (Wang et al., 1993) for the analysis of reaction rate coefficients. Two dominating channels leading to the formation of protonated naphthalenes were predicted for temperatures below 700 K and at a pressure of 20 Torr. When the temperature rises, the rate coefficients of these channels decrease, and also other reactions become more feasible. An analysis of the microcanonical rate constants and the collision stabilization rates allowed for an explanation of the observed trends. The process is governed by the competition for the hot reaction products between the exit reaction channels and the reverse of the entrance channel.

The rate coefficients calculated by Wang *et al.* (1993) deviate significantly from those predicted by the Langevin model, which assumes that for fast ion–molecule reactions, rate constants are nearly independent of temperature. Under typical combustion conditions of low pressure and high temperature (above 1500 K), the calculated rate constants are three orders of magnitude lower than the Langevin limit. Insufficient collision stabilization of the energized products was suggested by the authors as a main reason for such discrepancies.

4. CONCLUSIONS

A vast number of chemical reactions leading toward soot formation can be modeled by computational quantum-chemical *ab initio* (nonempirical) or semiempirical studies. *Ab initio* calculations prove to be a valuable method for the investigation of the earlier stages of the soot formation process. Molecular structures and properties of reactants and products could be predicted using accurate, nonempirical calculations. Also, reaction mechanisms are well characterized by such predictions.

Approximate predictions could be made for larger systems using different schemes of semiempirical theories. However, in a number of cases, such predictions have proved to be wrong, and these types of calculations should be replaced by rigorous, nonempirical calculations when computational resources allow.

Two investigations of the reaction schemes between positive hydrocarbon ions and acetylene, based on *ab initio* calculations for the $C_3H_3^+$ reactions and semiempirical, AM1 predictions for the reactions of $C_8H_7^+$ ions, respectively, were reported. Both studies contribute to the understanding of the ionic mechanism of soot particle formations. However, more theoretical, as well as experimental, explorations are necessary in order to judge which of the two possible reaction pathways, ionic or radical, is dominant. The significant improvement of computer resources in the last few years makes such computational investigation nowadays feasible for systems with ca. 30–40 atoms (*ab initio* methods) and a few hundred atoms (semiempirical schemes).

ACKNOWLEDGMENTS.   The author thanks Drs. Edet Archibong and Andrzej Nowek for reading the manuscript and their helpful suggestions, Mrs. Liqun Wong and Ms. Sherrie Hankins for assistance in preparing the manuscript, and Mrs. Shirisha Reddy for help with the figures. This work was supported by the U.S. Department of Energy through the LBL/JSU/AGMEF consortium.

REFERENCES

Anderson, W. P., Edwards, W. D., and Zerner, M. C., 1986, Calculated spectra of hydrated ions of the first transition-metal series, *Inorg. Chem.* **15**:2728.

Bacon, A. D., and Zerner, M. C., 1979, An intermediate neglect of differential overlap theory for transition metal complexes: Iron, cobalt, and copper chlorides, *Theor. Chim. Acta* **53**:21.

Brown, R. C., and Eraslan, A. N., 1988, Simulation of ionic structure in lean and close-to-stoichiometric acetylene flames, *Combust. Flame* **73**:1.

Calcote, H. F., 1981, Mechanisms of soot nucleation in flames: A critical review, *Combust. Flame* **42**:215.

Calcote, H. F., and Keil, D. G., 1988, Ionic mechanisms of soot formation in flames; Ion molecule reactions in sooting acetylene–oxygen flames, *Combust. Flame* **74**:131.

Calcote, H. F., Olson, D. B., and Keil, D. G., 1988, Are ions important in soot formation?, *Energy Fuels* **2**:494.

Cameron, A., Leszczyński, J., Zerner, M. C., and Weiner, B., 1989, Structure and properties of $C_3H_3^+$ cations, *J. Phys. Chem.* **93**:139.

Cone, C., Dewar, M. J. S., and Landman, D., 1977, Gaseous ions: MINDO/3 study of the rearrangement of benzyl cation to tropylium, *J. Am. Chem. Soc.* **99**:372.

Cox, S. R., and Williams, D. E., 1981, Representation of the molecular electrostatic potential by a net atomic charge model, *J. Comput. Chem.* **2**:304.

Dewar, M. J. S., and Haddon, R. C., 1973, MINDO/3 study of cyclopentadienyl (1^+) and cyclopentadienyl (1^-) ions, *J. Am. Chem. Soc.* **95**:5836.

Dewar, M. J. S., and Landman, D., 1977, Gaseous ions: MINDO/3 study of the rearrangement of toluene and cycloheptatriene molecular ions and the formation of tropylium, *J. Am. Chem. Soc.* **99**:2446.

Dewar, M. J. S., and Thiel, W., 1977, Ground states of molecules. 39. MNDO results for molecules containing hydrogen, carbon, nitrogen, and oxygen, *J. Am. Chem. Soc.* **99**:4907.

Dewar, M. J. S., Zoebisch, E. G., Healy, E. F., and Stewart, J. J. P., 1985, Development and use of quantum mechanical molecular models, *J. Am. Chem. Soc.* **107**:3902.

Eyler, J. R., Oddershede, J., Sabin, J. R., Diercksen, G. H. F., and Gruner, N., 1984, Excitation energy of linear propargylium ($C_3H_3^+$): Can this ion be detected by laser-induced fluorescence in flames?, *J. Phys. Chem.* **88**:3121.

Feng, J., Leszczynski, J., Weiner, B., and Zerner, M. C., 1989, The reaction $C_3H_3^+ + C_2H_2$ and the structural isomers of $C_5H_5^+$, *J. Am. Chem. Soc.* **111**:4648.

Frenklach, M., Clary, D. C., Gardiner, W. C., Jr., and Stein, S. E., 1985, Representation of multistage mechanisms in detailed computer modeling of polymerization kinetics, in *Twentieth Symposium (International) on Combustion*, p. 887, The Combustion Institute, Pittsburgh.

Gaydon, A. G., and Wolfhard, H. G., 1979, *Flames: Their Structure, Radiation, and Temperature*, 4th ed., John Wiley & Sons, New York.

Gerhardt, P., and Homann, K. H., 1990, Ions and charged soot particles in hydrocarbon flames: Positive aliphatic and aromatic ions in ethyne/oxygen flames, *J. Phys. Chem.* **94**:5381.

Gioumousis, G., and Stevenson, D. P., 1958, Reactions of gaseous molecular ions with gaseous molecules: Theory, *J. Chem. Phys.* **29**:294.

Glassman, I., 1989, Soot formation in combustion processes, in *Twenty-Second Symposium (International) on Combustion*, p. 295, The Combustion Institute, Pittsburgh.

Harris, S. J., and Weiner, A. M., 1985, Detection of incipient soot particles in a premixed ethylene flame, *Annu. Rev. Phys. Chem.* **36**:31.

Harris, S. J., Weiner, A. M., and Blint, R. J., 1988, Formation of small aromatic molecules in a sooting ethylene flame, *Combust. Flame* **72**:91.

Harrison, A. H., Haynes, P., McLean, S., and Meyer, F., 1965, The mass spectra of methyl substituted cyclopentadienes, *J. Am. Chem. Soc.* **87**:5099.

Hayhurst, A. N., and Jones, H. R. N., 1987, Ions and soot in flames, *J. Chem. Soc., Faraday Trans.* **83**:1.

Hayhurst, A. N., and Kittelson, D. B., 1978, The positive and negative ions in oxy-acetylene flames, *Combust. Flame* **31**:37.

Hehre, W. J., and Schleyer, P. v. R., 1973, Cyclopentadienyl and related $(CH)_5^+$ cations, *J. Am. Chem. Soc.* **95**:5837.

Hehre, W. J., Radom, L., Schleyer, P. v. R., and Pople, J. A., 1986, *Ab Initio Molecular Orbital Theory*, John Wiley & Sons, New York.

Herbst, E., Adams, N. G., and Smith, D., 1984, Ion-molecule synthesis of C_3O, *Astrophys. J.* **285**:618.

Hirst, D. M., 1990, *A Computational Approach to Chemistry* (J. P. Simons, ed.), Blackwell Scientific Publications, Cambridge.

Holmes, J. L., and Lossing, F. P., 1979, The reactivity of $[C_3H_3^+]$ ions: A thermochemical study, *Can. J. Chem.* **57**:249.

Homann, K. H., 1985, Formation of large molecules, particulates, and ions in premixed hydrocarbon flames: Progress and unresolved questions, in *Twentieth Symposium (International) on Combustion*, p. 857, The Combustion Institute, Pittsburgh.

Homann, K. H., and Wagner, H. G., 1967, Mechanisms of C formation in premixed flames, in *Eleventh Symposium (International) on Combustion*, p. 371, The Combustion Institute, Pittsburgh.

Keil, D. G., Gill, R. J., Olson, D. B., and Calcote, H. F., 1985, Ionization and soot formation in premixed flames, in *Twentieth Symposium (International) on Combustion*, p. 1129, The Combustion Institute, Pittsburgh.

Kohler, H. J., and Lischka, H., 1979, Theoretical investigations on carbocations: Structure and stability of $C_3H_5^+$, $C_4H_9^+$ (2-butyl cation), $C_5H_5^+$, $C_6H_7^+$ (protonated benzene), and $C_7H_{11}^+$ (2-norbornyl cation), *J. Am. Chem. Soc.* **101**:7863.

Kollmar, H., Smith, H. O., and Schleyer, P. v. R., 1973, CNDO calculations of isomeric $(CH)_5^+$ cations, *J. Am. Chem. Soc.* **95**:5834.

Kwiatkowski, J. S., and Leszczynski, J., 1992, *Ab initio* post-Hartree–Fock studies on molecular structure and vibrational IR spectrum of formaldehyde, *Int. J. Quantum Chem., Quantum Chem. Symp.* **26**:421.

Kwiatkowski, J. S., and Leszczyński, J., 1993, Harmonized infrared spectrum of formaldehyde: Experiment and theory, *J. Mol. Spectrosc.* **157**:540.

Lepp, S., Dalgarno, A., Van Dishoeck, E. F., and Black, J. H., 1988, Large molecules in diffuse interstellar clouds, *Astrophys. J.* **324**:553.

Leszczyński, J., and Goodman, L., 1994, Coupled cluster prediction of the formaldehyde harmonic force field, *J. Computational Chem.*, submitted.

Leszczyński, J., Wiseman, F., and Zerner, M. C., 1988, Toward an ionic mechanism of soot particle formation: Reactions between acetylene and tautomeric forms of the $C_3H_3^+$ ions, *Int. J. Quantum Chem., Quantum Chem. Symp.* **22**:117.

Lias, S. G., Liebman, J. F., and Levin, R. D., 1984, Evaluated gas phase basicities and proton affinities of molecules: Heats of formation of protonated molecules, *J. Phys. Chem. Ref. Data* **13**:695.

Lipkowitz, K. B., and Boyd, D. B. (eds.), 1990. *Reviews in Computational Chemistry*, Vol. I, VCH, New York.

Lipkowitz, K. B., and Boyd, D. B. (eds.), 1991. *Reviews in Computational Chemistry*, Vol. II, VCH, New York.

Lipkowitz, K. B., and Boyd, D. B. (eds.), 1992. *Reviews in Computational Chemistry*, Vol. III, VCH, New York.

Lipkowitz, K. B., and Boyd, D. B. (eds.), 1993. *Reviews in Computational Chemistry*, Vol. IV, VCH, New York.

Lipkowitz, K. B., and Boyd, D. B. (eds.), 1994. *Reviews in Computational Chemistry*, Vol. V, VCH, New York.

Lossing, F. P., 1972, Free radicals by mass spectrometry: Ionization potentials and heats of formation of C_3H_3, C_3H_5, and C_4H_7 radicals and ions, *Can. J. Chem.* **50**:3973.

Maclagan, R. G. A. R., 1992, The proton affinity of C_3H_2, *J. Mol. Struct. (Theochem)* **258**:175.

Masamune, S., Sakai, M., and Ona, H., 1972a, Nature of the $(CH)_5^+$ species: Solvolysis of 1,5-dimethyltricyclo[2.1.0.0.]pen-3-yl benzoate, *J. Am. Chem. Soc.* **94**:8955.

Masamune, S., Sakai, M., Ona, H., and Jones, A., 1972b, Nature of the $(CH)_5^+$ species: Direct observation of the carbonium ion of 3-hydroxyhomotetrehedrane derivatives, *J. Am. Chem. Soc.* **94**:8956.

McCreary, D. A., and Freiser, B. S., 1978, Gas phase photodissociation of $C_7H_7^+$, *J. Am. Chem. Soc.* **100**:2902.

Miller, J. A., and Melius, C. F., 1992, Kinetic and thermodynamic issues in the formation of aromatic compounds in flames of aliphatic fuels, *Combust. Flame* **91**:21.

Miller, W. J., 1973, Electron attachment kinetics in flames: Dissociative attachment to boric acid (HBO_2), in *Fourteenth Symposium on Combustion*, The Combustion Institute, Pittsburgh.

Olson, D. B., and Calcote, H. F., 1981, A comparison of polycyclic aromatic hydrocarbon in soot produced from various sources, in *Eighteenth Symposium (International) on Combustion*, p. 453, The Combustion Institute, Pittsburgh.

Ozturk, F., Baykut, G., Moini, M., and Eyler, J. R., 1987, Reactions of the propynylium isomer manifold with acetylene and diacetylene in the gas phase, *J. Phys. Chem.* **91**:4360.

Pople, J. A., Beveridge, D. L., and Dobosh, P., 1967, Approximation self-consistent MO theory: Intermediate neglect of differential overlap, *J. Chem. Phys.* **47**:2026.

Radom, L., Hariharan, P. C., Pople, J. A., and Schleyer, P. v. R., 1976, Molecular orbital theory of the electronic structure of organic compounds: Structures and stabilities of $C_3H_3^+$ and C_3H^+ cations, *J. Am. Chem. Soc.* **98**:10.

Raghavachari, K., Whiteside, R. A., Pople, J. A., and Schleyer, P. v. R., 1981, Molecular orbital theory of the electronic structure of organic molecules: Structures and energies of C_1–C_3 carbocations including effects of electron correlation, *J. Am. Chem. Soc.* **103**:5649.

Ridley, J. E., and Zerner, M. C., An intermediate neglect of differential overlap technique for spectroscopy: pyrole and the azines, *Theor. Chim. Acta* **32**:111.

Saunders, M., Berger, R., Jaffe, A., McBride, J. M., O'Neill, J., Breslow, R., Hoffman, J. M., Jr., Perchonok, C., Wasserman, E., Hutton, R. S., and Kuck, V. J., 1973, Unsubstituted cyclopentadienyl cation: A ground state triplet, *J. Am. Chem. Soc.* **95**:3017.

Schiff, H. I., and Bohme, D. K., 1979, An ion–molecule scheme for the synthesis of hydrocarbon chain and organonitrogen molecules in dense interstellar clouds, *Astrophys. J.* **232**:740.

Smith, D., and Adams, N. G., 1978, Molecular synthesis in interstellar clouds: Radioactive association reactions of methyl (+) ions, *Astrophys. J.* **220**:L87.

Stein, S. E., 1983, Structure and equilibria of polyaromatic flame ions, *Combust. Flame* **51**:357.

Stewart, J. J. P., 1990, Semiempirical molecular orbital methods, in *Reviews in Computational Chemistry*, Vol. I (K. B. Lipkowitz, and D. B. Boyd, eds.), pp. 45–78, VCH, New York.

Stohrer, W. D., and Hoffman, R., 1972, Electronic structure and reactivity of strained tricyclic hydrocarbons, *J. Am. Chem. Soc.* **94**:1661.

Takada, T., and Ohno, K., 1979, Calculations on the electronic structure of the cyclopropenyl cation, *Bull. Chem. Soc. Jpn.* **52**:334.

Tanner, S. D., Goodings, J. M., and Bohme, D. K., 1981, Hydrocarbon ions in fuel-rich, CH_4–C_2H_2–O_2 flames as a probe for the initiation of soot: An experimental approach, *Can. J. Chem.* **59**:1760.

Tse, R. S., Michand, P., and Delfau, J. L., 1978, Initial pyrolysis ions in hydrocarbon flames, *Nature (London)* **272**:153.

Turner, B. E., 1989, Detection of interstellar C_4D: Implication for ion–molecule chemistry, *Astrophys. J.* **347**:L39.

Wang, H., Weiner, B., and Frenklach, M., 1993, Theoretical study of reaction between phenylvinyleum ion and acetylene, *J. Phys. Chem.* **97**:10364.

Westmoreland, P. R., Dean, A. M., Howard, J. B., and Longwell, J. P., 1989, Forming benzene in flames by chemically activated isomerization, *J. Phys. Chem.* **93**:8171.

Wong, M. W., and Radom, L., 1989, Multiply charged isoelectronic analogues of $C_3H_3^+$: Cyclic or open chain?, *J. Am. Chem. Soc.* **111**:6976.

Zerner, M. C., 1991, Semiempirical molecular orbital methods, in *Review in Computational Chemistry*, Vol. II (K. B. Lipkowitz and D. B. Boyd, eds.), pp. 313–359, VCH, New York.

Zerner, M. C., Loew, G. H., Kirchner, R. F., and Mueller-Westerhoff, U. T., 1980, An intermediate neglect of differential overlap technique for spectroscopy of transition-metal complexes: Ferrocene, *J. Am. Chem. Soc.* **102**:589.

Chapter 7

Clean Combustion Utilizing Fluidized-Bed Boilers

Károly Reményi

1. INTRODUCTION

The relations between energetics, the environment, and firing technology are extremely broad. The most important problems of environmental pollution caused by the production of energy are as follows:

1. Atmospheric contamination
   - Sulfur dioxide
   - Nitrogen oxides
   - Increase in carbon dioxide concentration
   - Greenhouse effect
2. Contamination of waters
   - with chemical wastes
   - heat contamination
3. Treatment of radioactive wastes from power plants
   - Wastes from nuclear power plants
   - Other radioactive wastes from power plants
4. Disposal of slag and fly ash from coal-firing power plants.

KÁROLY REMÉNYI • Institute for Electric Power Research, H-1051, Budapest, Hungary.

Combustion Efficiency and Air Quality, edited by István Hargittai and Tamás Vidóczy. Plenum Press, New York, 1995.

1.1. Energy Situation

Power plants all over the world utilize both fuels of low-grade quality and better quality by-products of sorting. A fundamental condition for economical use of fuel is efficient operation of the boiler based on high-level theoretical knowledge of firing technology. Frequent and significant changes in the grade of the fuel make operation especially difficult. Insofar as the types of fuels used in power plants is concerned, a shift in favor of hydrocarbons, similar to the world tendency, had been characteristic until the beginning of the 1970s in Hungary. Following the rise in the price of oil, the reduction in the share of oil products became a fundamental aspect in the development of energetics. It caused serious difficulties for both oil-burning and coal-firing power plants.

A few words about perspectives on the world energy situation are cited here from a report of the International Energy Agency:

> The impact of natural gas use and environmental concerns on coal use will apply mainly in the advanced economies of the world. In the developing regions a more important constraint on the level of coal use will be the difficulty of financing the desired levels of new coal-fired power station capacity.

Such difficulties are hardly new, but they are set to grow in the 1990s. The power station construction programs of the 1970s and 1980s absorbed up to 30% of total public investment in developing regions. They now often account for a substantial portion of national debt; in many countries, the revenues from electricity sales have barely covered operating costs and fuel, let alone capital costs.

Future power station building plans are even more ambitious than in the past two decades. Over the period 1991 to 2000 the combined plans of the developing countries total at least 550 GW for all types of generating capacity. At an overall system cost of around U.S. $2 billion per gigawatt, the cost of the plans thus exceed U.S. $1000 billion, or U.S. $100 billion per year.

It is unlikely that this level of investment capital will be available from within the developing countries or from international loan or aid agencies. Even with moves in an increasing number of countries to raise electricity tariffs and restrict their electricity supply industries to improve cost-effectiveness, many power station construction programs will be significantly curtailed. This will affect all types of power-generating capacity. However, coal-fired projects seeking international funding could be particularly affected, because of the "green" criteria being adopted by an increasing number of international funding agencies.

1.2. Recent Forecasts

The combined effect of the various factors will be to reduce the previously anticipated rate of growth in power station capacity and use. This is already apparent in some recent forecasts. For instance, in 1988 127.6 million tonnes of coal equivalent (Mtce) was projected for the year 2000. In 1991, the projection was for only 1036.8 Mtce.

Assessing the effect of the various constraints on world coal use against such a fast-moving background requires detailed analysis. The prospects for coal use in individual countries need to be considered within the context of overall electricity demand and supply and against a background of common global assumptions on issues such as economic growth and oil and gas prices. The past performance of particular countries and utilities in matters such as financing and constructing new capacity also needs to be taken into account. In this study the position in over 120 countries was analyzed, and in most cases coal user prospects were assessed at the level of individual power plants.

1.3. Capacity Prospects

It seems likely that by 2000 world coal-capable capacity will reach 979 GW. This is 145 GW higher than at the end of 1988. However, if replacement coal-fired capacity is taken into account, the gross capacity commissioned in the 1990s will exceed 200 GW. Adding in repowered capacity could take the total to 250 GW. Much of the additional capacity—over 100 GW—will be built in Asia. China, India, Japan, South Korea, and Taiwan will be prominent. Much of the replacement and repowered capacity will be built in Europe (IEA Coal Research, 1992a).

1.4. Coal Demand Prospects

Total coal use in power stations is expected to reach 2967 million tonnes in 2000. This is well below the 3500 million tonnes or so anticipated at one time. Nevertheless, the net growth from 2641 million tonnes in 1989 means that world coal use in power stations will grow by over 80,000 tonnes a day throughout the 1990s. Not surprisingly, much of the growth will be in Asia. Over 250 million tonnes of additional coal will be burned in 2000 compared with 1989.

1.5. Coal Trade Prospects

The seaborne steam coal trade is expected to increase to 248 million tonnes in 2000, compared with 113 million tonnes in 1989. The annual growth rate is thus substantially higher than for overall power station coal use. The Pacific Rim and Europe are the main growth areas. However, there is a significant difference between the two regions. In Asia, the growth is primarily based on "new" coal use, whereas in Europe much of the increase in traded coal use represents the replacement of locally produced coal. In the European Economic Community, coal provides 20% of the total energy supply and almost 40% of electricity production.

Current estimates indicate that natural gas could double its share of world power generation by the year 2000. This growth potential has been stimulated by a combination of falling prices and the efficiency of modern, gas-based, combined-cycle technology. From an environmental viewpoint, the need to reduce sulfur and carbon dioxide emissions and the threat—if not the reality—of carbon taxes has resulted in a definite move, at least at the planning stage, away from coal and toward gas.

In the longer term, however, the prospects for coal are likely to be enhanced by the commercial development of cleaner and more efficient coal technologies, many of them currently at the demonstration stage, and by the geographical concentration and finite nature of gas resources (IEA Coal Research, 1992b).

1.6. Aging Coal-Fired Power Stations

Many countries have substantial numbers of aging coal-fired power stations, each of which, at some stage, will give rise to the question: Do we refurbish or replace? The decision is influenced by factors as varied as local and national fuel sources, government policy, the demand for power, environmental considerations, design, and the availability of new and retrofit technologies.

An important attraction of refurbishment is that it allows utilities to take advantage of existing site licenses and so shorten the overall time scale required to bring new capacity into service. While refurbishment may be limited to the addition of emission control equipment to enable an existing facility to comply with new regulations, it can encompass complete replacement of major plant components to take advantage of improved technologies.

Greater advantages can often derive from a more radical approach to refurbishment. Combined-cycle generation, for instance, can exploit the relative competitiveness of natural gas and considerably improve plant efficiencies by using a natural gas turbine in conjunction with a coal-fired plant. Fluidized-bed combustion offers reduced emissions and, because of its more compact design, the potential for increasing plant capacity.

1.7. Coal Cleaning

During the past 20 years, consumers have recognized the economic and environmental consequences of using coal containing unwanted mineral matter and impurities. The growing demand for a cleaner and more consistent product has stimulated research into advanced cleaning, and four proven technologies now exist for the production of ultraclean coals. Carbon recovery levels of at least 80–90% can be achieved, together with very low ash and sulfur contents.

The tendency of modern mining methods to produce coals containing a greater proportion of ash and fines has led to the gradual modernization of conventional coal cleaning plants, but the traditional location, adjacent to the mine, has largely been retained. Modern deep-cleaning techniques are carried out on finely ground coal. The cleaned product can be handled in its dry form but is usually transported more safely as a slurry, its stability being ensured by chemical additives. The cost of these additives and the need for more varied transport facilities is stimulating the development of point-of-use cleaning.

1.8. A Positive Approach to Environmental Impact Assessment

Coal-related projects can affect the environment ecologically, visually, and through emissions, waste disposal, or noise. Environmental considerations have long been a part of the planning process in many countries, but the formal implementation of environmental impact assessment (EIA) is relatively new to most.

2. FLUIDIZED-BED FIRING

We will emphasize the method of fluidization firing among the procedures developed for the purpose of energetic use of low-grade coals by means of an environment-saving technology. Its propagation, however, has not been as extensive in recent decades as had been anticipated based on preliminary assumptions. The reason for this—in addition to the increase in power demands during waiting—may be found in limitations with regard to the application of this method with existing equipment.

The fluidized bed was invented in the 1920s in Germany as a chemical processing technique. The turbulent mixing and close contact of materials within a fluidized bed were found to accelerate chemical reactions. Fluidized-bed combustion was derived from the gas–solid fluidization principle used in processes in the petrochemical and metallurgical industries and has become much more widely used over the past few years for coal combustion. Its success can largely be attributed to its outstanding combustion efficiency and environmental acceptability.

From the combustion viewpoint, this technology enables the use of a very wide range of solid fuels with great operational flexibility and exceptional thermal efficiency. However, its main advantage is that, simply through the addition of lime or limestone in the bed, it enables 90–95% sulfur retention without the need for flue gas treatment. If multistage combustion is used, nitrogen oxide emission levels can also be reduced considerably.

The fluidized-bed technique has been incorporated into a number of combustion processes, suitable for use in different types of boilers with varying levels of performance. Beds of this type have a maximum temperature of about 850°C. Low-temperature beds can be divided into two main groups:

1. *Bubbling fluidized bed boilers*. In these boilers, air injected at 2–3 m/s fluidizes the bed of solids. The fluidized bed, which is usually about 1 m thick, is homogenized by evenly distributed air injected through a grate at the base of the bed.
2. *Circulating fluidized bed boilers*. In circulating bed boilers, gas velocity (5–10 m/s) is far greater than the lower fluidization limit. The bed particles are continuously evacuated and recycled, and the bed is cooled by the recycling of cooled ash. The circulation rate is controlled by adjusting bed pressure loss. The lime or limestone is injected directly into the bed. The solids circulation rate enables high sulfur retention, up to 90% (Ca/S ratio: 1–2:1).

A 500-kW fluidized-bed test plant was built in Alexandria, Virginia, in 1965. It provided much of the design data for a 30-MW prototype unit at the Mononga-hela Power Company in Rivesville, West Virginia, built in the mid-1970s. The first commercially successful fluidized bed was an industrial-size atmospheric unit (10 MW) built with federal funds on the campus of Georgetown University in Wash-ington, D.C., in 1979. The unit still operates today (Alain and Hamelin, 1988).

An *atmospheric fluidized-bed combustor* performs roughly the same func-tions as a conventional boiler in driving a steam turbine, except with far less emissions. A *pressurized fluidized-bed combustor*, because of the increased energy in the high-pressure gases exiting the boiler, can drive also a gas turbine, an arrangement known as a *combined cycle*.

Fluidized-bed combustion is ideal for low-grade fuels, and this is the reason for their widespread use in new power plants. The advantages of the fluidized-bed technology are as follows:

- Stable combustion in spite of wide variations in the particle size, moisture content, ash content, and heating value of the fuel
- Possibility of using low-volatile fuels with high ash content
- Possibility of firing different fuels simultaneously with the same combus-tion equipment (the bed)
- Rapidity of load changes
- Possibility of efficient control of SO_x and NO_x emissions without expensive equipment

Fluidized-bed combustion has been undergoing constant development over the past 15–20 years. The breakthrough occurred with the development of the circulating technology in 1977–1979. The first units were a success, largely because of the utilization of reactive but heterogeneous low-grade fuels. The size of fluidized-bed boilers today ranges from 15 kW to 450 MW in thermal output. The smallest boilers were operated automatically in private houses for several years.

The benefits from environmental protection provide the greatest driving force for the spread of fluidized-bed firing. In this respect, a consensus has not been reached as to the relative merits of the usual steam boilers built with facilities for separation of harmful matter (SO_2 and NO_x) and boilers with fluidization firing. In addition, elimination of the environmental contamination effects of older equipment is still a problem, and it requires additional building in of separating equipment.

The main targets in the development of fluidized-bed firing were the following:

- implementation of stable firing with different and extremely fluctuating coal grades (7–12 MJ/kg, 35–50% ash content)
- combustibility of the mining wastes (coal sludge)
- utilization of the benefits of fluidized-bed firing with respect to the protection of air purity (reduction of SO_2 and NO_x emissions)
- possibility for application with existing boilers

In the development of fluidization firing, the following basic tasks had to be achieved:

- starting of fluidized-bed firing
- feeding of coal
- appropriate dosing and distribution of air
- removal of a part of ash from the layer

For starting of fluidized-bed firing, we have used an oil-firing flue gas generator (which heated the primary air to about 400°C) with an oil torch directed onto the layer from above; however, it could be omitted on the basis of experiments with the cooling circuit.

A combined fluidized-bed–cold furnace, vortex, pulverized coal-firing technology developed at the Institute for Electric Power Research (VEIKI) in Hungary can be utilized both with new equipment and in older facilities with moderate material and financial expense. The main feature of the procedure is that the lower part of the furnace is constructed as a bed suited for fluidization firing, and there is the possibility of feeding additional fuel through appropriate holes made in the furnace walls. In the case of new equipment, the designer is free to decide as to the construction, while in the case of older facilities the lower part of the furnace and part of the earlier coal grinding mills are transformed.

This technology has been in operation for almost six years in a boiler generating 48 tonnes of steam per hour at the Tatabánya thermal power plant in Hungary. The advantage of the procedure is that the fuel is fed into the fluid bed and the different parts of the furnace after appropriate preparation (drying, grain size, etc.). Thermal performance of the fluid bed can reach two to three times (4–5 MW/m²) the usual value. It is less sensitive to the fuel grade (coals of low grade, coals of different ash and humidity content, products of coal washing, wastes, etc.), even by comparison with the usual fluid firing methods. A substantial part of the auxiliary equipment in older facilities can be used directly or with small modifications.

In terms of environmental control, this technology brings important benefits without further measures. Sulfur capture from coals with high sulfur content increases in parallel with the important growth of sulfur self-capture of the ash and can be increased by the use of appropriate dosing additives. Reduction in nitrogen oxides can be achieved by means of implementation of multistage, flexible air feeding and by employing appropriate ratios of the fed fuel. Thus, the reduction of nitrogen oxides can be ensured on the basis of a multistage firing principle.

The fundamental idea behind this technology is that more efficient firing of low-grade coals can be achieved, with important environmental benefits not only with new boilers, but also by refurbishment of existing facilities. The technology as originally elaborated (Fig. 1) was patented in Hungary and was by the United States and other countries, too. The fundamental requirement in order to achieve

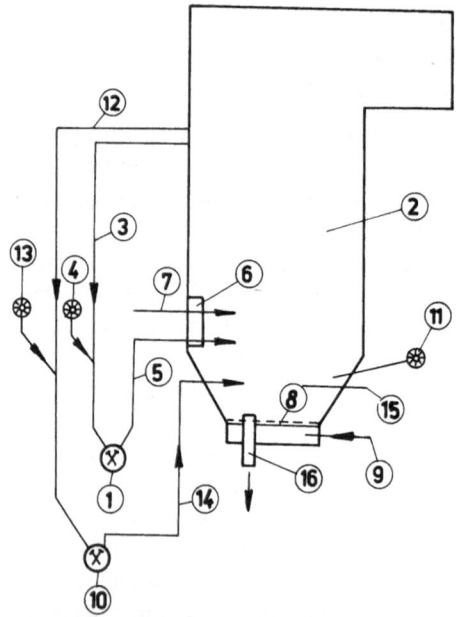

Figure 1. Hybrid fluidized-bed combustion system developed by VEIKI. 1, Fan mill with separator; 2, combustion chamber; 3, flue gas resuction; 4, coal feeder; 5, PC ducts; 6, coal burner; 7, secondary air duct; 8, wind box; 9, main air ducts; 10, fan mill without separator; 11, spreader; 12, flue gas resuction; 13, coal feeder; 14, PC duct; 15, second air; 16, bottom ash removal.

the lowest possible cost is to keep as much of the old equipment as possible; only minor transformations should need to be performed.

Taking into account the above aspects, the following are the most important modifications that occur during the transformation of existing equipment:

- The lower part of the furnace is removed and replaced with an air distributor suited for fluidization firing and a bed including the nozzle system; removal of the excess bed material must be ensured (Fig. 2).

Figure 2. Lower part of the furnace for hybrid fluidization firing.

- A part of the existing mills is transformed so that it will provide rough materials with a maximum grain size of 6–8 mm because its performance is increased, too.
- For the purpose of regulating the bed temperature, one regulating circuit is inserted into the boiler control.
- For the purpose of boiler starting, starting oil (or gas) burners are arranged at the appropriate points of the furnace wall.
- A new fluidization fan is erected.

Following laboratory experiments, a 48 t/h boiler was transformed in 1986 at the Tatabánya thermal power plant. In 1990 and 1991, two 100 t/h boilers each at the Ajka thermal power plant were transformed; in 1992, two other boilers at this plant were modified.

The transformations could be performed within a very short period of time, practically during one general overhaul. The results of these transformations have been very favorable. The problems of firing stability with extremely poor-quality coal have ceased, and an essentially lower minimum loading could have been achieved without oil support. The boiler efficiency has improved. Sulfur capture reached very high values without further additives, due to the limestone content of the ash. Feeding of further additives is also possible, if necessary. The volume of nitrogen oxides was significantly cut.

3. ENVIRONMENTAL AND ECONOMIC BENEFITS OF HYBRID-FLUID FIRING

The advantages of hybrid-fluid firing include the following:

1. SO_2 emission and NO_x generation are significantly reduced due to the low firing temperature, even without the use of additives.
2. It can be favorably matched with existing boilers, at relatively low cost, and it can be implemented in a short period of time.
3. It is suited for burning coals of high ash content and of varying grade, without oil support.
4. The boiler ensures stable firing even at low loading (at 40% of the nominal loading).
5. The firing conditions improve and the tendency for slagging is reduced.

4. TRANSFORMATION OF THE BOILERS AT THE AJKA THERMAL POWER PLANT WITH RESPECT TO FIRING TECHNOLOGY

The boiler transformed at the Ajka thermal power plant is a facility with natural circulation. It is a two-draft, membrane wall equipment, with an air heater of the Ljungström type. The original pulverized coal firing is a direct injection system, in which 4 pulverized coal firings of ventilator mills connect to one-one corner burners, supplied with a Raymond classifier. The main data for the boiler are:

- Steam-generating capacity: 100 t/h
- Steam pressure: 72.5 bar
- Steam temperature: 500°C
- Feed water temperature: 190°C
- Temperature of air preheating: 310°C
- Temperature of the released flue gas: 160°C

The raw coal data are:

- Calorific value: 8–10 Mj/kg
- Humidity: 18–22%
- Sulfur: 3.2–4.0%
- Ash: 35%
- CaO content in the ash 30–40%
- Ca/S mole ratio: 2.5–2.9

In the case of combined fluidization and pulverized coal firing (Fig. 2), lower furnace temperatures and exit temperatures develop as compared to those in a pulverized coal-firing boiler. Distribution of the thermal flow is also changed in the furnace. That is why, when transforming a pulverized coal firing boiler, thermal calculations have to be performed for the boiler. This will show whether the surfaces are appropriate or need modification. In the case of the boilers at the Ajka power plant, the surfaces could have been left unchanged, because the exit steam temperature could have been achieved with reduced water injection. Of course, the proportion of fluidization and "cold" pulverized coal firing will slightly affect the results of the thermal calculations for a particular boiler.

New regulating circuits have been established for the hybrid-fluid firing equipment. These are:

- layer temperature regulator
- layer mass regulator
- modified boiler loading
- modified firing regulator

Regulation of the temperature of the fluidized layer is performed by changing the volume of the blow air. Coal loading of the mills without a classifier separator is adjusted by means of manual intervention. The regulated characteristic is the temperature, which is measured by the nine thermocouples arranged in the main and auxiliary layers. The intervening component is the catch before the high-pressure (fluid) ventilator. When the layer temperature is reduced, the catch will open.

Regulation of the layer mass is performed by the draining catches. The regulated characteristic is the layer mass, calculated from the static pressure measured in the air distribution boxes of the main and auxiliary layers and the volume of flowing air. When the layer mass reaches the upper limit value, the draining valves open, and at the lower limit they close.

Loading control of the boiler is changed so that only the mills with a classifier participate in regulation. When the mill with a classifier does not operate, then there is no regulation of loading. In the case of the operation of mills with air classifier separation, the number of revolutions of the coal feeders is adjusted by means of manual intervention. This solution, however, cannot be considered final.

Firing control is essentially the regulation of air volume in proportion to the volume of the produced fresh steam. The regulated characteristic is the rotation regulator of the original pressure ventilator of the boiler.

5. MEASUREMENT RESULTS

In order to evaluate the success of the first hybrid-fluid firing system, we have performed measurements of the boiler efficiency and of the harmful emissions. The efficiency of the boiler calculated by an indirect method was 85.8% at a nominal loading of 100 t/h. This value represents about a 2% improvement in efficiency with hybrid-fluid firing.

The extent of sulfur adsorption as sulfur dioxide (measurements performed between May 27 and May 30, 1991) relative to the total sulfur entering with the coal without feeding of limestone, calculated for 6% O_2, is presented in Table I. The extent of sulfur adsorption without limestone addition is determined decisively with the firing method by the total sulfur (S) entering with the coal, as well as by the calcium content of the carbon ash (Ca), and it is favorable in the case of the coal employed at the Ajka power plant. That is why it was very reasonable to change the method of firing, because, by means of hybrid-fluid firing, this specificity of the coal can be better utilized than in the case of pulverized coal firing. The extent of sulfur adsorption previously measured in the case of the

Table I.   Extent of Sulfur Dioxide Capture without Feeding of Limestone

Mill position[a]	Steam production (t/h)	SO_2 (mg/m_N^3)	NO_x (mg/m_N^3)	Sulfur adsorption (%)
1-2-4	102.5	5724	546	59.7
1-3-4	94.3	4155	680	70.4
1-4	73.4	5022	689	70.9

[a]During the measurements, only mill no. 2 had a classifier.

pulverized coal-firing boilers was in the range of 15–25% relative to the total sulfur entering with the coal.

6. OPERATING EXPERIENCE

At the beginning, one mill without air classifier separator and two mills with air classifier separators were operated at the nominal loading of the boiler. At lower loading, one mill with air separation was kept in operation in addition to that without air classifier separation. Overheating temperature of the fresh steam was kept at the stipulated value with reduced water injection.

We have concluded from the measurement results that one mill without an air classifier separator would not be able to evenly distribute the coal on the surface of the fluidized layer, so zones with a significant amount of excess air and air-deficient zones will develop. Above a certain loading of the mill, CO content indicative of improper burning appeared in the flue gas.

The limit of loading of the mill without classifier determined the relative proportions of fluidization and coal powder firing. The share of coal powder firing had a higher value than expected. Slag deposits found on the pipes of the furnace confirmed this fact, too. After adjustment of the firing, these problems have been eliminated.

Fluidization firing is firing at low temperature. The temperature in the layer is maintained at around 850–860°C. This requires a thermal balance between the released coal heat and the thermal volume of the blown air of 310°C, on the one hand, and the thermal volume discharged toward the boiler tubes and the thermal volume taken away by the flue gas and the solid matter, on the other hand. The target is to raise the volume of coil heat released in the layer at a layer temperature of 860°C so that the boiler efficiency is not reduced.

In order to ensure an improvement in the share of fluidization firing, we have taken the following measures:

- We allowed for the operation of two counter-blow mills without air classifier separation by means of removing the air classifier separator.
- We increased cooling of the fluidized layer by means of removing the thermo-resistant concreting from the membrane walls above the air distribution; these walls have been made previously by other concepts. In this way, while maintaining the layer temperature of 860°C, we could release larger coal heat volumes in the fluidized layer.
- After performing the modifications, two counter-blow mills without air separation and one with air classifier separation were operated at the nominal loading of the boiler. In this case the volume of heat released by means of fluidization firing was 60–70% in the case of maximum loading, depending on the quality of the coal.

Circulation of coal within the previously mentioned furnace plays some role in cooling of the fluidized layer and in even distribution of the furnace temperature. Leaving the fluidized layer, which extracts heat, and on returning to the layer, it cools the layer. Dust circulation is realized in a larger circle; for example, when eco fly ash is returned to the layer, it has a stronger effect than it had previously.

In the case of hybrid-fluid firing, where "cold" coal powder firing and a slightly higher flame temperature occur in the loading conditions near the nominal boiler loading, our endeavors to release more heat in the fluidized layer and to reduce the share of pulverized coal firing are reasonable. By means of recirculation of the layer material and eco fly ash within the furnace, we can reduce the unfavorable effect exerted on SO_2 and NO_x formation by the pulverized coal flame. Fly ash recirculation is also useful in the case of those kinds of coal where the Ca/S mole ratio is favorable, as in the case of the coal employed at the Ajka power plant or when, in the case of limestone feeding, we are able to better utilize the material capable of adsorbing SO_2 by means of recirculation. By increasing the residence time, the combustible losses of fly ash will be smaller.

In summary, based on the operating experience acquired with the boilers transformed during the past six years, we can state that the hybrid-fluid system developed at VEIKI can be safely adapted for higher performances as well. The environmental benefits are significant, but they can still be improved by further experiments. There is an important base of experience for the selection of the feeding points, for air distribution, and for regulation. Even when the separate starting flue gas generator is omitted, a safe starting technology is available. The sulfur adsorption provided by the system can be increased, if necessary, by means of using additives, and the reduction in nitrogen oxides is significant. Fluidization can be especially favorable to use with firing coals with high humidity because dried fuel will be fed into the furnace. The hybrid system utilizes the whole furnace space well, and it allows firing at a much lower furnace temperature than with pulverized coal firing even above the fluid bed.

REFERENCES

Alain, M., and Hamelin, M., 1988, Air Pollution Prevention and Measurement, French Ministry of Environment.

IEA (International Energy Agency) Coal Research, 1992a, Power Station Coal Use: Prospects to 2000, Gemini House, London, January 1992.

IEA Coal Research, 1992b, Perspectives on Seaborne Steam Coal Trade, Gemini House, London, November 1992.

Chapter 8

Aspects of Catalyzed Coal Liquefaction

Sol William Weller

1. BACKGROUND

Among the many coal conversion processes are those that produce liquid fuels from solid coal. It is not easy to do this, and one underlying problem is apparent even from a casual inspection of the mass or elemental formulas of coal and some other common fuels. A typical value for the carbon/hydrogen mass ratio of a bituminous coal is about 15; for gasoline, the value is about 6, and for natural gas, 3. In terms of an elemental formula, coal might be $CH_{0.8}$, gasoline CH_2, and natural gas CH_4. To make liquid fuel from coal requires either addition of hydrogen or loss of carbon.

A distinction is usually made between indirect and direct liquefaction: Indirect processes require gasification of coal first to make "synthesis gas," with subsequent catalyzed production of hydrocarbons (e.g., Fisher–Tropsch synthesis), methanol, or other oxygenates; direct (or primary) processes convert coal directly to liquids by reaction with hydrogen under elevated pressure and temperature. This chapter is concerned with the direct hydroliquefaction of coal, origi-

SOL WILLIAM WELLER • Department of Chemical Engineering, State University of New York at Buffalo, Buffalo, New York 14260.

Combustion Efficiency and Air Quality, edited by István Hargittai and Tamás Vidóczy. Plenum Press, New York, 1995.

nated by Friedrich Bergius in the early years of the 20th century. In particular, it is concerned with selected studies of the role of catalysts in the Bergius process, the kinetics and mechanism of liquefaction, and the economic prospects for "oil from coal."

The history of the Bergius process is interesting. After receiving his doctorate in 1907, Bergius spent a year working with Nernst and with Haber, just at the time that they were developing the foundation for the high-pressure synthesis of ammonia from the elements. Bergius began his own high-pressure research in 1909, first on the dissociation pressure of CaO_2, next on converting cellulose into a coal-like substance, and finally on the hydroliquefaction of coal (Bergius, 1932). His first patents on coal conversion were issued in 1913. Ultimately, the patents and the process were taken over by I.G. Farben, who in 1927 built the first commercial plant at Leuna; the feedstock was brown coal. Imperial Chemical Industries (I.C.I.) entered into an agreement with I.G. Farben to exchange information, and the first plant in England, with bituminous coal as feed, was built in 1935 by I.C.I. at Billingham. The collaboration was later extended to petroleum companies in the Netherlands and the United States.

The story of catalysts in the Bergius process is also interesting and still incomplete. Although Bergius initially believed that his process could not be specifically catalyzed, he did add to the feed about 5 wt % of "Luxmasse," an alkaline, titaniferous iron oxide, as a sink for the H_2S generated during coal liquefaction. It developed subsequently that Luxmasse does catalyze coal liquefaction and that the presence of sulfur compounds is actually helpful to the catalytic action. It should be mentioned that pyrite, often present in coal mineral matter, is also a catalyst. The initial catalyst favored by I.G. Farben for the liquefaction of brown coal was molybdenum oxide in very low concentration, along with some sulfuric acid to neutralize calcium humates, which otherwise deposited in the preheater tubes as "caviar" and caused plugging. Ultimately, I.G. Farben switched to ferrous sulfate (in larger amounts) impregnated on the coal as a cheaper and more available catalyst. For bituminous coal, I.C.I. preferred to use tin oxides as catalysts, noting that incremental enhancement of liquefaction occurred if compounds capable of generating HCl were added with the tin.

In terms of the following discussion of catalyst research, it is both chastening and salutary to note that key results have been recognized for over 60 years. These include the outstanding effectiveness of molybdenum and tin compounds as catalysts; the lesser though real effectiveness of iron compounds; the importance of catalyst dispersion (exemplified by impregnation on coal); and the promoting effect (or synergism) of sulfuric acid addition with Mo catalysts and of HCl addition with Sn catalysts.

Some mention of terminology commonly used in this field is appropriate:

- "%Conversion" or "%liquefaction" is defined as

    $$100 \times [(\text{maf coal in}) - (\text{maf insoluble residue out})]/(\text{maf coal in})$$

- Oil: Soluble in n-alkanes, usually n-hexane or n-pentane
- Asphaltene: Soluble in benzene, insoluble in n-hexane or n-pentane
- Asphaltol (or pre-asphaltene): Insoluble in benzene, soluble in more powerful solvents such as pyridine.

Catalysts have been used in the direct hydroliquefaction of coal in two modes, once-through or fixed bed. The earlier and more common once-through mode involves adding unsupported catalyst (powdered or impregnated on coal) to the feed and disposing of the catalyst with other mineral matter in the product stream, sometimes with attempts to reclaim and recycle all or part of the catalyst. In fixed-bed operation the metal-containing catalyst is usually supported on alumina pellets of high surface area, which remain in the reactor during continuous operation. Although molybdenum compounds tend to be the catalysts of choice in fixed-bed as well as in once-through operation, this chapter will be concerned mainly with the once-through mode.

2. CATALYST STUDIES

As pointed out in Section 1, by 1945 industrial practice in Germany and Britain had indicated that catalysts based on Mo, Sn, and Fe are to be preferred in coal liquefaction, that good catalyst dispersion is important for optimum catalyst effectiveness, and that some instances of catalyst synergism exist. What was not available was the research evidence that had led to these conclusions. Aside from the (mostly unknown) catalyst screening that must have taken place, information was lacking as to (a) the chemical fate of the catalysts and their possible recovery and (b) the chemical role played by different catalysts in the various steps of the coal-to-oil conversion.

Weller and Pelipetz (1951) addressed the question of catalyst dispersion by a comparison of catalyst added as dry powder, ball-milled with coal, or impregnated on coal from water solution, in each case without added vehicle. Ammonium molybdate was the best catalyst tested, but only if impregnated on coal; addition of some sulfuric acid showed incremental effectiveness. Ferrous sulfate was reasonably effective if impregnated on coal. Tin catalysts in combination with ammonium chloride (or as stannous chloride) were effective even without impregnation. Shibaoka et al. (1980) later showed, by optical and electron microscopy, that stannous chloride is reduced under reaction conditions, and Sn and Cl migrate into the coal particles. If sufficient sulfur is present, the final form of tin is stannous sulfide. The rank ordering Mo > Sn > Fe, found in previous autoclave testing, was confirmed in pilot plant runs by Clark et al. (1952) on coal impregnated with catalyst.

Catalysts based on tin compounds receive relatively little attention in present-day coal research. Iron has been and remains an important catalyst for reasons of cost, but molybdenum compounds seem to be the focus of much catalyst research because of their potential effectiveness in very low concentration. The experiments of Weller and Pelipetz, which showed large increases in liquefaction for ammonium molybdate on impregnation, were performed with dry coal (no added solvent). They do not prove that impregnation of Mo catalysts on coal is essential. Schlesinger *et al.* (1962) carried out experiments with ammonium molybdate at concentrations as low as 0.01%. Their work showed that, in the presence of a recycle vehicle oil and with good mixing, at 450°C and high hydrogen pressure the maximum difference in yield of light oil was only 4% in a comparison of impregnated versus powdered ammonium molybdate catalyst.

An alternate approach to achieving good catalyst dispersion was introduced by Hawk and Hiteshue (1965). Instead of impregnating coal with a water-soluble catalyst, they studied oil-soluble metal naphthenates without impregnation when vehicle was present (but with impregnation from hexane solution in "dry hydrogenation" experiments). Autoclave tests were performed with commercially available naphthenates of Cr, Co, Cu, Fe, Mo, Ni, Sn, and Zn; of these, not surprisingly, Mo naphthenate was the most effective. Hawk and Hiteshue advanced the following rationale for their study:

> These products are quite expensive as presently manufactured, and it was well understood before the project began that if required in any but minuscule amounts, the cost would be prohibitive in a liquid-fuel-from-coal process. On the favorable side, the successful application of oil-soluble catalysts to coal hydrogenation would be sufficiently novel that the results would be a worthwhile contribution to chemical knowledge.

The Dow Coal Liquefaction Process, developed during the 1970s, also used Mo in low concentration but took a different process approach (Moll and Quarderer, 1979). In an effort to avoid the cost of catalyst impregnation and drying, Dow introduced into the feed stream a water-in-oil microemulsion of ammonium molybdate at very low Mo concentration (ca. 0.01%). The process depended on recovering and recycling part of the catalyst from the overflow of a hydroclone used for solids separation. In one example, the Mo concentration increased from 275 ppm in fresh feed to 615 ppm in the total feed after 10 passes through the reactor and hydroclone. Addition of CS_2 was found helpful in the treatment of low-sulfur, low-ash coal, and molybdenum disulfide was found, by microprobe analysis, in the solids recovered from the liquefaction products.

It is remarkable that in the cases of the three metal catalysts most used in coal liquefaction—molybdenum, tin, and iron—the form of the metal that seems to be most active is the sulfide. In the case of iron, the active species is pyrrhotite; even ferrous sulfate is converted to this compound under reaction conditions (Artok *et al.*, 1992). In traditional hydrogenation catalysis with group VIII metals such as platinum, sulfur is well known as a poison.

With the activity of Mo sulfide in mind, attention has turned to the study of ammonium tetrathiomolybdate (ATTM) rather than ammonium molybdate as catalyst precursor. ATTM can decompose under reaction conditions directly to the disulfide. Derbyshire *et al*. (1986) studied coal impregnated with ATTM from water solution, with subsequent drying but in the absence of added oil vehicle ("dry catalytic hydrogenation"). They found, *inter alia*, that water removal after impregnation under vacuum at 25°C resulted in much better dispersion than water removal at 110°C. As is so often the case, details of experimental procedure turn out to be important.

A tungsten analog of ATTM, ammonium tetrathiotungstate (ATTW), has been studied by Garcia and Schobert (1989, 1990) for the hydrodesulfurization of a high sulfur (ca. 10%, largely organic) lignite. They found ATTM to be very effective for total sulfur removal and suggested that a sequential mechanism is the principal process in the removal of organosulfur as hydrogen sulfide.

In the last few years, even more unusual catalyst precursors have been studied. Hirschon and Wilson (1992) used $(C_6H_5)_2Mo_2(\mu\text{-}SH)_2(\mu\text{-}S)_2$, impregnated on coal from tetrahydrofuran (THF) solution, as a Mo precursor. From a comparison of tetralin and *n*-hexadecane as vehicle they concluded, among other things, that a good donor solvent is not required when a good catalyst precursor is used. Hayward and Schobert (1993) have reviewed research with a number of precursors, including metal carbonyls and various metal organic compounds, as well as their own new experiments with thia-crown ethers of Fe, Co, Ni, and Cu with BF_4^- as anion. They did not discuss these complexes as possible acid catalysts.

By contrast, Wender and his colleagues have focused on superacid catalysts. Pradhan *et al*. (1991a,b) examined the catalytic activity of sulfated (Fe_2O_3, SnO_2, or Mo/Fe_2O_3 (all said to be superacids) when added as powders. The sulfated Mo/Fe oxide was the most active of these. They postulated that the sulfate group helps to prevent sintering of the catalyst, thus aiding accessibility of dissolved constituents to the small catalyst particles. Fraenkel *et al*. (1991) studied the combination of a mild superacid (triflic acid, CF_3SO_3H) with iodine. The acid is presumed to enhance the depolymerization of coal to asphaltenes; the major effect of iodine, a known catalyst, is to hydrogenate and hydrocrack asphaltenes to oils.

3. KINETICS AND MECHANISM

In the 1940s investigators at the U.S. Bureau of Mines began a survey of catalysts for the hydrogenation of coal. The research that dealt specifically with the kinetics and mechanism of liquefaction was described in a group of three papers published in 1950–51. The first of these (Weller *et al*., 1950) considered the mechanism of coal and asphalt (modern terminology, asphaltene) hydrogenation in an attempt to rationalize the effects of different catalysts. The other two papers (Weller *et al*., 1951a,b) were early efforts to describe quantitatively the rates at which coal is converted to asphaltene, oil, and gas.

The focus in the 1950 paper was on the synergistic catalyst combination tin–halogen acid, a very effective catalyst system for the liquefaction of bituminous and sub-bituminous coal. This combination also proved very effective for the conversion of asphaltene isolated from the products of a coal hydrogenation pilot plant run on Bruceton (Pittsburgh seam bituminous) coal. Figure 1, extracted from more extensive data in the paper, illustrates the effects of tin, hydrochloric acid, and ammonium chloride, in an atmosphere of hydrogen or helium, on the conversion of benzene-soluble asphaltene either to lighter products ("progressive reactions") or to benzene insolubles (now termed "regressive" or "retrogressive" reactions, since the objective is the production of oil, not new insoluble solids.)

Interpretation of such results led the authors to a descriptive theory of coal hydrogenation. For the case of asphaltene, in the authors' words:

> Asphalt is assumed to give either gas, by a noncatalytic reaction, or reactive fragments of some kind (possibly free radicals). The cleavage to give reactive fragments is catalyzed by halogen acid-producing substances (hydrochloric acid, ammonium chloride, etc.). These reactive fragments may recombine to produce asphalt again; they may polymerize still further to produce benzene-insoluble material; or they may be stabilized with addition of hydrogen to produce an oil. This hydrogenation stabilization is assumed to be catalyzed especially well by tin.

The descriptive theory led to some predictions: Tin, not a splitting catalyst, in the absence of HCl should behave as no catalyst; phosphoric acid, a polymerization catalyst, should produce especially large quantities of benzene insolubles (the result is not shown in Fig. 1, but the prediction is fulfilled); and HI, which is both a splitting and a hydrogenation catalyst, should result in a relatively high conversion of asphalt and a relatively small production of benzene insolubles (it did). The authors noted the similarity of response by coal and asphalt to hydrogenation catalysts, and they suggested a similar reaction scheme for coal:

> It is postulated that coal . . . is thermally split to form reactive fragments, the splitting being catalyzed by halogen acids. The fragments either polymerize to form benzene insoluble products or are stabilized by the addition of hydrogen to form soluble products.

Further:

> Many of the separate elements of the theory are not unique, nor is it supposed that all the details of coal hydrogenation will be covered by the crude picture presented. It is hoped, however, that the hypotheses will permit a correlation of the major steps in the over-all hydrogenation and of the role played by the separate catalyst constituents, and that they will provide a framework for the design of future experiments.

The liquefaction description advanced in 1950 was essentially a linear one, in the sense that it proposed conversion of coal to asphaltene, then of asphaltene to oil, by consecutive reactions, with side reactions to gas and water. As later research has shown, this model does not cover all the details of coal hydrogena-

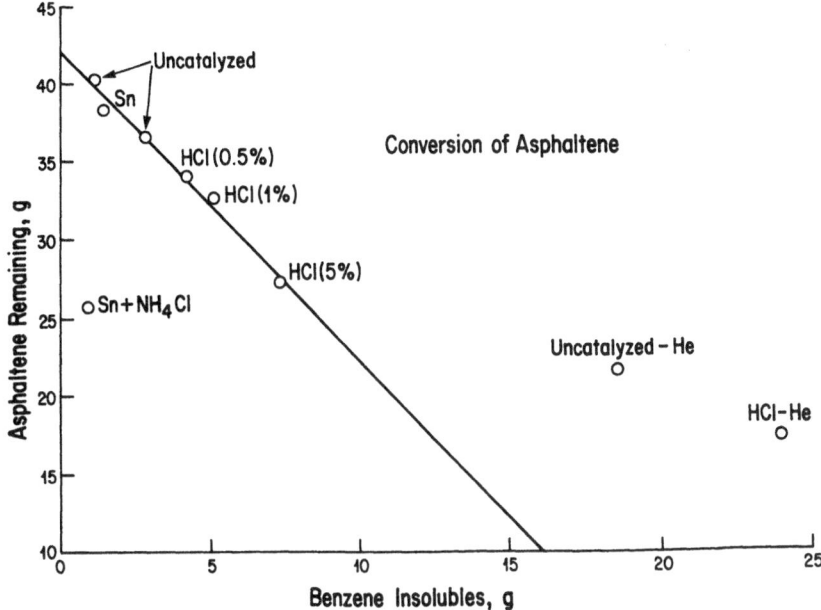

Figure 1.   Effect of catalysts (tin, ammonium chloride, hydrochloric acid) on the conversion of benzene-soluble asphaltenes to hexane-soluble oil + gas ("progressive" reactions) or to benzene-insoluble material ("regressive" reactions), in hydrogen or in helium. Data taken from Weller *et al.* (1950).

tion. However, a number of the major postulates remain in the current canon of the liquefaction process. These include the notion of thermally generated (and possibly produced by an acid-catalyzed splitting reaction) "reactive fragments," which may be free radicals. In turn, these reactive species may be stabilized by addition of hydrogen (progressive reaction) or may polymerize to generate insoluble products (regressive or retrogressive reactions).

The second paper of this group (Weller *et al.*, 1951a) presented batch autoclave data for the hydrogenation of crude asphaltene, isolated from pilot plant runs on coal, as a function of temperature and nominal time at temperature. The catalyst was 1 wt % SnS + 0.5 wt % NH$_4$Cl. Plots of log "asphalt remaining" versus time were linear for temperatures of 400–440°C, suggesting an integral rate expression $A = A_0 \exp(-k_2 t)$, where A_0 and A represent the weight of asphalt originally present and that present at time t, respectively, and k$_2$ is an apparent first order rate constant.

The third paper (Weller *et al.*, 1951b) presented rate data on the hydrogenation of hand-picked anthraxylon (a glossy constituent of banded coal). Overall conversion of the coal obeyed the rate expression $C = C_0 \exp(-kt)$, and plots of asphaltene produced versus nominal time at temperature were generally consistent with the linear scheme coal → asphaltene → oil. The authors added this caution: "This scheme does not imply that asphalt is the first intermediate product; in these experiments, however, it is the earliest intermediate which is isolated and measured." Figure 2 illustrates the findings at 400°C. At the bottom of the figure is the note that at 430°C the conversion of asphaltene to oil occurs at a rate ten times lower than that for the conversion of coal to asphaltene. This result is consistent with the idea, widely accepted, that the bonds broken initially in coal conversion are much weaker than those necessary for the asphaltene-to-oil conversion.

Neavel (1976) made important contributions to the methodology and to the interpretation of kinetic studies by (a) introducing the use of miniature autoclaves, or tubing bombs, to study coal hydrogenation at short times at temperature and (b) showing that solvents more powerful than benzene—e.g., pyridine—can dissolve

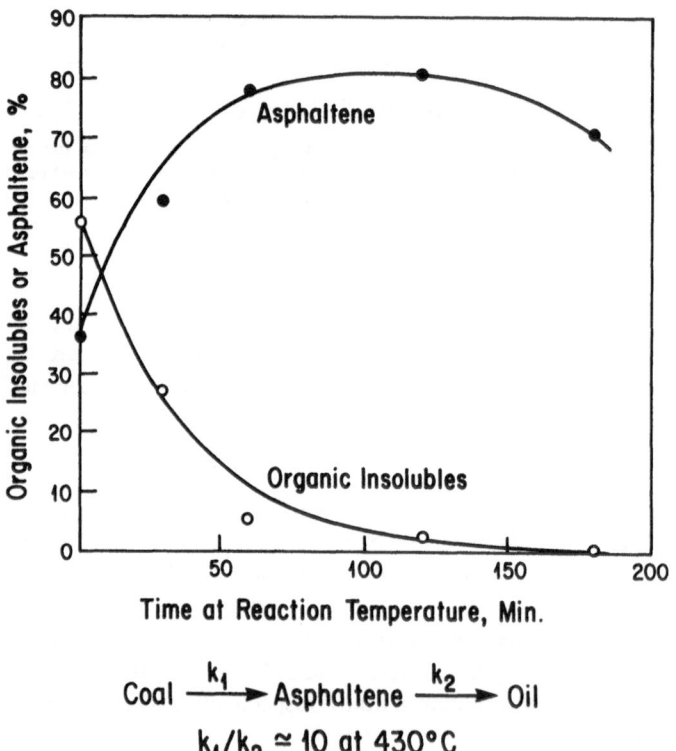

Figure 3. Hydrogenation of Bruceton anthraxylon at 400°C: organic insolubles remaining, or asphaltene produced, as a function of time at reaction temperature. Asphaltene is taken as the intermediate in the reaction scheme. Data from Weller *et al.* (1951b).

some of the benzene-insoluble material in the product. Neavel found that in less than 5 minutes at 400°C the product from a mixture of tetralin (a H-donor solvent) and vitrinite-enriched Illinois bituminous coal was more than 90% soluble in pyridine, though only 40% was soluble in benzene. This led to the concept that asphaltene (classically defined as benzene soluble, hexane insoluble) is *not* the first isolable intermediate product in the coal-to-oil conversion. Something that is benzene insoluble but pyridine soluble is formed at an earlier stage. Neavel used the term "asphaltols" for this material; the term "pre-asphaltenes" is also in wide use and may be a more self-explanatory name. Neavel also found that in a non-H-donor solvent (dodecane or naphthalene), regressive reactions ("repolymerization") occurred which led to the formation of benzene-insoluble material.

Figure 3 represents an effort to incorporate the pre-asphaltene concept into the sequential model of Weller *et al.* (1951b). In this diagram the possibility that the fast conversion of coal to pre-asphaltenes can be catalyzed is left indeterminate; the question has not been much studied. The splitting reactions 1 and 3 and the regressive reactions 5 and 6 can occur thermally but may be catalyzed by acids. The stabilization reactions 2 and 4 are presumably catalyzed by metal compounds such as tin, molybdenum, or iron oxides and sulfides.

Squires (1978) has provided a summary of research which indicates that oil may sometimes be produced from coal without intermediate formation of pre-asphaltenes or asphaltenes. In this case the sequential route for progressive reactions leading from coal to oil, illustrated in Fig. 3, would be accompanied by a path for coal to oil parallel to the sequential path. It is accepted that there is a diversity of bond strengths in coal. Squires suggested the term "scissile bonds" for those linkages in coal which are easily broken. These might be hydrogen bridges,

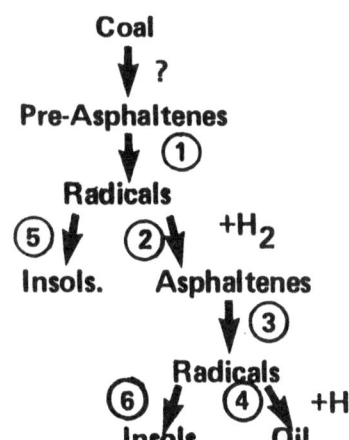

Figure 3. Modified linear reaction scheme for the stepwise conversion of coal to pre-asphaltenes to asphaltenes to oil ("progressive" reactions) or to polymerized, benzene-insoluble material. The intermediate reactive fragments are labeled "radicals" in this drawing.

ether linkages, or C–C bonds such as that joining two benzyl groups in dibenzyl. Vernon (1980) has pointed out that the bond strength for the central C–C bond in dibenzyl is only 235 kJ/mol, much less than the value of 480 kJ/mol for the bond joining the two phenyl groups in diphenyl.

The participation of free radicals as necessary intermediates in coal liquefaction is often assumed (e.g., Curran et al., 1967; Wiser, 1968). Weller et al. (1950) were ambivalent about this, suggesting that their "reactive fragments" might (or might not) be free radicals. The reactive species resulting from bond cleavage are not necessarily radicals since, depending on the nature of the original linkage and of the catalyst, both heterolytic splitting and homolytic splitting of bonds are possible. Acid-catalyzed splittings of hydrogen bonds or ether linkages are plausible candidates for a heterolytic mode of attack.

Experimentally established rate expressions for the conversion of a given coal under given conditions have important engineering value. However, there are basic problems in the interpretation of kinetic data. One is a generic question: Can reaction mechanism be deduced from the form of an empirical, global rate expression? Weller (1956, 1975, 1992) has discussed this question in the context of heterogeneous catalysis, arriving at the conclusion that the answer is "yes" only for the special case of an elementary reaction. A separate problem arising in coal liquefaction is identifying the meaning of "concentration" of coal or asphaltene or oil in a mass action formulation of reaction rate.

In their work with coal and asphaltene, Weller et al. (1950, 1951a,b) found that their limited kinetic data were consistent with successive first-order reactions. Their rate expressions used weight (or weight percentage) of coal, asphaltene, or oil to represent a concentration, in the sense that concentration is used in the law of mass action. Other investigators, also using weights to represent concentrations, have found rate data that were better fitted by second-order kinetics or by more complex expressions. An example is the work of Gun et al. (1979). On the basis of a classical initiation–propagation–termination free-radical mechanism, Gun et al. arrived at the overall rate expression

$$d[P_m]/dt = b_1[C] + b_2[C]^2 + b_3$$

where $[P_m]$ denotes gas and benzene-soluble products, $[C]$ represents the percentages of organic matter in coal and benzene-insoluble intermediates, and b_1, b_2 and b_3 are constants.

The law of mass action underlies the formulation of rate expressions for chemical reactions. The mass action principle was proposed in 1862–67, simultaneously by investigators in France, England, and Norway, to quantify the rate of elementary reactions in single-phase systems between identifiable chemical species. The mass action rate expressions involve concentrations (moles/unit volume) of individual species, not weight fractions, and the expressions in general do not describe the rates of nonelementary reactions.

The reactions occurring during coal liquefaction are complex, not elementary, and "concentrations" of species in the mass action sense are not available. One may ask, for example, what is the concentration (moles/unit volume) of unreacted coal or of partially depolymerized and hydrogenated coal? Weller *et al.* (1951a,b) implicitly equated concentrations in mass action formulations to weight fractions of unknown species that are lumped together on the basis of common solvent solubility. This procedure causes no distress if the objective is the development of a useful, empirical description of the progression of reactions. There is a temptation, however, to deduce mechanistic details from the form of a kinetic expression. The temptation should be resisted unless there is more direct evidence for a mechanism than is available from overall rate laws. Gorin (1981) has expressed this conclusion well: "The rather good fit of the equations to experimental data does not in itself prove that the mechanism is valid; they should be regarded primarily as a convenient semiempirical method of correlating the data. They should be useful also for scaleup purposes and engineering design."

4. ECONOMICS AND ENVIRONMENT

4.1. Economic Considerations

Although coal liquefaction was practiced on a large scale in Germany, England, and Japan through the end of World War II, the cost of liquid fuels from coal in a free, peacetime economy has been notoriously difficult to pin down. The number is very sensitive to the assumptions made in the calculation, e.g., the costs of capital and labor, the return on investment, and the by-product credit that may be taken. The difficulty was appreciated as far back as the late 1940s, at which time the U.S. Bureau of Mines established some common guidelines (or assumptions) for simultaneous cost calculation by several groups. In particular, the costs were to be based on a 6% return on total investment after a 50% income tax, for the production of gasoline from a Wyoming coal.

Estimates were made by the Bureau, by Ebasco Services (a major consulting firm), and by the National Petroleum Council (NPC). The various estimates were intended both for demonstration of consistency (or inconsistency) between estimates and for comparison with the cost of gasoline from a petroleum refinery. The results, given by Wu and Storch (1968), are illuminating.

The final comparative costs, published in 1952–3, were 19.1 cents per gallon (¢/gal) by the Bureau of Mines estimate, 21.8 ¢/gal by Ebasco, and 34.8 ¢/gal by NPC. The major reason for the large difference between the Bureau and NPC was the lesser by-product credit allowed in the NPC estimate. With respect to by-product credit, NPC pointed out that the market for by-product chemicals could not support a large-scale hydrogenation industry. At the time of these estimates, the cost of gasoline at an Oklahoma petroleum refinery was 10.6 ¢/gal.

The Bureau of Mines estimate of 19.1 ¢/gal of gasoline from coal was 80%

higher than the cost of gasoline from petroleum at the time. A similar differential may persist 40 years after these estimates. An unsubstantiated estimate of the cost of synthetic crude from coal, heard by the author in 1993, was that it could be as low as $30/barrel. In late 1993 petroleum crude was selling for about $17/barrel on the international market. Based on these numbers, the estimated cost of synthetic crude from coal would be about 75% higher than the cost of petroleum crude. Although the estimates are very rough, they illustrate the continuing large cost disadvantage for synthetic crude. Since the unfavorable economic comparison of coal liquefaction with petroleum seems to persist, the term "receding break-even point," suggested for the phenomenon more than a decade ago (Haggin, 1981), may still apply. One factor contributing to this result was mentioned by Wu and Storch (1968) in their review, "Hydrogenation of Coal and Tar":

> In the conventional hydrogenation process part of the aromatic hydrocarbons and the aromatic compounds containing oxygen and nitrogen are converted into lower priced gasoline with expensive high-pressure hydrogen. This is contrary to economic principles in peacetime . . . In conclusion, the existence of giant petroleum-refining facilities, the high construction cost of the coal hydrogenation plant, and the continuous discovery of petroleum reserves makes gasoline from hydrogenation of coal and tar unable to compete with petroleum in the United States at the present time.

The by-product credit question, namely, whether manufacture of chemicals from coal would provide a profitable venture, can be partially answered by examining the experience of one company, Union Carbide. Having decided to use coal as a primary chemical feedstock, in 1952 Carbide started operation of a 300 ton/day pilot plant in West Virginia mainly to produce chemicals—benzene, phenols, heterocyclics, etc.—by the high-pressure hydrogenation of West Virginia coal. The plant was closed four years later, in 1956, for economic reasons (Stranges, 1983). Problems of separation of individual chemicals from the product stream, in sufficiently high purity to satisfy market demands, proved to be formidable and expensive.

During the 1970s many approaches to "clean fuels from coal" were developed by American companies, with substantial government assistance in the research and development. The industrial and governmental spending on fossil fuel research peaked in 1979–80 but fell sharply during the 1980s. A striking example of the importance of political/economic considerations was the experience of the Gulf Oil Corporation in promoting the development of the SRC-I and SRC-II (SRC = solvent-refined coal) processes for making clean boiler fuel by high-pressure hydrogenation of coal. The SRC process was originally developed by the Spencer Chemical Company, later acquired by Gulf. Gulf built a pilot plant in Fort Lewis, Washington, which began operation in 1974. In 1978 the Consolidated Edison Co. in New York City successfully burned 4500 barrels of SRC-II fuel oil from the Fort Lewis plant to generate electricity. The Department of Energy (DOE) then awarded Gulf a contract to design, build, and operate a demonstration plant. DOE also agreed to share technology with West Germany and Japan. In 1981, however, a year before construction was to begin, DOE discontinued its support of the project. The new national administration had adopted a policy calling for private industry to play a larger role in the development of synthetic fuels from coal (Stranges, 1983); it was reported that the President intended to ask Congress to dismantle DOE. After this, the German, Japanese, and American backers of the demonstration plant finally decided to terminate the entire project (Williams, 1982).

4.2. Environmental Concerns

Production of synthetic liquid fuels from coal introduces its own set of environmental problems. During hydroliquefaction, much of the sulfur and nitrogen contained in the coal is converted to hydrogen sulfide and ammonia, respectively. Although these gaseous contaminants and the methods for their control are well known, the amounts would be formidable for a coal liquefaction plant of commercial size. More insidious is the occurrence of a range of monocyclic and polycyclic aromatic hydrocarbons (PAH), along with heterocyclic analogs, in the liquid product. Many of the PAHs are known carcinogens; even benzene and toluene are now subject to severe restrictions in the United States.

REFERENCES

Artok, L., Schobert, H. H., and Davis, A., 1992, Temperature-staged liquefaction of coals impregnated with ferrous sulfate, *Fuel Process. Technol.* **32**:87.

Bergius, F., 1932, Chemical reactions under high pressure, in *Nobel Lectures, Chemistry, 1922–1941*, pp. 244–279, Elsevier, Amsterdam.

Clark, E. L., Hiteshue, R. W., and Kandiner, H. J., 1952, Producing fuel oil from coal, *Chem. Eng. Prog.* **48**:15.

Curran, G. P., Struck, R. T., and Gorin, E., 1967, Mechanism of the hydrogen-transfer process to coal and coal extract, *Ind. Eng. Chem.* **6**(2):166.

Derbyshire, F., Davis, A., Lin, R., Stansberry, P. G., and Terrer, M. T., 1986, Coal liquefaction by molybdenum catalyzed hydrogenation in the absence of solvent, *Fuel Process. Technol.* **12**:127.

Fraenkel, D., Pradhan, V. R., Tierney, J. W., and Wender, I., 1991, Liquefaction of coal under mild conditions: Catalysis by strong acids, iodine, and their combination, *Fuel* **70**:64.

Garcia, A. B., and Schobert, H. H., 1989, Comparative performance of impregnated molybdenum-sulfur catalysts in hydrogenation of Spanish lignite, *Fuel* **68**:1613.

Garcia, A. B., and Schobert, H. H., 1990, Catalytic hydrodesulfurization of a high organic sulfur Turkish lignite: Amount, form, and mechanism of sulfur removal, *Fuel Process. Technol.* **26**:99.

Gorin, E., 1981, Fundamentals of coal liquefaction, in *Chemistry of Coal Hydrogenation*, 2nd suppl. vol. (M. A. Elliott, ed.), pp. 1846–1915, John Wiley & Sons, New York.

Gun, S. R., Sama, J. K., Chowdhury, P. B., Mukherjee, S. K., and Mukherjee, D. K., 1979, A mechanistic study of hydrogenation of coal, *Fuel* **58**:176.

Haggin, J., 1981, Synfuels and coal conversion: What cost?, in *Chem. Eng. News* **1981** (June 27):43.

Hawk, C. O., and Hiteshue, R. W., 1965, Hydrogenation of Coal in the Batch Autoclave, U.S. Bureau of Mines Bulletin 622.

Hayward, N., and Schobert, H. H., 1993, Thia crown ether complexes as catalyst precursors for direct hydrogenation, *Energy Fuels* **7**:326.

Hirschon, A. S., and Wilson, R. B., 1992, Highly dispersed coal liquefaction catalysts, *Fuel* **71**:1025.

Moll, N. G., and Quarderer, G. J., 1979, The Dow coal liquefaction process, *Chem. Eng. Prog.* **75**(10):46.

Neavel, R. C., 1976, Liquefaction of coal in hydrogen-donor and non-donor solvents, *Fuel* **55**:237.

Pradhan, V. R., Herrick, D. H., Tierney, J. W., and Wender, I., 1991a, Finely-dispersed iron, iron–molybdenum, and sulfated iron oxides as catalysts for coprocessing reactions, *Energy Fuels* **5**:712.

Pradhan, V. R., Tierney, J. W., Wender, I., and Huffman, G. P., 1991b, Catalysis in direct coal liquefaction by sulfated metal oxides, *Energy Fuels* **5**:497.

Schlesinger, M. D., Frank, L. V., and Hiteshue, R. W., 1962, Relative Activity of Impregnated and Mixed Molybdenum Catalysts for Coal Hydrogenation, U.S. Bureau of Mines Report of Investigations 6021.

Shibaoka, M., Ueda, S., and Russell, N. J., 1980, Some aspects of the behaviour of tin (II) chloride during coal hydrogenation in the absence of a solvent, *Fuel* **59**:11.

Squires, A. M., 1978, Reaction paths in donor solvent coal liquefaction, *Appl. Energy* **4**:161.

Stranges, A. N., 1983, Synthetic petroleum from coal hydrogenation: Its history and present state of development in the United States, *J. Chem. Educ.* **60**:617.

Vernon, L. W., 1980, Free radical chemistry of coal liquefaction: Role of molecular hydrogen, *Fuel* **59**:102.

Weller, S. W., 1956, Analysis of kinetic data for heterogeneous reactions, in *AIChE J.* **2**:59.

Weller, S. W., 1975, Kinetic models in heterogeneous catalysis, *ACS Adv. Chem. Ser.* **148**:26.

Weller, S. W., 1992, Kinetics of heterogeneous catalyzed reactions, *Catal. Rev.-Sci. Eng.* **34**(3):227.

Weller, S. W., and Pelipetz, M. G., 1951, Coal hydrogenation catalysts: Studies of catalyst distribution, *Ind. Eng. Chem.* **43**:1243.

Weller, S., Clark, E. L., and Pelipetz, M. G., 1950, Coal hydrogenation catalysts: Mechanism of coal liquefaction, *Ind. Eng. Chem.* **42**:334.

Weller, S., Pelipetz, M. G., and Friedman, S., 1951a, Kinetics of coal hydrogenation: Conversion of asphalt, *Ind. Eng. Chem.* **43**:1572.

Weller, S., Pelipetz, M. G., and Friedman, S., 1951b, Kinetics of coal hydrogenation: Conversion of anthraxylon, *Ind. Eng. Chem.* **43**:1575.

Williams, R. O., Jr., 1982, The crossroads in energy policy, *ASTM Standardization News* **1982**(April):12–25.

Wiser, W. W., 1968, A kinetic comparison of coal pyrolysis and coal dissolution, *Fuel* **47**:475.

Wu, W. R. K. and Storch, H. H., 1968, Hydrogenation of Coal and Tar, U.S. Bureau of Mines Bulletin 633.

Chapter 9

Incineration of Waste Solvents Containing Chlorinated Hydrocarbons: Some Critical Remarks

Zoltán Adonyi and Sándor Kántor

1. INTRODUCTION

Although environmental factors have had an impact on industrial production throughout history, their role has only recently become widely known owing to local, regional, and global environmental problems. The nature of the relation between production and environmental problems is highly empirical, as evidenced by the relatively undeveloped description and management system of these problems (Adonyi, 1993). An outstanding example of this is that industrial production is often blamed for generating toxic wastes, but the inducting role of the consumption structure is hardly taken into consideration. The result is the low efficiency of environmental activities and a lack of methodology for licensing environmental technologies (Gauweiler, 1992).

The technologies for the elimination of toxic wastes depend on the following critical aspects:

• The crucial goal of industrial production is to make profit.

ZOLTÁN ADONYI AND SÁNDOR KÁNTOR • Department of Chemical Technology, Budapest Technical University, H-1521 Budapest, Hungary.

Combustion Efficiency and Air Quality, edited by István Hargittai and Tamás Vidóczy. Plenum Press, New York, 1995.

- Environmental regulations are becoming more restrictive.
- The number of alternative possibilities for problem solving is on the increase, leading to a wide range of new results in research and development.

A new important aspect is the profit-oriented and occasionally aggressive waste management industry, which is in some cases helped but in several cases impeded by the local population and authorities.

The increasing concentration of halogenated hydrocarbons in the atmosphere, the high stability of halogenated hydrocarbons, and the possibility of the emission of highly toxic compounds during their incineration impart an outstanding importance to this group of compounds in the frame of toxic waste management. Incineration techniques can be considered the main technology for the destruction of these compounds (e.g., Lee *et al.*, 1986; Wiley, 1987) because research into alternative technologies has not led to industrially applicable solutions.

In this chapter we will give an overview of the present regulations concerning toxic waste incineration and point out the problems involved in carrying out experiments to assist decision makers in the regulation process. Some new information gained by the reinterpretation of previously published experimental data is also presented.

2. PROBLEMS IN THE MODELING OF HALOGENATED WASTE SOLVENT INCINERATION

The high toxicity of a group of combusted chlorinated hydrocarbons is well known (U.S National Institute of Health and Human Services, 1991) and results in restrictive operating requirements for waste incinerators. A group of hydrocarbons containing chlorine and several of their combustion products (see, e.g., Öberg *et al.*, 1985; Visalli, 1987) are among the most stable and toxic chemicals that have ever been produced. Probably, this is the reason why no contractor was found for a long time in the United States for any polychlorinated biphenyl (PCB) incineration pilot-plant research.

Nowadays, high-temperature incineration has become a widely used and accepted technique for the elimination of chlorinated hydrocarbons including PCBs (e.g., Oppelt, 1987; Hunt *et al.*, 1984) in spite of huge efforts spent on finding alternatives (Meardon *et al.*, 1989). The commonly used type of toxic waste incineration systems contain a rotary furnace, an afterburner, and a flue gas scrubber (Wiley, 1987).

The first Hungarian law concerning the storage and treatment of wastes that are considered hazardous to the environment and human health was passed in 1981 (Hungarian Council of Ministers, 1981). Ten years later, the Ministry for Environment and Regional Policy (KTM) promulgated the first emission standards for waste incineration (KTM, 1991). According to these standards, if the waste contains any chlorine:

- Combustion temperature must be 1150°C (±50°C).
- Residence time must reach 2 s at 1150°C.
- Oxygen content of stack gas cannot be lower than 6%.
- If the nominal capacity of the waste incinerator is more than 1000 kg/h, then equipment for the continuous measurement of and collection of data on the oxygen, carbon monoxide, hydrogen chloride, hydrogen fluoride, sulfur dioxide, nitrogen oxides, organic carbon, and particulate matter content of the flue gas must be installed.

A comparison of the Hungarian emission standards and the German requirements concerning the flue gas of waste incinerators is presented in Table I. It illustrates the environmental regulations in countries that have recently established or updated their environmental policy.

A different approach is exemplified by the planning and operation requirements of the U.S. Environmental Protection Agency (EPA). The following requirements must be met by incinerators fed by chlorinated hydrocarbons (U.S. EPA, 1982):

Table I. Emission Standards for Toxic Waste Incinerators in Hungary and in Germany

	Maximum concentration in stack gas on 30 min. average	
	Hungary (mg/Nm^3)	Germany (TA-LUFT/86) (mg/Nm^3)
Particulate matter	30	30
Carbon monoxide	100	100
Sulfur dioxide	200	100
Nitrogen oxides (denoted by NO_2)	400	500
Hydrogen chloride	50	50
Hydrogen fluoride	2	2
Organic carbon (denoted by C_1)	20	20
Hg	0.1	—
Total of Cd and Tl	0.1	0.2[a]
Total of As, Te, Se, Ni, and Co	1	1
Total of Pb, Cr, Cu, V, Sn, Mn, and Sb	1	5[b]
Total polychlorinated dibenzodioxins and polychlorinated dibenzofurans in 2,3,7,8-TCDD toxicity equivalent	0.1 ng/Nm^3	10 ng/Nm^3 (0.1 mg/Nm^3)[c]

[a]Including Hg.
[b]Not including V and Sn.
[c]Beginning in 1996.

1. The incinerator must achieve a destruction and removal efficiency (DRE) of 99.99% for each principal organic hazardous constituent (POHC). POHCs have been listed in the Resource Conservation and Recovery Act (RCRA) since 1980.
   This list is updated semiannually in the Code of Federal Regulations (40 CFR 261).
2. If the feed waste reaches more than 0.5% chlorine concentration or the hydrogen chloride emission in the stack gas is greater than 1.8 kg/h, at least 99% of the hydrogen chloride in the stack gas must be removed.
3. An incinerator burning hazardous waste must not emit particulate matter in excess of 180 mg/dry standard cubic meter (dscm) corrected to 7% O_2 in the stack gas.
4. The feeding of chlorine content fuel should be stopped when one of the measured parameters is out of its required range for a given period of time.

PCB incineration must meet some additional requirements (U.S. EPA, 1979):

(a) Combustion temperature at 1200°C or 1600°C (±100°C).
(b) Residence time of 2 s at 1200°C or 1.5 s at 1600°C.
(c) At least 3% oxygen, for a combustion, temperature of 1200°C, or 2% oxygen, for a combustion temperature of 1600°C, in the stack gas.
(d) Combustion efficiency (CE) calculated from the stack gas CO_2 and CO concentration as $[C_{CO_2}/(C_{CO_2} + C_{CO})] \times 100\%$, shall be at least 99.9%.
(e) The flue gas emission should not exceed 0.001 g per kilogram of PCB introduced into the incinerator. This requires 99.9999% elimination (combustion and flue gas scrubbing) efficiency.

The U.S. EPA (1983) has ranked the POHCs by their heat of combustion per unit mass ($\Delta H_c/g$). This list is based on the presumption that the higher the heat of combustion, the easier the compound is to incinerate. Results of studies have indicated that this ranking is not consistent with the relative gas-phase thermal stabilities of numerous POHCs (Taylor et al., 1990).

Carrying out experiments to provide the scientific data required by decision makers in working out regulations for planning and operation of waste incinerators is a difficult task because of the high toxicity of the investigated compounds. It was necessary, for instance, in the case of PCBs to choose some less toxic model compounds with similar thermal properties. Such model compounds may be selected on the basis of the following considerations:

• They must be registered highly toxic materials.
• They should provide a wide range of calorific values—covering those of a large percentage of possible incoming wastes—so that the incineration behavior of waste mixtures can be modeled.
• Their thermal stability, including that at the temperature of incineration,

and their self-ignition and evaporation behavior should be representative of that of the incinerated chemicals.
- The molecular structures of the model compounds—e.g., aromatic and chlorinated—should be representative of those of all possible incinerated chemicals.

The Energy and Environmental Research Corporation (EERC) worked out a systematic model for studying PCB combustion (Kramlich *et al.*, 1984). Based on considerations similar to those listed above, chloroform, 1,2-dichloroethane, benzene, chlorobenzene, and acrylonitrile were used as model molecules. The measured kinetic constants of the examined compounds suggest that the combustion efficiency for a residence time of 1 s and at a significantly higher temperature than that at which 99.99% destruction of chlorinated compounds is obtained should meet present incineration requirements. This would indicate a great opportunity for waste incineration, for example, in cement kilns, where the temperature can reach 1450°C. Unfortunately, further examination of the kinetic constants makes this result doubtful. It is a widely accepted idea in technology development that the reduced temperature $T_{act} = E_{ap}/R$, where E_{ap} is the apparent activation energy, and R is the gas constant, and the preexponential factor A characterize the examined chemicals. These factors are considered as independent in the mathematical description of the chemical process. However, there is a strong linear relationship between the reported values of $\ln A$ and T_{act} (Fig. 1). The regression equation for the all of the investigated molecules is

$$\ln A = 0.9195 * T_{act} + 5.2675 \qquad r = 0.9955$$

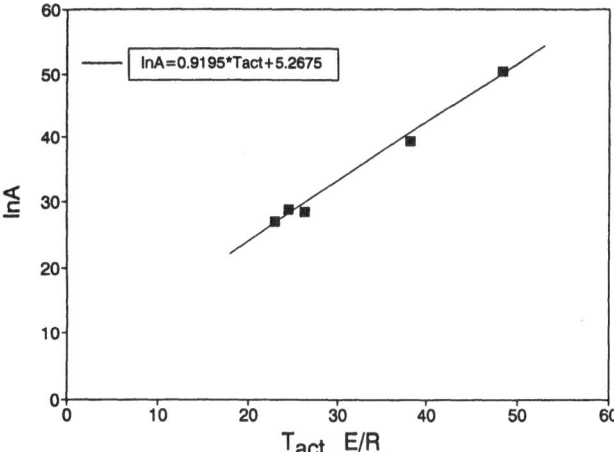

Figure 1. Relation between the logarithm of the preexponential factor, ln A, and the reduced temperature, T_{act}.

It shows that the so-called kinetic compensation effect is a result of the confidence level of the data evaluation method and the experiments. Therefore, the calculated kinetic constants characterize the conditions of the experiment rather than the chemical reactions. It is obvious from this observation that conclusions originating from formal kinetic constants alone are not helpful in the identification of adequate operating parameters for toxic waste incineration.

3. SOME OBSERVATIONS ABOUT THE RELATIONSHIP BETWEEN THERMODYNAMIC CALCULATIONS AND THE PRACTICE OF TOXIC WASTE INCINERATION

3.1. Incineration in the Presence of an Oxygen Excess

The relation between the results of laboratory experiments and the industrial practice of toxic waste incineration is dynamic due to fast-changing environmental regulations. Real incinerator facilities do not have a homogeneously mixed feed, therefore the role of evaporation and diffusion is more important and the reaction mechanism is more complex than would allow for a direct relation between experimental results in the laboratory and industrial practice. Problems similar to those in kinetic experiments have appeared in theoretical equilibrium calculations. A further problem is that thermodynamic data are not available for calculations of the formation of polychlorinated dibenzofurans (PCDFs) and polychlorinated dibenzodioxins (PCDDs) in PCB incineration. Therefore, indirect methods are necessary to solve this problem.

Several calculation methods and models have been presented in the past decade. In spite of the fact that sometimes 200–300 different chemical reactions were included in these models, some important factors were neglected. For instance, the kinetic factors in the equilibrium calculation were disregarded by Ackerman *et al.* (1981). For example, the water–gas reaction $H_2 + CO_2 \rightarrow H_2O + CO$ is a slow chemical reaction, and therefore the real concentration of CO during experiments would not be as high—and the calculated combustion efficiency would not be as low—as indicated by the thermodynamic calculations. Putting this consideration into the models, the calculations give lower CO concentration and higher combustion efficiency.

In spite of this, the inadequate turbulence that frequently occurs in practical realizations of toxic waste incineration make the combustion efficiency significantly lower than that calculated for the case of ideal mixing. One of the best examples of this is incineration in cement kilns, where the high measured CO concentration—up to 9000 ppm (Kálmán and Varga, 1988)—indicates inadequate turbulence in the rotary furnace and the presence of oxygen-deficient conditions.

3.2. Incineration in an Oxygen-Deficient Atmosphere (Pyrolysis)

Another way of calculating the incineration of chlorinated hydrocarbons uses the conditions of pyrolysis as a starting point. Both the process and the products are characterized by the lack of oxygen. The model discussed above was suited to simultaneous calculation of 200 compounds including alkenes, chlorinated benzenes, and phenol (Shih, 1975).

Under pyrolysis conditions and in the case of chlorine concentrations up to 10%, the main products were CO, ethylene, acetylene, methane, benzene, styrene, chlorobenzene, hydrogen, and hydrogen chloride. When the chlorine concentration was in the range of 50 to 70%, the main products were CO, hydrogen chloride, acetylene, benzene, styrene, and mono-, di-, tri- tetra, penta-, and hexachlorobenzene. Later, the same products were detected in several experiments, verifying the results of these calculations (e.g., Taylor and Dellinger, 1988).

Based on these calculations and measurements, it can be declared that pyrolysis conditions during incineration of waste solvents containing chlorinated hydrocarbons are favorable for the production of chlorinated benzenes, which are intermediates in the formation of PCDDs and PCDFs (Graham *et al.*, 1986). Therefore, the thermal destruction of any chlorinated hydrocarbons can lead to the formation of PCDDs and PCDFs even in the high oxygen content of flue gas if pyrolysis conditions occur as a result of inadequate planning or operation of the technology. The oxygen-deficient conditions can also easily arise in high-temperature industrial processes where the quality of the main product is of equal importance to the efficiency of destruction of the incinerated toxic wastes. This phenomenon can explain the wide range of concentrations of, for example, dioxin in waste incineration emission.

We should mention that this is just one reason for the wide spread of the data obtained by analytical measurements at industrial toxic waste incinerators. Although work on the development of reliable sampling methods for concentrations less than 1 ng/m^3 concentration has been conducted in the past few years (e.g., Rappe and Buser, 1989), the results of the measurements of different PCDDs and PCDFs at waste incinerators significantly depend on the location and the method of sampling (Lützke, 1987).

4. RELATIONSHIP BETWEEN CO CONCENTRATION AND EMISSION OF PRODUCTS OF INCOMPLETE COMBUSTION

The effect of CO concentration on the concentration of products of incomplete combustion (PIC) of waste incinerator flue gas is controversial. Although several investigations were not able to find a significant correlation between CO concentration and the POHC or PIC (e.g., U.S. EPA, 1989), the CO concentration in stack gases is used as a gross indicator of combustion problems (U.S. EPA, 1990).

We have reexamined the data from 39 measurements (Staley *et al.*, 1989) dealing with the problem of CO concentration and have employed a different approach from that originally applied for the statistical evaluation of these data. We have found that a strong linear relationship between the measured POHC emission and CO concentration is obtained by relating average POHC and average PIC/POHC to CO concentration instead of evaluating this relationship for individual POHC and PIC compounds. Data in the range between 100 and 600 ppm (without the data for 500 ppm) gives the following equation:

$$POHC = 7.83 \times 10^{-8} \times C_{CO} - 5.69 \times 10^{-6} \qquad r = 0.9791$$

This relationship between POHC and CO concentration is plotted in Fig. 2.

This observation supports the practice of automatically stopping toxic waste feeding into waste incinerators utilized with afterburners when the CO concentration in the flue gas reaches 50 ppm or the oxygen content is lower than 4.5% (Carboni and White, 1981). Moreover, the flue gas CO concentration of a commercially available mobile incinerator developed by EPA's Office of Research and Development is between 1.3 and 7.7 ppm (Freeman and Olexsey, 1986)! In view of these results and the problems mentioned, the planned practice in Hungary of operating cement kilns at up to 8000 ppm CO flue gas concentration (e.g., Aerne and Sólyom, 1992) is disquieting.

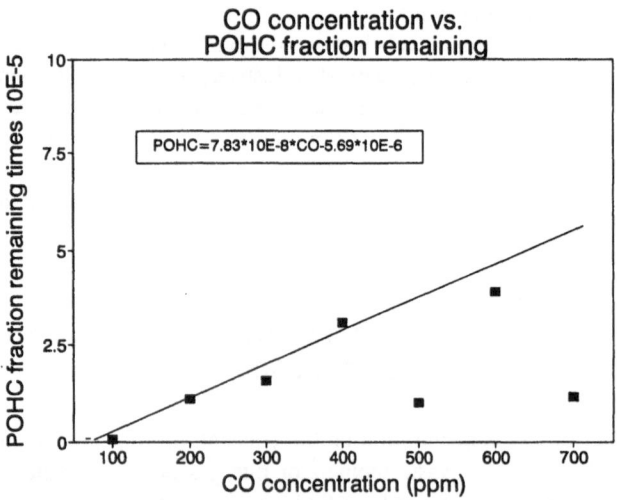

Figure 2. Relationship between the average POHC fraction remaining and the CO concentration of flue gas.

5. CONCLUSIONS

The technologies used for hazardous waste management are based on the presently available results of scientific research and development. The most developed and widely used technique for the elimination of chlorinated hydrocarbons is incineration in special toxic waste incinerators and in high-temperature industrial processes. It is our expectation that this technology cannot be replaced in the near future. Comparing current operating practices with the results of relevant kinetic and thermodynamic calculations, it is evident that the continuous review and development of these technologies and the systems used to model them is necessary and should take into account the scientific and practical imperfections that have been discussed in this chapter. In addition, it must be taken into consideration that the social elements of the environmental problems have the same importance as the knowledge gaps in scientific understanding and methodology.

REFERENCES

Ackerman, D. G., Scinto, L. L., Bakshi, P. S., Delumyea, R. G., and Johnson, R. J., 1981, Guidelines for the Disposal of PCBs and PCB Items by Thermal Destruction, Final Report, TRW, Inc., Redondo Beach, California, EPA-600/2-81-022.

Adonyi, Z., 1993, Thoughts about ideas, methods and ethics protecting the humanity, Paper presented at the conference "Development of the Earth and Life—Protection of the Earth and Life," Budapest, June 3–4, 1993 [in Hungarian].

Aerne, B., and Sólyom, L., 1992, Proposal for alternative fuel use, Lábatlan Cement Ltd., Lábatlan, Hungary [in Hungarian].

Carboni, J. V., and White, J. M., Jr., 1981, Environmental Assessment of Proposed Entitlement Benefits for the Energy Systems Company, Policy Planning and Evaluation Inc., McLean, Virginia, Contract DOE/RG/10089-T-1, March 10, 1981.

Freeman, H. M., and Olexsey, A. R., 1986, A review of treatment alternatives for dioxin wastes, *J. Air Pollut. Control Assoc.* **36:**67–75.

Gauweiler, P., 1992 (Staatsminister für Landesentwicklung und Umweltfragen in Bayern), *Entsorgungs Praxis* **6:**395.

Graham, J. L., Hall, D. L., and Dillinger, B., 1986, Laboratory investigation of thermal degradation of a mixture of hazardous organic compounds. 1, *Environ. Sci. Technol.* **20:**703–710.

Hungarian Council of Ministers, 1981, Government Order, 56/1981 (XI.18).

Hunt, G. T., Wolf, P., and Fennelly, P. F., 1984, Incineration of polychlorinated biphenyls in high-efficiency boilers: A visible disposal option, *Environ. Sci. Technol.* **18:**171–179.

Kálmán, J., and Varga, A. T., 1988, Destruction of Waste Solvents in Cement Kilns, Technology Development Report, Manuscript, Budapest, February 1988 [in Hungarian].

Kramlich, J. C., Heap, H. P., Pohl, J. H., Poncelet, E., Samuelsen, G. S., and Seekei, W. R., 1984, Laboratory-Scale Flame-Mode Hazardous Waste Thermal Destruction Research, Energy and Environmental Research Corporation, Irvine, California, EPA-600/2-84-086.

KTM, 1991, Order of the Minister of Environment and Regional Policy, November 1991.

Lee, C. C., Huffman, L., and Oberacker, D. A., 1986, An overview of hazardous/toxic waste incineration, *J. Air Pollut. Control Assoc.* **36:**922–929.

Lützke, K., 1987, Emissionmessungen von PCDD and PCDF, *VDI-Ber.* **634**:95.

McDonald, L. P., 1977, Burning Waste Chlorinated Hydrocarbons in a Cement Kiln, Technology Development Report, Water Pollution Control Directorate, EPS4-WP-77-2.

Meardon, J. A., Lyanders, S., Rees, J. T., and Shealy, S., 1989, Evaluation of alternative processes for final treatment of hazardous waste effluents, *Environ. Prog.* **8**:62–71.

Öberg, T., Aittola, J. P., and Berström, J. G. T., 1985, Chlorinated aromatics from the combustion of hazardous waste, *Chemosphere* **14**:215–221.

Oppelt, E. T., 1987, Incineration of hazardous waste, *J. Air Pollut. Control Assoc.* **37**:558–587.

Rappe, C., and Buser, H. R., 1989, Chemical and physical properties, analytical methods, sources and environmental levels of halogenated dibenzodioxins and dibenzofurans, in *Halogenated Biphenyls, Terphenyls, Naphthanes, Dibenzodioxins and Related Products*, Elsevier, Amsterdam.

Shih, C. C., 1975, Thermal Degradation of Military Standard Pesticide Formulations for U.S. Army Medical Research and Development Command, TRW Report No. 24768-6019-R4-00, March 1975.

Staley, L. J., Richards, M. K., Huffman, G. L., Olexsey, R. A., and Dellinger, B., 1989, On the relationship between CO, POHC and PIC emission from a simulated hazardous waste incinerator, *J. Air Pollut. Control Assoc.* **39**:321–327.

Taylor, P. H., and Dellinger, B., 1988, Thermal degradation characteristics of chloromethane mixtures, *Environ. Sci. Technol.* **22**:438–447.

Taylor, P. H., Dellinger, B., and Lee, C. C., 1990, Development of a thermal stability based ranking of hazardous organic compound incinerability, *Environ. Sci. Technol.* **24**:316–328.

U.S. EPA, 1979, Polychlorinated Biphenyls (PCBs). Manufacturing, Processing, Distribution in Commerce, and Use Prohibitions, *Federal Register* **44**:31514.

U.S. EPA, 1982, Standards Applicable to Owners and Operators of Hazardous Waste Management Facilities, *Federal Register* **47**:27516–27535.

U.S. EPA, 1983, Guidance Manual for Hazardous Waste Incinerator Permits, Prepared by Mitre Corporation for the U.S. EPA Office of Solid Waste and Emergency Response, NTIS PB84-100577.

U.S. EPA, 1989, Review of OSW's Proposed Controls for Hazardous Waste Incinerators: Products of Incomplete Combustion, U.S. Environmental Protection Agency, Science Advisory Board, October, 1989.

U.S. EPA, 1990, Standards for Owners and Operators of Hazardous Waste Incinerators and Burning of Hazardous Wastes in Boilers and Industrial Furnaces; Proposed and Supplemental Proposed Rule, Technical Corrections, and Request for Comments, 55 FR 82, U.S. Environmental Protection Agency, April 27, 1990.

U.S. National Institute of Health, 1991, TOXNET, N.L.'s Toxicology Data Network, November 13, 1991.

Visalli, J. R., 1987, A comparison of dioxin, furan and combustion gas data from test programs at three MSW incinerators, *J. Air Pollut. Control Assoc.* **37**:1451–1463.

Wiley, S. K., 1987, Incinerate your hazardous waste, *Hydrocarbon Process.* **8**:51–54.

Chapter 10

Concentrations of Combustion Particulates in Outdoor and Indoor Environments

Alfred H. Lowrey, Lance A. Wallace, Sándor Kántor, and James L. Repace

1. INTRODUCTION

Human exposure to combustion particulates is a risk imposed on society justified by the need for the beneficial uses of energy. The respiratory system is the major route for this exposure in the form of airborne suspensions (aerosols) of these particles (U.S. EPA, 1982). Like Prometheus (Fig. 1), we find ourselves chained to the rock of the essential benefits of combustion while at the same time suffering the deterioration of the internal human systems necessary for life support. It is common knowledge that particle exposure, particularly from combustion, is a source of lung disease. In broad terms, outdoor particulates exhibit a bimodal size distribution (U.S. EPA, 1986a) consisting of *fine* particles

ALFRED H. LOWREY • Laboratory for the Structure of Matter, Naval Research Laboratory, Washington, D.C. 20375.   LANCE A. WALLACE • Atmospheric Research and Exposure Assessment Laboratory, U.S. Environmental Protection Agency, Warrenton, Virginia 22186.   SÁNDOR KÁNTOR • Department of Chemical Technology, Budapest Technical University, H-1521 Budapest, Hungary. JAMES L. REPACE • Office of Research and Development, Exposure Assessment Division, U.S. Environmental Protection Agency, Washington, D.C. 20460.   This work was performed by the authors in their private capacity. No official support or endorsement by the Naval Research Laboratory or the Environmental Protection Agency or any other federal agency is intended or should be inferred.

Combustion Efficiency and Air Quality, edited by István Hargittai and Tamás Vidóczy. Plenum Press, New York, 1995.

Figure 1. Legend of Prometheus.

(less than 2.5 μm in diameter, with peak size concentration about 0.9 μm) and *coarse* particles (> 2.5 μm in diameter with peak concentrations in the size range of 10–20 μm). Figure 2 illustrates this distribution, indicating some of the common constituents in each size range. The coarse particles include reentrained surface dust, salt spray, and particles formed by mechanical processes such as crushing and grinding. Particles from combustion general fall into the fine range and occur in two size categories: condensation nuclei and accumulation mode. The condensation nuclei are generally considered to range in size from 0.005 to 0.05 μm in diameter and result from cooling condensation of vapors or plasmas produced by high-temperature processes in combustion. Accumulation mode

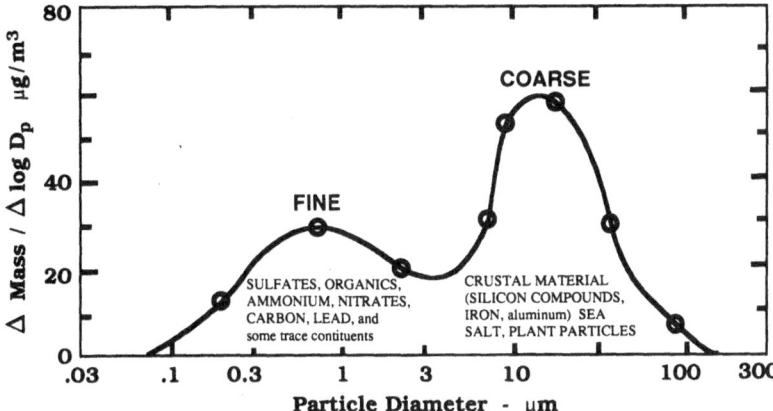

Figure 2. Representative example of typical bimodal mass distribution (measured by impactors) and chemical composition of an urban aerosol.

particles generally range from 0.5 to 2.0 μm in diameter and form principally by coagulation or are grown through vapor condensation of short-lived particles originally in the nuclei mode. Particles in the accumulation mode normally do not grow into the size range of the coarse mode (U.S. EPA, 1986a). It is well known that fine particles evade the natural defenses of the human respiratory tract. Often this fraction of exposure is designated respirable suspended particulates (RSP) to distinguish this feature from exposure to the total quantity of suspended particulates (TSP). In 1987, the U.S. Environmental Protection Agency (EPA) introduced new annual and 24-hour standards for particulate matter, using a new indicator, PM_{10}, that includes only those particles with an aerodynamic diameter smaller than 10 μm (U.S. EPA, 1991). An additional measure, $PM_{2.5}$, was also developed that includes only particles smaller than 2.5 μm in diameter.

Public awareness of the problems of particulate exposure has had enormous impact (Telliard, 1992). As early as the year 1257, Queen Eleanor of England left the city of Nottingham to reside in Tutbury because of smoke. In the year 1273 the use of coal was prohibited in London because it was considered that clouds of smoke were prejudicial to health. The worst of the recorded smog disasters in London occurred in 1952, 1956, and 1962 with estimated mortalities of 4000, 1000, and 750, respectively, which produced major changes in regulatory policy (Parker, 1978). Emphasis has been on trying to achieve the proper balance between the essential need for combustion and the control measures necessary to manage the by-products. This was evident when the interest in energy conservation affected ventilation practices and thrust the indoor airshed into prominence as a major environment for human particulate exposure. The problems of particulates have generated significant amounts of research on exposure in both outdoor and indoor environments, resulting in vast amounts of information and significant new perceptions about managing the inevitable risks.

This chapter will present simple ideas which concern managing exposure to combustion particulates. The problem is widespread and complex and, in detail, is often not subject to simplistic analysis, particularly given the intricacies of the cultural or social context. These complexities have been reflected in the vastly different worldwide responses to the problems of air pollution (Schulze, 1993). However, certain observations and methods of analysis have been found to have comprehensive significance for understanding particulate exposure in both indoor and outdoor environments. They are the subject of this chapter. It is intended that the simple concepts will provide a framework for providing references to more complex research literature and methodologies. Basic principles of risk analysis, with some illustrative examples, are also included for the purpose of providing perspective on managing human exposure to combustion particulates. The purpose of this chapter is to provide a current view about the balance between the rock of our need for combustion and the vulture's beak of the impact on our health.

2. CHARACTERIZATION OF PARTICLES

An aerosol is defined as a gaseous suspension of fine particles. They are formed either by conversion of gases to particulate matter, disintegration of liquids or solids, or resuspension of powdered material. A great variety of words have been used to signify aerosols; dust, smoke, haze, fume, fog, mist, and smog are common with somewhat different popular definitions. The negative associations of many of these words are indicative of the problems which motivate the continuing research efforts. For atmospheric aerosols, material directly suspended in the form of particles is labeled *primary* while material resulting from gas–particle conversion is called *secondary*. Experimental sampling techniques are a well known and difficult problem in dealing with aerosols (U.S. EPA, 1982) and are closely correlated with the nature of the information available. This is particularly significant for aerosols of biological material (Cox, 1987). It is worth noting, for example, that the well-known variances in exposure estimates derived from differing sampling methodologies (Wallace *et al.*, 1991; Willeke and Baron, 1993; Wick *et al.*, 1994) have led to the suggestion of a new analytical method blending the techniques of individual hygiene with those of environmental monitoring by measuring an individual's actual exposure using personal monitoring techniques. Personal monitors consistently find higher levels of particles than observed by stationary samplers, suggesting the concept of a personal cloud of particles associated with this individual (Pellizzari *et al.*, 1993). Particle size, concentration, and chemical composition are usually the most important factors in determining the effects on human health (Friedlander, 1977).

2.1. Physical Characteristics

The distribution of particle sizes usually encountered closely approximates a lognormal distribution (NAPCA, 1969). In this distribution, the familiar bell-shaped probability curve appears when particle size frequency is plotted against the logarithm of the particle size. Figure 3 shows this graph for a lognormal distribution chosen to have an arbitrary median particle size at 1 μm. Figure 4 shows the same distribution plotted against the direct particle size with the values of several other weighted average diameters indicated in comparison to the arbitrary 1-μm size median. The particle size is one of the important features of the physical configuration because it is a major determinant for particle motion and its impact on human health. Table I shows the size range of various biological particles in the atmosphere (Wick *et al.*, 1994) and the approximate region of their deposition in the human respiratory system.

Figure 5 illustrates the respiratory deposition regions as a function of particle size. Thus, RSP are generally more of a concern because of the direct relationship between concentrations of particles this size and effective human dose. Other important characteristics are the particle shape, structure, and density (U.S. EPA, 1982). Table II gives various particle shapes associated with some of the possible sources.

The density is important because it affects motion and behavior. The density for particles of regular geometry is usually associated with the bulk density of the

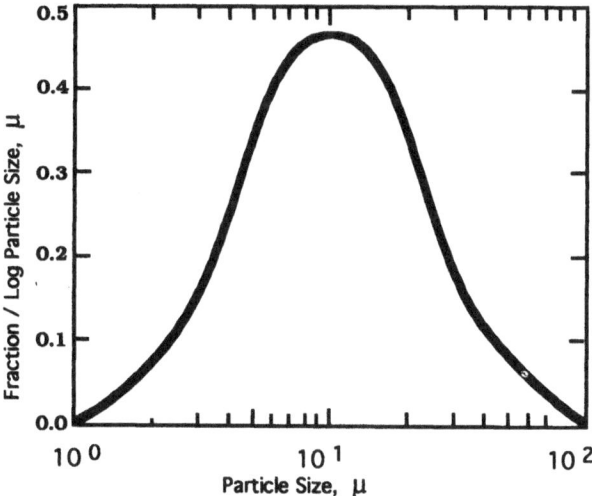

Figure 3. Particle size distribution, plotted logarithmically, for idealized lognormal distribution with a median diameter of 1 μm. With a logarithmic plot, this is symmetrical in contrast with the same distribution displayed in Fig. 4.

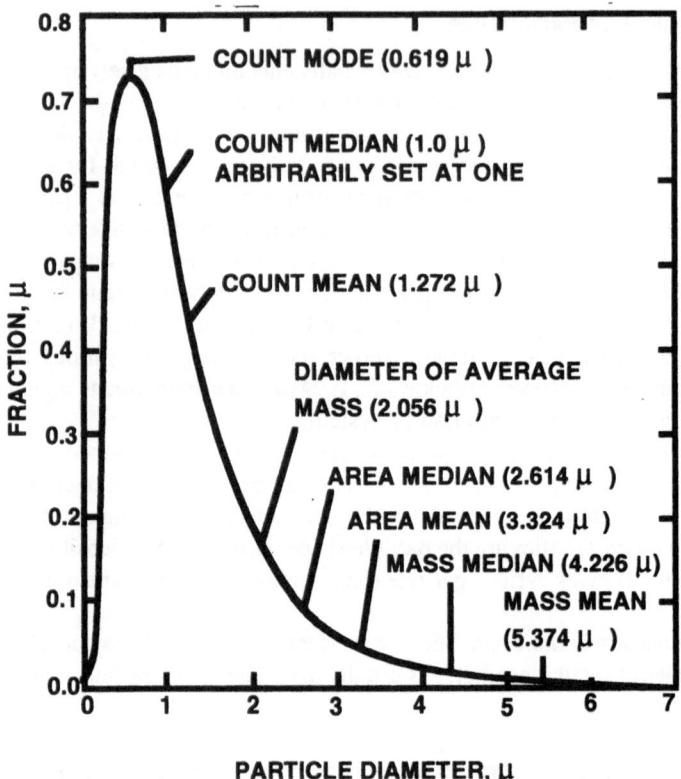

Figure 4. Idealized lognormal distribution of particles having a count median diameter of 1 μm and showing numerical values of several other relative weighted diameters.

Table I. Biological Particles in the Atmosphere and Their Diameters (Wick *et al.*, 1994)

Types[a]	Diameter (μm)	Deposition region
Viruses	0.015–0.45	Tracheobronchial
Bacteria	0.3–10	Tracheobronchial
Algae	0.5–10,000	Pulmonary
Fungus spores	1–100	Pulmonary
Protozoa	2–10,000	Pulmonary
Moss spores	6–30	Nasopharyngeal
Fern spores	20–60	Nasopharyngeal
Pollen	10–100	Nasopharyngeal
Plant fragments, minute seeds, insects, spiders, mites, etc.	>100	Nasopharyngeal

[a]Many of the very small particles are carried on fragments of organic matter or "rafts," airborne dust, etc., which are of large diameter, or by insects.

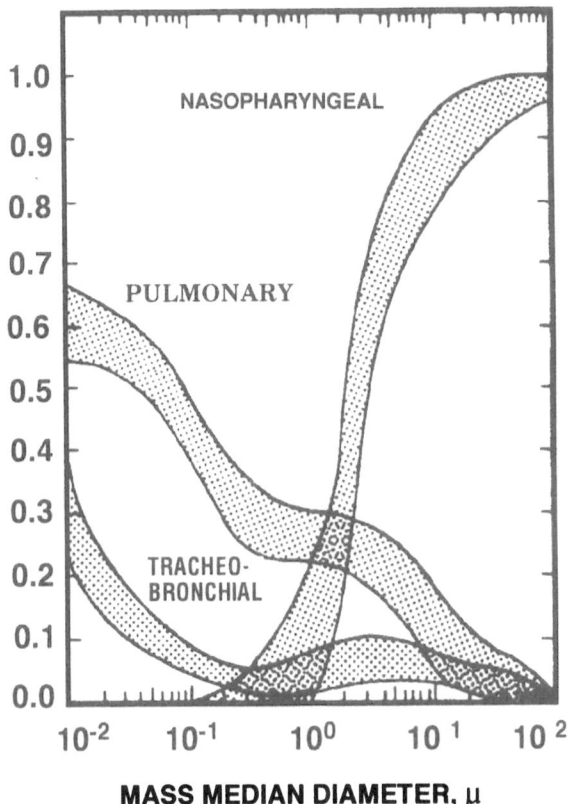

MASS MEDIAN DIAMETER, μ

Figure 5.   Deposition region in the human respiratory system as a function of particle diameter.

material of origin. However, many particles are agglomerates of smaller particles of various compositions. The structure of these particles is often important because they include air-filled voids. Because many particles are irregular or agglomerates and have unknown density, it is common to represent the shape, structure, and density of particles in terms of dynamically equivalent spheres of unit volume. These are characterized by their *aerodynamic diameter*, defined as the diameter of a sphere of unit density that attains the same terminal velocity at low Reynolds number in still air as the actual particle under consideration (U.S. EPA, 1982). This allows the approximate use of the Navier–Stokes equation to model the motion of the particles in the aerosol for problems ranging from atmospheric dispersion to deposition in the human respiratory system. Figure 6 shows the settling velocity for spherical particles having unit density (NAPCA, 1969). Stokes' law applies over a considerable range of particle sizes but

Table II.   Particle Shapes and Some
Possible Source Types (U.S. EPA, 1982)

Shape	Source type
Spherical	Smoke, pollen, fly ash
Irregular	Cinder, biological
Cubical	Mineral
Flakes	Dander, mineral
Fibrous	Lint, plant fiber
Condensation flocs	Carbon, smoke, fume

requires corrections at extremes of the curve. The particle surface characteristics
are related to these size and shape factors and are known to be important for
sorption, nucleation, and adhesion (NAPCA, 1969). Figure 7 shows frequency
plots for particle number, surface, and volume for the Pasadena smog aerosol,
illustrating differences in the distribution of these properties. Extensive computational modeling approaches have been developed for simulating atmospheric
transport, transformation, and deposition using such techniques as Gaussian
diffusion (U.S. EPA, 1982). Extensive computational modeling efforts have also
been devoted toward understanding pollution transport in the indoor airshed

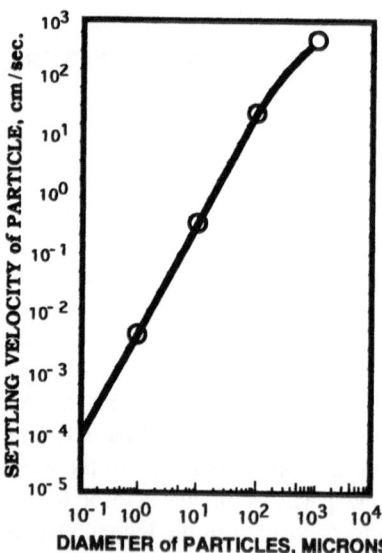

Figure 6.   Settling velocities in still air at 0°C and 760 mm pressure for particles having a density of
1 g/cm^3.

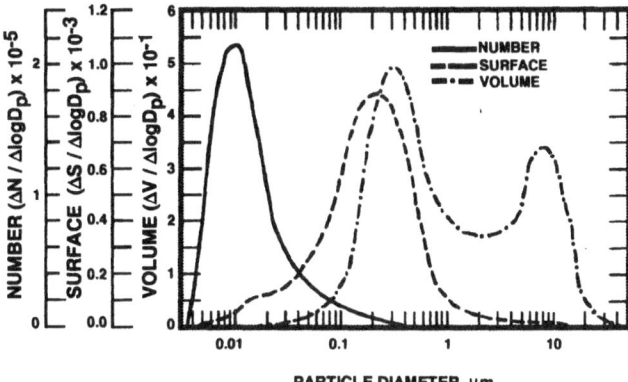

Figure 7. Frequency plots of particle number, surface, and volume distribution for 1969 Pasadena smog aerosol.

(Austin *et al.*, 1992; Axley, 1988; Sparks, 1988). A continuing problem is the limited experimental validation of these models (Willeke and Baron, 1993; Wick *et al.*, 1994).

2.2. Chemical Characteristics

Understanding the chemical composition of the particulates has direct significance for balancing the human health risks with the benefits associated with the source. Table III gives a breakdown of the sources of 24-hour average particulate samples from Los Angeles area smogs in 1972 and illustrates the chemical complexity of major sources contributing to urban aerosols.

Chemical composition is related to particle sizes as roughly indicated in Fig. 2. Table IV gives the predominant particle sizes for various chemical species.

Particulate organic matter has often been fractionated by means of acid–base extractions followed by column chromatography. Table V shows the general proportions of broad classes of organic compounds for organic particulates collected in Detroit around 1977.

Indoor particulates tend to be biological materials or organic combustion products with a much higher probability of being in the respirable range than those found outdoors (NRC, 1981). Major indoor aerosols have been classified into six types: bioaerosols (plant and animal), mineral, combustion, home/personal care products, and radioactive aerosols produced by attachment of radon decay products to existing particles (Owen *et al.*, 1992).

The importance of understanding the physical and chemical characteristics of the particulates is directly related to decisions about risk management and control.

Table III.   24-Hour Source Breakdown for Particulate
Samples from Los Angeles Area Smogs[a]

Source	Micrograms per cubic meter		
	Pasadena	Pomona	Riverside
Sea salt	0.7	5.7	1.3
Soil dust	19.8	15.1	28.5
Auto exhaust	5.1	7.2	3.9
Cement dust	1.4	3.3	2.3
Fly ash	0.1	0.2	0.1
Diesel exhaust	1.4	1.9	0.9
Industrial and agricultural	4.7	6.6	20.5
Tire dust	0.5	0.7	0.4
Aircraft	1.3	1.8	7.4
SO_4^{2-}	2.9	19.	5.9
NO_3^-	4.9	36.4	12.9
NH_4^+	2.3	16.3	5.7
Organics	29.6	29.3	24.8
Water	12	18	—[b]

[a]Adapted from Friedlander 1977.
[b]Unknown.

The respirable particles present a more significant impact on potential human dose since the lungs are the primary entry for particulate-based pollutants. There are some sources over which we have little control, such as sea spray. Combustion, on the other hand, is almost always human-generated, with extensive opportunities for informed control. There are different human environments which are characterized by different particulate sources, with a variety of opportunities for understanding and modifying total human exposure, which is a primary determinant of risk. In short, atmospheric aerosols present complex problems of physical and chemical characterization, but relatively simply analysis and classification are useful tools for managing the resulting risks.

Table IV.   Correlations of Chemical Content with Particle Size[a]

Chemical content	Particle size
Sulfates, organic condensate, lead, arsenic, selenium, hydrogen ion, NH_4^+ salts, soot	Normally fine
Iron, titanium, magnesium, potassium, phosphate, silicon, aluminum	Normally coarse
Chloride, nitrate	Normally bimodal
Nickel, tin, vanadium, antimony, manganese, zinc, copper	Variable

[a]U.S EPA, 1982.

Table V. Composition of the Organic
Fraction of Airborne Particles Collected
in Detroit[a] (U.S. EPA, 1982)

Aliphatic hydrocarbons	48.3%
Aromatic hydrocarbons	3.6%
Neutral oxidized hydrocarbons	20.8%
Acidic compounds	14.8%
Basic compounds	0.55%
Insolubles	10.8%

[a]Average benzene-soluble organics, 3 μg/m^3.

3. PRINCIPLES OF RISK ANALYSIS

It is difficult to communicate the significance of particulate exposures merely through the use of tables of concentrations and graphs of trends. This is because significance involves judgment and trade-offs to evaluate what levels of such exposures are acceptable. We are surrounded by the hazards of modern life and by the many communications about these hazards. The messages about the risks associated with these hazards are often controversial generally because there is enough uncertainty in the underlying knowledge to allow contradictory conclusions by different experts whose research comes from different points of view. Messages about significance and risk necessarily compress technical information, and this can lead to misunderstanding, confusion, and distrust (NRC, 1989). For the purpose of bridging the intricate balance between science and policy, the Government of the United States of America implemented the process known as risk assessment (NRC, 1983). The purpose was to institute mechanisms that best foster a constructive partnership between government and science, to ensure that government regulation rests on the best available scientific knowledge, and to preserve the integrity of scientific data and judgments in the unavoidable collision of contending interests that accompany most important regulatory decisions.

Because risk is ubiquitous, judgments and decisions are made in many ways (CEQ, 1989). For the purpose of this chapter, a few basic definitions are given here:

- *Hazard* is a source of *risk* and refers to a substance or action that can cause harm. The likelihood of harm from exposure distinguishes *risk* from a *hazard*.
- *Risk* is defined as the possibility of suffering harm from a *hazard*.
- *Risk Management* uses information from risk assessment and risk analysis, together with information about technical resources, social, economic, and political values, and control or response options to determine means of reducing or eliminating a risk.

For the purpose of determining the significance of combustion particulate levels in outdoor and indoor environments, the process of risk analysis will be defined in terms of four basic steps:

- Hazard identification
- Risk assessment
- Determining the significance of the risk
- Communication of risk information

This outline provides a framework for bridging the gap between science and policy. However, in general, there is no single correct approach for analyzing risks nor do all risk analyses follow these steps. In fact, the interactive nature of the process may change the sequence (CEQ, 1989).

Hazard identification is the process of ascertaining whether there is a potential risk associated with a suspected hazard. Many technical analyses of environmental and health risks define this potential for risk as a combination of the following factors:

- Probability of occurrence of an event
- Probability that a hazard will be released by an event
- The probable quantity, concentrations, transport, and fate of hazardous materials released into the environment at the time of the event.
- The probability of exposure of individuals, populations, or ecosystems to hazardous materials released into the environment
- The probability of adverse human health or environmental effects from exposure to hazardous materials released into the environment

Integral to the process are scientific inferences based on available evidence. For carcinogenic risk, such evidence falls into four general classes: epidemiological data, animal-bioassay data, data on *in vitro* effects, and comparisons of molecular structure/chemical similarity (NRC, 1983). The scientific judgments about such evidence are an integral part of the process and strongly influence the probability factors for the projected risk.

Risk assessment is the process of estimating the severity and likelihood of harm to human health or the environment occurring as a result of exposure to a substance or activity that, under plausible circumstances, can cause harm. Four distinct but related techniques are encompassed here: source/release assessment, exposure assessment, dose–response assessment, and risk characterization.

Source/release assessment estimates the amounts, frequencies, and locations of the introduction, release, or escape of risk agents from specific sources into occupational, residential, or outdoor environments.

Exposure assessment provides quantitative data on individuals, populations, or ecosystems that are, or may be, exposed to a risk agent; data on the concentrations of the risk agent; and information about the duration and other characteristics of the exposure.

Dose–response assessment provides quantitative data on the specific amounts of a risk agent that may reach organs or tissues of exposed individuals or populations and attempts to estimate the percentage of the exposed populations that might be harmed or injured and to identify, where relevant, particular characteristics of populations potentially at risk from the agent. The important characteristics of the dose–response assessment may be determined from information about exposures and risks of the average individual and exposures and risks of the most exposed individuals.

Risk characterization integrates the above information with the complex political, economic, and social factors into a comparative risk statement that includes one or more quantitative estimates of risk, often including sensitivity studies and quantitative values based on alternate assumptions. This is essential for determining the significance of risk.

Determining the significance of risks is essential because it is impossible to eliminate all risks associated with human activities. Risk assessment techniques provide a means of organizing relevant information and estimating adverse health or environmental consequences. In doing so, they may convey a level of precision that does not reflect the tentative nature of underlying assumptions and uncertainties that characterize this process. A balance may be achieved by using the analytical tools and by considering what constitutes a level of acceptable risk. This proves difficult in principle because different people perceive risks differently, and this has prompted a great deal of controversy (NRC, 1989; CEQ, 1989). In practice, the U.S. federal government has adopted a 1 in a million (1×10^{-6}) lifetime risk as a *de minimis* standard below which generally no regulation is needed and a 3 in 10,000 (3×10^{-4}) lifetime risk as a *de manifestis* level above which regulation is always needed (Repace and Lowrey, 1993).

4. THE CONCEPT OF TOTAL EXPOSURE

The development and application of the techniques of risk analysis have demonstrated the need to understand the total human exposure to a pollutant. Because of the complexities of human activity patterns (Szalai, 1972; Ott, 1988), it is clear that ambient outdoor concentrations of pollutants such as combustion particles are an unreliable indicator of the actual dose inhaled by the average person (Repace *et al.*, 1980). Figure 8 shows the particulate levels measured by a piezobalance in the vicinity of James Repace for a 24-h period, October 16, 1979. Figure 9 shows the total exposure percentages for indoor, outdoor, and transit environments relative to the time budget spent in these environments. It is also important to consider the uncertainties of sampling: the use of personal monitors located on an individual show higher concentrations than are found with stationary samplers, suggesting an effect associated with a personal cloud which moves with the individual (Pellizzari *et al.*, 1993; Wallace *et al.*, 1991). The general lack of

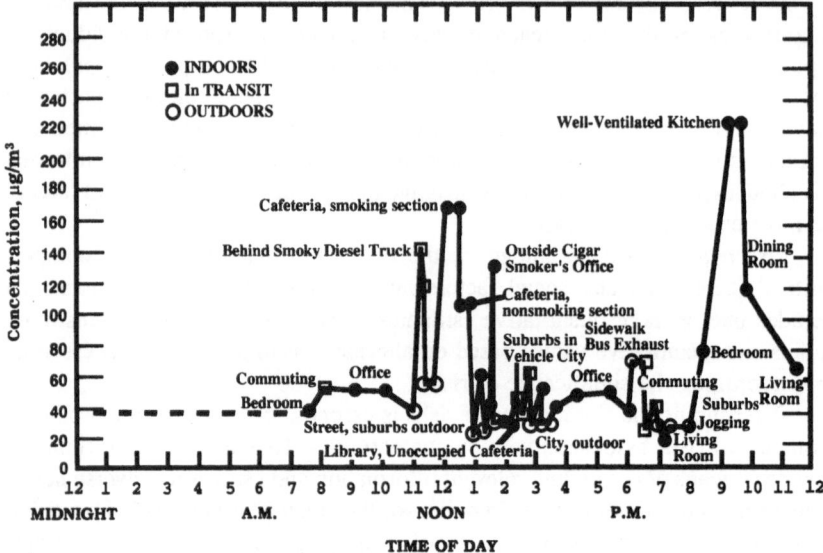

Figure 8. Personal exposure of James Repace to respirable particles (as measured by a piezobalance) over a 24-h period.

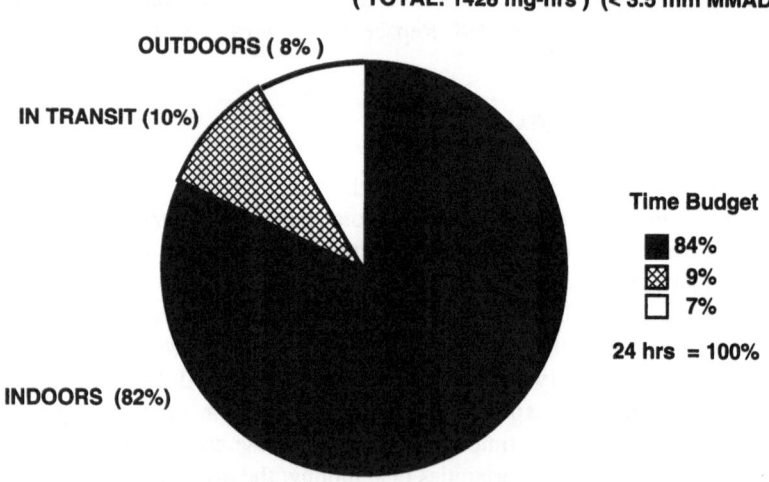

Figure 9. Total particulate exposure percentages for indoor, outdoor, and transit environments derived from time and concentration measurements shown in Figure 8.

specific location exposure data is a major contributor to uncertainty because detailed representative data are very limited and concentration levels are very location specific (Jenkins *et al.*, 1992). Total exposure is the integrated sum of these exposures over the time that the individual is exposed. This exposure is crucial for understanding the true risks of environmental pollutants (Rodricks, 1992; Lappe, 1991). This concept of total exposure is essential for understanding the emphasis on the indoor environment, especially for combustion particulates.

5. OUTDOOR LEVELS OF PARTICULATES

Atmosphere particles in the size range of 1–10 μm tend to have a composition characteristic of local sources and soil: in maritime locations, the bulk of airborne sea salt will be found in particles of this size; in urban locations, industrial and combustion processes generally contribute the bulk of these particles (NAPCA, 1969). National PM_{10} emission estimates for 1985 are given in Table VI (U.S. EPA, 1991).

Table VI illustrates the complex problems of interpreting experimentally observed levels of particulates. These complexities are reflected in various regional measures of particulates shown in Table VII (U.S. EPA, 1991). The highest values are consistently found in region IX, which includes California, Hawaii, Arizona, and Nevada, encompassing the Los Angeles smog basin, Pacific islands,

Table VI. National PM_{10} Emission Estimates for 1985 (U.S. EPA, 1991)

Source category	Million metric tons/year
Transportation	1.3
Fuel combustion	1.2
Industrial processes	2.4
Solid waste	0.2
Miscellaneous	0.8
Total from above categories	5.9
Fugitive emissions	
Agricultural tilling	7.4
Burning	0.7
Construction	12.2
Mining and quarrying	0.4
Paved roads	5.9
Unpaved roads	17.3
Wind erosions	3.8
Total fugitive emissions	47.7

Table VII. U.S. Regional Average Particulate Levels ($\mu g/m^3$) for 1989[a]

Region	TSP level Geometric mean	PM$_{10}$ level Arithmetic mean	90th percentile
I	40	24	41
II	42	32	53
III	46	33	56
IV	41	32	50
V	52	33	57
VI	48	32	53
VII	57	33	56
VIII	46	32	56
IX	74	45	72
X	49	33	61

[a]U.S. EPA, 1991.

and the wind-blown southwest deserts. It is of interest to note that, based on data through 1988, failure to meet the U.S. National Ambient Air Quality Standards (NAAQS) for particulates occurs not only in urban areas such as Los Angeles, Denver, and Cleveland but also in agricultural areas such as the central California valley and timber areas in northern Maine (U.S. EPA, 1991).

Table VII also illustrates important changes in the NAAQS for particulates. On July 1, 1987, the U.S. EPA promulgated new annual and 24-hour standards for particulate matter, using the new indicator PM$_{10}$, which includes only those particles with aerodynamic diameter smaller than 10 μm. The original TSP standards were an annual geometric mean of 75 $\mu g/m^3$ and a 24-hour concentration of 260 $\mu g/m^3$, not to be exceeded more than once a year. The new PM$_{10}$ standards specify an expected annual arithmetic mean not to exceed 50 $\mu g/m^3$ and an expected number of 24-hour concentrations greater than 150 $\mu g/m^3$ *not to exceed one*. The PM$_{10}$ particles appear to represent essentially all the particulate emissions from transportation sources and most of the emissions in the other traditional categories. The trends in the former NAAQS for TSP show a documented 20% decrease in national average TSP concentrations between 1978 and 1982, despite difficulties arising from changes in sampling technique. This trend is attributed to installation of control equipment by electric utilities, resulting in a decrease in particulate emissions *despite increased fuel consumption*, and also to reduced activity in industrial processes, for example, in the iron and steel industries (U.S. EPA, 1991). Smaller increases in regional values have been attributed to forest fires and drought conditions. Figure 10 shows idealized effects on particle size distribution from various types of particle emissions (U.S. EPA, 1982). The complexity of these standards, however, is illustrated by the problem of

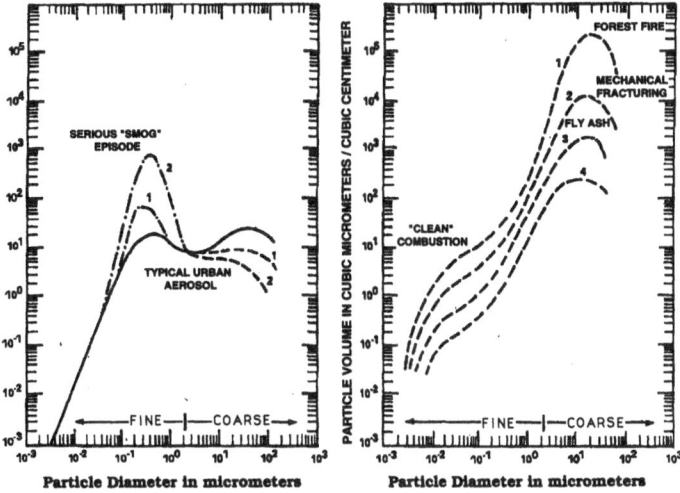

Figure 10.   Relative impact on idealized size distributions for typical events impacting outdoor air quality.

relating the 90th percentile values in the table to the 24-hour concentration standard. The 90th percentile value is a reasonable peak concentration for temporal comparisons (and is well below the 150-μg/m^3 maximum concentration); however, this standard is exceeded by maximum concentrations at 11% of the reporting sites (U.S. EPA, 1991).

5.1. Recent Examples

5.1.1. Los Angeles

In 1987, Congress mandated that the U.S. EPA undertake a study of exposure to particles. EPA's Atmospheric Research and Exposure Assessment Laboratory (AREAL) joined with California's Air Resources Board to sponsor a study in the Los Angeles Basin (Pellizzari *et al.*, 1993). The full study was carried out in Riverside, California, in the fall of 1990. The city of Riverside was selected because it was known to have highly variable outdoor PM$_{10}$ concentrations and is typical of southern California communities. A wide range of outdoor concentrations facilitated determination of the contribution of outdoor levels to indoor and personal exposure levels. The main goal of the study was to estimate the frequency distribution of exposures to PM$_{10}$ particles for all nonsmoking Riverside residents aged 10 and above, based on a probability sample of 178 residents. A second major objective was to estimate the frequency distribution of concentrations of PM$_{10}$ and

PM$_{2.5}$ particles in residences and nearby outdoor areas (e.g., backyards). Particles in the PM$_{2.5}$ range are less than 2.5 μm in diameter and represent the fine mode of the size distribution corresponding to RSP. A central site monitor measured coarse particles and PM$_{10}$ size particles. Figure 11 shows levels of 12-hour samples for a 48-day period beginning September 22, 1990. Table VIII shows exposures (weighted to represent nonsmoking population aged 10 years or more) to particulates in outdoor and indoor environments and effective concentrations observed through the use of personal monitors.

One of the important results of this study was that the comparisons between measurements at the central site and measurements at individual locations, including those obtained by the use of personal monitors, showed good correlation. This suggests that a single monitor may be a good predictor of ambient particle levels in areas several miles away. The most important observation of the study is the excess measured personal exposures compared with the indoor and outdoor measurements using stationary monitors. This suggests a "personal cloud" of particles, possibly due to reentrainment from clothes, carpets, or furniture. Overnight *indoor* concentrations were also correlated with outdoor concentrations, although with an increased scatter due to variations in level of indoor activity. The personal concentrations, perhaps due to the "personal cloud," did not correlate well with outdoor concentrations, with $r^2 = .16$ (Pellizzari *et al.*, 1993).

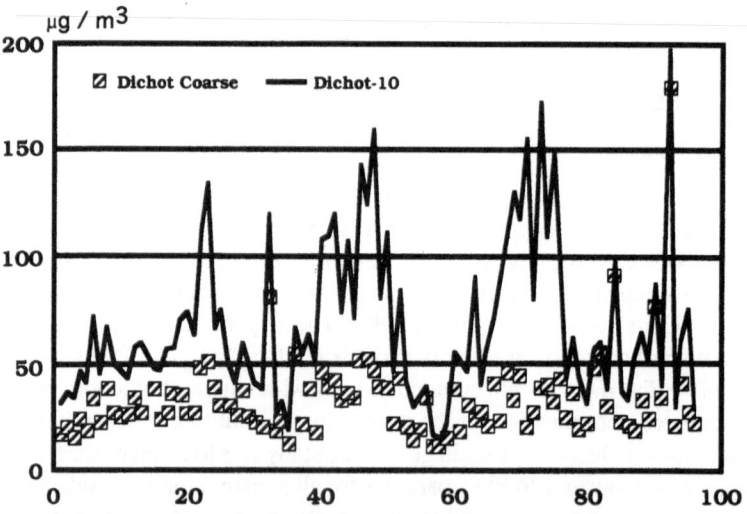

Figure 11. Particle total exposure assessment methodology (PTEAM) measurements of 12-h average coarse and PM$_{10}$ particle concentrations over a 48-day period in Riverside, California.

Table VIII.   Population-Weighted[a] Concentrations ($\mu g/m^3$) of Particulates[b]

Sample type	N	Median	Arithmetic mean	Percentile 90th	Percentile 98th
Daytime PM10					
Personal	171	130	150	260	380
Indoor	169	82	95	180	240
Outdoor	165	83	94	160	240
Overnight PM10					
Personal	168	66	77	140	190
Indoor	163	52	63	120	160
Outdoor	162	74	87	170	210
Daytime PM2.5					
Indoor	173	34	48	100	170
Outdoor	167	36	49	100	170
Overnight PM2.5					
Indoor	166	26	36	83	120
Outdoor	161	35	51	120	160

[a]Personal samples were weighted to represent nonsmoking population of 139 Riverside residents aged 10 or above. Indoor and outdoor samples were weighted to represent 61,500 homes with at least one nonsmoker aged 10 or above.
[b]Pellizzari et al., 1993.

5.1.2. Budapest

Because of the scarcity of information on levels of indoor and outdoor air pollution in Eastern European countries, measurements of RSP levels were made outdoors and in various indoor environments in the city of Budapest, Hungary. This is a busy national capital city with high traffic and a well-publicized problem of inefficient combustion engines for transport. Data were collected between September 6 and September 15, 1991, primarily in the downtown area. Preliminary results have been presented at the 6th International Conference on Indoor Air Quality and Climate, Indoor Air '93, in Helsinki, Finland (Lowrey et al., 1993). Measurements were made with a portable TSI model 8510 piezobalance. This measures mass concentrations in the range of 10 $\mu g/m^3$ to 10 mg/m^3. For these experiments, a velocity filter selected respirable particles with a 50% cut diameter of 3.5 μm. Table IX gives outdoor RSP levels at various locations.

Within the limitations of the experiment, the locations are meant to be representative rather than statistically significant, and the values measured may be compared with those from a similar study in Washington, D.C., in 1980 (Repace

Table IX. Measurements at Outdoor Locations in Budapest

Location	RSP ($\mu g/m^3$)	Sample time (min)	Comments
1. Buda Castle beer hall	28	6	Intermittent cooking smell
2. Buda Castle street	40	6	
3. Moscow square, by subway	43	6	Transient tobacco smell
4. Moscow square, away from street	45	8	10 m away from street edge
5. Nonsmoking apartment	45	6	Some cooking smell
6. Deak Ferenc square	45	8	Traffic center
7. Buda Castle overlook	47	8	
8. Downtown beer hall	50	6	Noticeable rise from bus
9. Kossuth square	50	4	Karoli Mihaly center
10. Downtown side street	50	6	
11. Moscow square, plaza	53	6	Interior by tram stop
12. Student restaurant	60	6	
13. Moscow square, by street	60	8	Curbside
14. Downtown intersection	68	10	
15. Moscow square	69	7	Friday, Sept. 13, 5:30 P.M.
16. Student pub	77	8	
17. Old Buda apartment complex	85	16	
18. Small park beside tram stop	85	8	Between buildings away from traffic
19. Outside student club, by river	86		
20. Small park close to large office	87	6	
21. Railroad station	93	6	Some drifting smoke
22. Small office	95	12	Some influence from traffic
23. Moscow square, bus stop	97	6	Some drifting smoke
24. Downtown bus stop	98	16	Park by major bus stop
25. Large office	102	10	
26. Bus stop alongside bridge	117	6	Below alongside of bridge
27. Roadway under bridge	118	8	On road along river under bridge
28. Margit Hid tram stop	150	6	Center of road

and Lowrey, 1980). These outdoor background levels in Budapest averaged 75 ± 29 $\mu g/m^3$ ($n = 28$). This value may be compared with values for Washington, D.C., of 38 ± 15 $\mu g/m^3$ ($n = 5$) in heavy traffic and 46 ± 13 $\mu g/m^3$ in other outdoor locations ($n = 15$). The values recorded by the piezobalance are not directly comparable to the EPA standards for the United States because of differences in particle sizes and sampling methodologies (U.S. EPA, 1991). However, the average value reported for Washington, D.C., in 1980 is similar to the highest arithmetic mean PM_{10} concentration for Washington, D.C., reported by EPA in 1989 (43 $\mu g/m^3$), consistent with the documented decline in the early 1980s. Thus, the value of 73 $\mu g/m^3$ measured in Budapest in 1989 is similar to

1989 values for cities such as El Paso, Texas, St. Louis, Missouri, and Fresno, California, while it is *significantly less* only compared to the 93-μg/m^3 value reported for the Riverside–San Bernardino, California, area (U.S. EPA, 1991).

The impact of automobile traffic on RSP background levels is evidenced by several measurements in Table IX. Transportation source emissions may be divided into two categories: engine-related emissions from vehicle exhaust and other highway vehicle-related particles from tire wear and clutch and brake lining wear (U.S. EPA, 1982). Engine-related particulate emissions are composed primarily of carbonaceous matter (including adsorbed organics), sulfates, and lead halides. Highway vehicle-related particulates are emitted at the rate of about 0.01–0.30 g/mi by gasoline engines and 0.5–3.0 g/mi by diesel engines. The three high outdoor measurements were made at sites near the Margit bridge during rush hour traffic on a Friday afternoon. The tram stop is located between lanes of traffic. By comparison, the two low measurements were made at locations in the Buda Castle area, which is above the general traffic patterns and airshed for most of the city, and were taken at weekend times with relatively little traffic. Some proximity measurements were also taken to assess distribution due to traffic. Measurement 18 was prompted by the large value observed at the tram stop. The pedestrian exit opens into a small pocket park with typical urban trees and plantings. It is surrounded on three sides by buildings and open to the street on the fourth. The measurement was made about 20 m from the road and is almost half that found in the middle of traffic. Similar proximity measurements were made at Moscow Square, which is a transfer stop for the subway, the bus system, and the tram system. This is the main road for traffic from Buda to Pest. The site arrangement is roughly triangular, with automobile and bus traffic on one side, tram lines both parallel with the road and also forming another side, and the third side consisting of a large hillside with the road to Buda Castle on top. The subway stop is located in about the middle along with associated vending stands. A series of measurements were made on a Saturday morning. The locations away from the street (3, 4, 11) were lower than those by the street and at the bus stop location at the street (13, 23). Measurement 15 was made in the center of the concession area on Friday during evening rush hour. This traffic-related variation is well correlated with measures of air quality in other cities (Bardeschi *et al.*, 1991) and supports the inference that the high RSP background level reported here is related to the inefficient combustion engines in vehicles. Most transportation is in vehicles not well designed for emission control, with much of the recent increase in traffic due to an influx of inefficient automobiles produced in Eastern Europe.

5.2. Outdoor Summary

In summary, a few simple observations have been shown to be true for levels of combustion particulates in the outdoor environment:

1. Combustion particulates are a major source of outdoor particulate levels, especially in urban environments.
2. They are not the only source, as demonstrated by the spread of areas in the United States with elevated levels of particulates.
3. The PM_{10} standard for the United States is an annual arithmetic mean of 50 $\mu g/m^3$. Most of the metropolitan statistical areas (MSA) in the United States are in compliance with this standard. For 1989, only 13 out of 340 MSAs exceeded this level. Reported PM_{10} levels ranged from 16 to 96 $\mu g/m^3$.
4. Control of emissions from combustion has significantly affected the ambient particulate levels with a 20% decrease between 1978 and 1982. This effect is greater than the fluctuations due to drought and forest fires.

6. INDOOR LEVELS OF PARTICULATES

The importance of the indoor environment as a source of polluted air has only been recently recognized (Nagda *et al.*, 1987). The first modern studies on indoor air quality were conducted in the mid 1960s (Biersteker *et al.*, 1965) and early 1970s (Yocum *et al.*, 1971); among the pollutants studied were total suspended particulates (TSP), sulfur dioxide, and carbon monoxide (Nagda *et al.*, 1987). The indoor airshed is the predominant source for human exposure to air pollution (Szalai, 1972; Coghlin *et al.*, 1989): urban dwellers in industrialized countries spend over 90% of their time indoors (Jenkins *et al.*, 1992), with 88% of their time spent in the home and workplace (Repace *et al.*, 1980). The measurements of an individual exposure to particulates over a 24-hour period (Repace *et al.*, 1980) demonstrate that higher concentrations of RSP may easily be found in indoor environments than are usually observed in outdoor environments.

Characterization of air pollution indoors is an undertaking of similar complexity as understanding the outdoor airshed. In 1988, the U.S. EPA listed 11 major indoor pollutants (OAR, 1988):

- Radon
- Environmental tobacco smoke (ETS)
- Biologicals
- Carbon monoxide
- Nitrogen dioxide
- Respirable particles
- Organic gases
- Formaldehyde
- Pesticides
- Asbestos
- Lead

These have been easily identifiable, given their obvious risks to human health. However, the chronic health problems of sick building syndrome (SBS) and multiple chemical sensitivity are as yet poorly understood and remain a major challenge (NRC, 1992; Hileman, 1991).

Combustion, principally the smoking of tobacco products (ETS), is the primary source of particles in the indoor environment (Repace and Lowrey, 1980; Owen *et al.*, 1992). Other types of combustion are known to contribute to particulate levels; however, their contributions are small. For example, homes in Suffolk Country, New York, using kerosene heaters had mean RSP levels of 56.3 $\mu g/m^3$ compared to levels of 34.2 $\mu g/m^3$ without this source; for similar homes in Apex, North Carolina, the corresponding values were 73.7 $\mu g/m^3$ compared to 56.1 $\mu g/m^3$ (Guerin *et al.*). Particulate levels indoors are partially determined by outdoor levels (Pellizzari *et al.*, 1993) but in the absence of smoking, these levels are low (Repace and Lowrey, 1980, 1982; Repace *et al.*, 1980; Repace, 1987). In the homes of asthmatics in southern California, mean mass concentrations were 42.5 $\mu g/m^3$, compared to 60.8 $\mu g/m^3$ observed in the outdoor environment (Colome *et al.*, 1992). Regressions on the data indicated that indoor concentrations are moderately correlated with the outdoor concentrations. These should be compared with typical particulate levels from tobacco smoke of between 100 and 200 $\mu g/m^3$ found in smoking environments (Repace and Lowrey, 1980) and with the observations that, on average, homes with smokers had RSP levels 30–35 $\mu g/m^3$ higher than nonsmoking homes (Spengler *et al.*, 1986). ETS is the overwhelming dominant source of particulate levels indoors; this has been observed in Washington (Repace and Lowrey, 1980), in highly industrialized regions of Poland (Pastuszka, 1993), in Japan (Dr. Hiroshi Nitta, private communication), and in Budapest (see Tables IX–XII). All the RSP levels measured with the piezobalance for indoor locations with smoking in both Washington, D.C., and Budapest exceed the NAAQS ambient PM_{10} standard of 50 $\mu g/m^3$. This may not be the case in China, where home heating using smoky coal produced indoor PM_{10} levels of 9500–24,000 $\mu g/m^3$, compared to a wood smoke level of 22,300 $\mu g/m^3$ and a smokeless coal level of 1800 $\mu g/m^3$ (Chuang *et al.*, 1992). Corresponding outdoor particulate background levels were 250–580 $\mu g/m^3$ for the smoky coal measurements and 50 $\mu g/m^3$ for the wood smoke experiment (background levels were not observed relative to the smokeless coal).

6.1. Representative RSP Levels in Budapest

For comparison with values reported for Washington, D.C. (Repace and Lowrey, 1980), RSP measurements were made in Budapest in September 1989 using a similar experimental design. The values found in the outdoors are given in Table IX and were used as the background measurements for the indoor experiments.

6.1.1. Measurements Made Indoors in the Absence of Smoking

The generally elevated background levels, compared to those in the United States (Repace and Lowrey, 1980), are also reflected in the indoor measurements reported in Table X, which were taken in the absence of smoking. The $100\text{-}\mu\text{g/m}^3$ level found in the school hall was at a location near a lounge used for smoking. The two lowest measurements are a hall in a university laboratory and one measurement in the nonsmoking apartment. The laboratory hall was a site in which smoking was forbidden and was used primarily by nonsmoking scientists.

6.1.2. Measurements Taken in the Presence of Smoking

Table XI gives the data from the field survey of indoor locations in the presence of smoking. The room volumes were estimated and the number of smokers averaged over the sampling time. This was a relatively mild change of season time, and many of the restaurants operated with windows and doors open to the street. The circulation of outdoor air clearly affected the distribution of smoke patterns in some locations. Smoker density is the dominant variable for indoor RSP as was previously demonstrated for the locations in Washington (Repace and Lowrey, 1980).

Figure 12 represents the data in Table XI and shows the indoor RSP levels as a function of smoker density. The clear elevation with higher density is present even with very different ventilation characteristics and room volumes. It was not possible to quantify the effects of ventilation, but low levels were clearly associated with outside infiltration such as in the tourist hotel coffee shop located beside

Table X. Measurements in Indoor Area without Visible Smoking

Location	Volume (m^3)	Number of people	RSP ($\mu\text{g/m}^3$)	Sample time (min)	Background ($\mu\text{g/m}^3$)	Comments
Cathedral	17,000	5	77	6	50	Some candles
Small office	72	1	82	8	94	Background outside at street
Laboratory hall	720	3	42	10	68	Background on small side street
School hall	800	5	100	4	—	School hall before smoking break
Nonsmoking apartment	96	1	64	10	—	
Nonsmoking apartment	96	1	54	10	45	Background outside on balcony
Nonsmoking apartment	96	1	70	6	—	RSP measured 3 hrs after controlled experiment

Table XI. Measurements in Indoor Areas in the Presence of Active Smoking

Location	Volume (m³)	Occupancy	Number of smokers	Smoker density (no./100 m³)	RSP (µg/m³)	Sample time (min)	Background (mg/m³)	Comments
Tourist hotel coffee shop	544	40	1.5	0.28	56	16	—	Doors/windows open, large breeze
Small beer hall	160	9	1.1	0.69	230	16	50	Doors/windows open
Tourist bar, Buda Castle	230	10	1	0.43	121	32	28	Smoking in other vaulted dining room/outside door open
Neighborhood restaurant	300	15	1.3	0.44	147	18	45	Background from apartment above
Downtown pastry shop	254	10	1.2	0.47	90	22	—	Intermittent smoking 5 m distant
Expensive bus tour restaurant	612	60	0	0	75	12	50	Smoking observed before/after readings
Wein stube	72	20	3.1	4.3	230	10	50	Door open/background from cathedral
Student restaurant	144	50	7	4.86	255	30	60	Little ventilation
Student pub	216	40	9	4.17	563	6	77	Downstairs/only one count for occupancy
Student pub 2	216	40	11	5.09	650	2	77	Proximity effect of two smokers at next table
Railroad station pub	109	4	0.9	0.86	97	16	70	Door open, in/out traffic
Large public office cafe	640	15	3.9	0.61	172	32	95	High vaulted ceiling/lots of traffic
Teachers' lounge	135	18	4	2.96	420	12	100	Open window/a background from outdoors
Small office 1	72	3	1	1.39	170	6	83	Window open during smoking
Small office 2	72	1	0	0	140	18	83	26-min decay after smoking
Smoking hall	22	5	0.4	1.67	132	8	50	Background from nonsmoking hall
Student club	4000	—	—	—	355	30	87	Crowded student hall/unknown occupancy
Student disco	4000	—	—	—	600	—	—	Smoke machine from band
Open-air railroad station	20000	—	—	—	144	10	93	Essentially enclosed open-air space by tracks

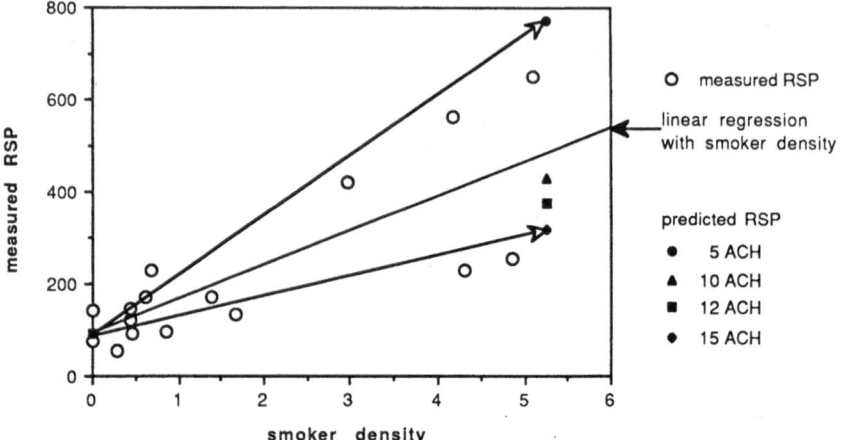

Figure 12. Measured RSP values, from Table XI, plotted as a function of smoker density (O). The linear regression of these data is shown by straight line. Predicted RSP values for a smoker density of 5.25/100 m³ are shown as a function of different air exchange rates (ach) by filled data points, with arrows bracketing the range from 5 ach to 15 ach.

the river with a large breeze coming through the open windows and door. The student club and the student disco were in a large-volume university building almost completely filled with people and no apparent outside ventilation, resulting in high RSP levels. The smell of tobacco smoke was also associated with the elevated level in the railroad station, which was an enormous high-ceiling space enclosed on three sides.

6.1.3. Controlled Experiments

To evaluate the effect of tobacco combustion on indoor space in Budapest, some controlled experiments were performed. Three experiments were performed in the nonsmoking apartment, which measured approximately 4 m × 12 m × 3 m (144-m³ volume). An outside background measurement registered 80 μg/m³, and the indoor measurements in the absence of tobacco smoke were 60 μg/m³ and 70 μg/m³. Figures 13–15 show the rise and decay of tobacco RSP generated from Sopianae multifilter cigarettes smoldered in the center of the living area (4 m × 5 m) about 2 m distant from the piezobalance. The cigarette was extinguished at the times indicated on the graphs, and the remaining measurements represent the decay in the closed space ventilated only by natural infiltration. Experiment 2 (Fig. 14) represents the contribution of a second cigarette ignited at the end of experiment 1 (Fig. 13). Additive behavior was observed, with each cigarette contributing about 100 μg/m³ RSP on top of the "background." The higher

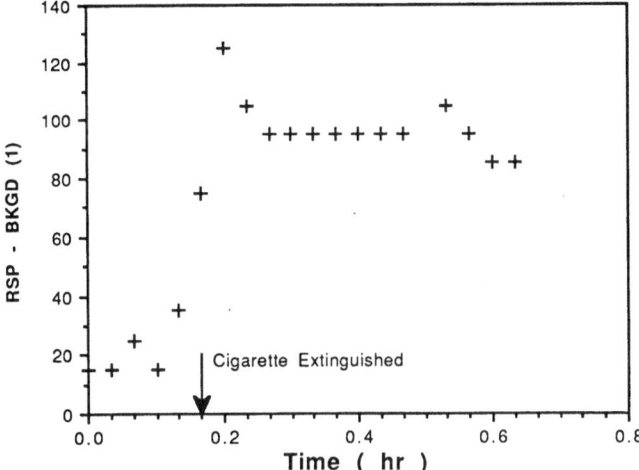

Figure 13. Controlled experiment with smoldered cigarette in nonsmoking apartment. The plot shows the measured RSP values minus the measured background RSP value from outside. The arrow indicates the time the cigarette was extinguished.

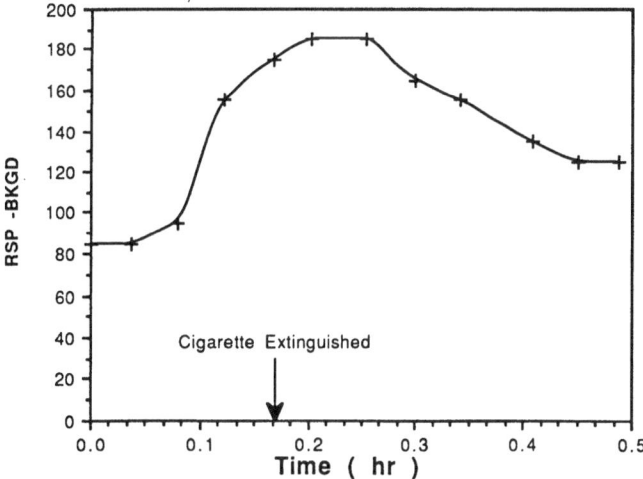

Figure 14. Ignition of second smoldered cigarette at the end of experiment in Fig. 13. This shows additivity of contributions of smoking cigarettes to RSP levels.

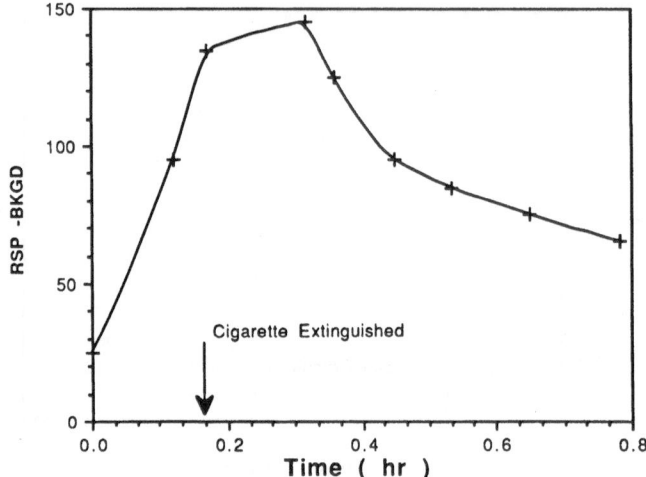

Figure 15.   Repeat of experiment in Fig. 13 late in the evening, showing different decay characteristics probably due to outside infiltration.

starting level for Fig. 14 represents the RSP level of $130 - 45 = 85$ $\mu g/m^3$ from the level in Fig. 13 which had not yet completely decayed. Experiment 3 (Fig. 15) represents a repeat of experiment 1 with measurements taken at close to 5-min intervals instead of from consecutive 2-min samples. The differences in decay rate represent fluctuations in infiltration and background levels; experiment 1 was performed at a time when the restaurant on the first floor of the apartment block was in full operation whereas experiment 3 was performed after it had closed.

It was possible to make limited measurements in a small office space under working conditions. The office measured approximately 6 m × 3.5 m × 3.5 m (74-m³ volume) and had a large open window at one end. The office contained two smoking workers although only one cigarette was consumed during the experiment. Figure 16 shows the rise and decay of smoke. The RSP level at the beginning of the experiment was 170 $\mu g/m^3$, which is presumably the residual RSP from previous smoking. With the natural ventilation resulting from the open window, the additional cigarette (also a Sopianae multifilter) contributed an additional 30 $\mu g/m^3$ RSP while it was being smoked. The levels then decayed to 110 $\mu g/m^3$ after an additional 16 minutes.

6.2. Analysis

6.2.1. Similarity of Cigarette Emissions

From an extrapolation of the $\ln(\text{RSP} - \text{BKGD})$ versus time curve (Fig. 16; for small office, natural smoking) we find $\ln A_0 = 5.5875$; thus, $A_0 = 267$ $\mu g/m^3$.

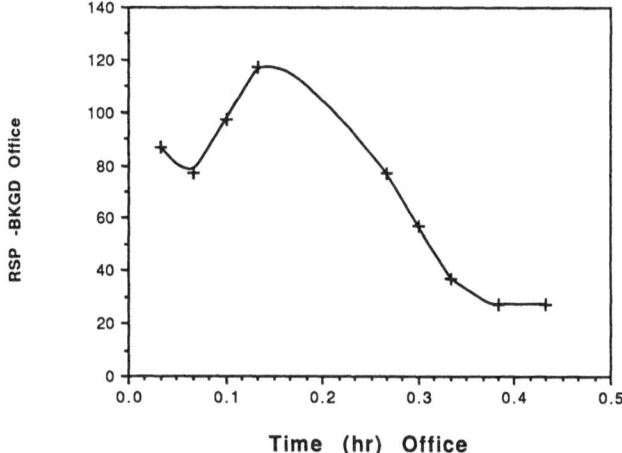

Figure 16. Controlled experiment in small office (74 m³) with naturally smoked cigarette. Single office window was wide open.

The slope of the regression line shows the air exchange rate for this open-window office, C_a = 5.5 air changes per hour. From Repace and Lowrey (1980), the equilibrium concentration which would result from continuous smoking is A_{eq} = $A_0/e^{t_b C_a} - 1$) = 364 µg/m³. From Repace (1987), C_0, the total RSP mass, emitted during smoking of this cigarette for a burning time t_b of 6 minutes is given by C_0 = $A_{eq}C_a t_b/n$, where n = 1 cigarette. Substituting the parameter values yields C_0 = 14.28 mg of tar. If the cigarette had been smoked for a standard 10 minutes, the total RSP emission would be C_0 = 23.8 mg for this Hungarian cigarette, in agreement with the values for sidestream emissions of American, British, and Canadian cigarettes (Repace, 1987).

6.2.2. Air Exchange Rates

The Hungarian ventilation standard MSZ 21875-79, titled Labor Safety Requirements of Heating and Ventilation of Workplaces, with amendment MSZ 21875-2 issued April 1991, governs standards for building construction. Ventilation rates are set at 30 m³/person-hr for office and light work conditions, 40 m³/person-hr for medium labor, and 50 m³/person-hr for heavy labor. There is a requirement to plan for the use of natural ventilation if possible. Air flow speeds are not to be greater than 0.25 m/s, with a maximum of 90% recirculated air for purposes of energy conservation and a limit of 0.5 mg/m³ concentration of dust. For the areas sampled in this experiment, 4 m appears to be a standard ceiling height, and, assuming an occupancy of 10 persons/100 m², the specified air exchange rate is about 0.75 ach, slightly less than that specified for offices by

ASHRAE 62–1989 (ASHRAE, 1989). The ASHRAE standard of 0.84 ach is a consensus standard for building ventilation published by the American Society of Refrigeration, Heating and Air Conditioning Engineers. As can be seen from Fig. 9, the bulk of the data suggests observed air exchange rates that lie between 5 and 15 ach. In contrast, Repace and Lowrey (1980) observed a range of 0.5 to 7.0 ach for indoor spaces in Washington, D.C. The difference appears to be from the contribution of infiltration, which adds greatly to the mechanical ventilation during this mild season of open windows. This reduces the RSP concentration to approximately ⅙ to ½₀ of the closed-window concentrations expected to be found in wintertime.

6.3. Indoor Summary

In summary, a few simple observations may be made about levels of particulates in the indoor air:

1. Indoor *background* particulate levels are loosely coupled to outdoor levels in the absence of indoor sources (Wallace *et al.*, 1991).
2. In the absence of combustion or deliberate aerosol generation such as humidifiers, indoor particulate levels can be less than outdoor background levels due to deposition and other aerosol decay mechanisms.
3. Combustion processes are commonly recognized as the principal indoor source of particulate matter, carbon monoxide, nitrogen oxides, and polycyclic aromatic hydrocarbons (PAH) (Guerin *et al.*, 1992).
4. Tobacco smoking (ETS) is the major source of indoor combustion particulates.

Indoor air pollution is complex with many sources of potential hazards. For indoor particulates, one classification has suggested six types: bioaerosols (plant and animal), mineral, combustion, home/personal care, and radioactive aerosols (Owen *et al.*, 1992). Experimental data contributing to an understanding of these sources are limited, since most reports include total mass or total number of particles, not merely those from a specified source. The recognition that, in urban environments, 90% of personal exposure originates in the indoors is recent (U.S. EPA, 1982), and the basic components of many indoor air problems are yet to be understood (NRC, 1992).

The above conclusions are a basic subset of the concerns about indoor air pollution and are supported by the following arguments. In the absence of specific identified sources of particulates, indoor levels can be significantly lower than outdoor levels when measured by similar techniques. This is exhibited in a recent study in the Los Angeles area, in which the indoor/outdoor ratios of fine particulate levels were found to range from 0.18 to 0.96 for a series of museums with widely varying ventilation characteristics (Ligoki *et al.*, 1993). Inefficient combustion of biomass used in cooking, such as cattle dung, wood, and coal, was found to dominate TSP levels in poorly ventilated homes in India, while more efficient combustion of kerosene and liquid propane gas produced TSP levels that were indistinguishable from the background (Raiyani *et al.*, 1993). In the United States, kerosene space heaters have been shown to contribute significantly to indoor RSP, with a geometric mean value of 52.5 $\mu g/m^3$ in homes using kerosene space heaters, compared to a background mean of 38.9 $\mu g/m^3$ in similar homes without this source of combustion particulates. Combustion of tobacco products is the major source of indoor combustion particulates (U.S. EPA, 1982; Repace and Lowrey, 1980; Spengler *et al.*, 1986). Tobacco smoke is found to cause dramatic increases in immediate levels of RSP, as shown in Fig. 11 and Table XI, and has been shown to be a substantial contributor to 24-hour and long-term average exposures relative to background (Spengler *et al.*, 1986, 1981).

7. SIGNIFICANCE OF COMBUSTION PARTICULATE CONCENTRATIONS

The significance of the simple observations about combustion particulates lies in understanding the associated human health risks and the compromises necessary for control of these risks. The benefits of combustion are built into the bedrock of human society, and thus our potential exposure to particulates and other byproducts of burning is ubiquitous. What has changed is our understanding of the complexities of exposure and the potential risks.

7.1. Social Response to Outdoor Particulate Levels

In certain industrialized societies such as the United States, the response to the health risks of outdoor particulates has been the implementation of controls over the emissions (U.S. EPA, 1982, 1991) and the establishment of bureaucracies to monitor these controls and generate new information to better understand the risks (Schulze, 1993). The NAAQS standards (U.S. EPA, 1991) were originally developed based on concerns about effects on vegetation, effects on visibility and climate, and toxic and carcinogenic effects on human health (U.S. EPA, 1991, 1982). With respect to cancer, one of the best studied and most feared of human diseases, regulatory criteria were established for control of this specific risk (U.S. EPA, 1986b). Table XII shows the airborne carcinogens regulated and the expected number of cancer deaths estimated for each (Repace and Lowrey, 1985a).

In the United States, the effect of controls on stationary and transportation sources has produced a well-known 20% decrease in outdoor particulate levels (U.S. EPA, 1991). The effects of combustion on the outdoor air continue to be a major concern; this problem, for example, is an integral part of the complex discussions on the North American Free Trade Association (NAFTA) treaty.

Table XII.   Relative Risks of Airborne Carcinogens[a]

Pollutant	Estimated annual mortality[b,c]
ETS (passive smoking)[d]	5000 LCDs
Vinyl chloride	<27 CDs
Radionuclides (worldwide impact of U.S. Department of Energy facilities)	17 CDs
Coke oven emisssions	<15 LCDs
Benzene	<8 CDs
Arsenic	<5 LCDs

[a]Repace and Lowrey, 1985a.
[b]CD, Cancer death; LCD, lung cancer death.
[c]Risks for radionuclides are best estimates, and other risks are upper bounds.
[d]In the United States, this is currently regulated at the state and local governmental levels and administratively controlled in federal buildings by individual federal agencies and by the General Services Administration.

7.2. Slowly Developing Social Response to Indoor Particulate Levels

For the indoor environment, a similar social response is slowly beginning to emerge (ICIAQ, 1992). The delayed recognition of these problems is partially due to the difficulties in obtaining health effect data for specific indoor pollutants, particularly over the long latency periods associated with diseases such as cancer. Risk assessment methodology has been essential in understanding the health problems associated with the indoor air (Repace and Lowrey, 1990).

7.3. Summary of Risk Arguments for Environmental Tobacco Smoke

Beginning with the four simple conclusions about indoor combustion particulates, Repace and Lowrey published a risk assessment in 1985 for lung cancer deaths from ETS (Repace and Lowrey, 1985a). They estimated 5000 excess lung cancer deaths a year in the United States among nonsmokers of age 35 and older. A brief summary of the ETS arguments is given below.

Hazard Identification. There is a variation in exposure to ETS as a function of lifestyle, and there is epidemiological evidence for the variation of risk with lifestyle. Mainstream smoke is a known carcinogen and has a chemical composition related to that of ETS. There is biological plausibility that ETS is a carcinogen.

Risk Assessment. Individual aggregate exposure is estimated to be 1.4 mg of tobacco tar per day. Epidemiological evidence established an age-adjusted risk ratio for lung cancer of 0.54 between Seventh Day Adventist (SDA) populations in southern California and a similar non-SDA population. Because of their religious beliefs and associated lifestyle, the SDA population best represents a true control group of people not likely to be exposed to ETS, a ubiquitous pollutant until recently. This risk ratio and age-adjusted mortalities were found to be in good agreement with other large studies in Japan (Hirayama, 1981) and the United States (Garfinkel, 1980).

Exposure–Response Relationship. Using the estimated aggregate exposure to tobacco tar of 1.4 mg/day and the epidemiological data from the SDA study, the calculated aggregate risk if 5 lung cancer deaths (LCDs)/100,000 at risk per 1-mg/day nominal exposure. The risk to the most exposed is approximately 10 times this based on about 14-mg/day nominal exposure. An alternative lower bound risk of 0.5 LCDs/100,000 at risk was estimated by extrapolation of dose–response studies in smokers; this procedure is known to have problems due to saturation effects of heavy tar dose in smokers (Jarvis and Russell, 1985).

Risk Characterization; Determining the Significance of the Risk. Table XII places the lung cancer risk from ETS in perspective to the risks of other regulated airborne carcinogens. It is important to understand that the large risk from ETS is directly related to the carcinogenic potency of tobacco tar and the vast number of people exposed. The following quotation comes from Indoor Air Facts No. 5, published by the U.S. EPA; this helps further explain the significance of the risk of ETS.

> In the United States, 50 million smokers annually smoke approximately 600 billion cigarettes, 4 billion cigars and the equivalent of 11 billion pipefuls of tobacco. Since people spend approximately 90% of their time indoors, this means that about 467,000 tons of tobacco are burned indoors each year. Over a 16-hour day, the average smoker smokes about two cigarettes an hour, and takes about ten minutes per cigarette. Thus, it takes only a few smokers in a given space to release a more-or-less steady stream of ETS into the indoor air. (OAR, 1989)

Risk Communication. Risk communication is now in the hands of the public health agencies, governmental bureaucracies, and the economic controversies that generate headlines in the press. In 1992, the U.S. EPA released its risk assessment on the Respiratory Health Effects of Passive Smoking (OHEA, 1992). Based on analysis of 30 epidemiological studies, EPA concluded that ETS is responsible for approximately 3000 excess lung cancer deaths in nonsmokers, in good agreement, to order of magnitude, with the estimates of Repace and Lowrey (1985a). Control strategies for ETS are based on source removal or separate ventilation. In 1985, Repace and Lowrey (1985b) published an indoor air quality standard for ETS of 0.75 $\mu g/m^3$ daily average RSP concentration based on the risk assessment. This was followed in 1993 by a nicotine standard of 7.5 nanograms of airborne nicotine per cubic meter of workplace air for a working lifetime of 40 years (Repace and Lowrey, 1993). This is enforceable because nicotine is uniquely associated with tobacco and ETS. These standards suggest that ventilation is not a feasible strategy for control of ETS, either economically or in terms of practical considerations. Removing the sources of ETS appears the only feasible control, a practice adopted by numerous state and local regulatory bodies and numerous federal facilities and recommended by the U.S. EPA, the U.S. Public Health Service, and the National Institute of Occupational Safety and Health.

8. CONCLUSIONS

Information has been presented that is useful for understanding air pollution from combustion. This Promethean problem is clearly complex with difficulties in sampling methodologies, confounding particulates from other sources, total exposures resulting from different microenvironments, and questions of relative effective human dose. However, simple ideas have proven useful for creating the proper social and economic balance between the risks and benefits of combustion. For the United States, annual average particulate levels in the outdoor air are less than 100 $\mu g/m^3$, with only 13 out of 340 MSAs having average levels higher than the annual PM_{10} standard of 50 $\mu g/m^3$. Particulate levels in the outdoor air have been lowered through emission controls on stationary and transportation sources; this has proven a successful strategy despite the confounding problems of fugitive emissions (U.S. EPA, 1991). Risk assessment techniques have led to the development of the understanding of the problems of total exposure (Repace *et al.*, 1980) and the significance of the indoor airshed as a source of combustion pollution. Given the quantities of tobacco consumed indoors (Repace and Lowrey, 1990; OAR, 1989) and the danger of its chemical constituents (Guerin *et al.*, 1992), ETS has been identified as the major indoor source for combustion pollution in most industrialized societies (Repace and Lowrey, 1980; Guerin *et al.*, 1992) and is possibly the primary combustion pollutant with respect to human exposure (Repace and Lowrey, 1985a). The lung cancer risk from ETS is larger than that from other regulated outdoor airborne carcinogens (Repace and Lowrey, 1985a,b). Removal of the source is an effective means of control for ETS (Repace and Lowrey, 1985b, 1993). Control of indoor particulate sources remains a new frontier.

REFERENCES

ASHRAE, 1989, Ventilation for Acceptable Indoor Air Quality, American Society of Heating, Refrigeration & Air Conditioning Engineers.

Austin, B. S., Greenfield, S. M., Weir, B. R., Anderson, G. E., and Behar, J. V., 1992, Modeling the indoor environment, *Environ. Sci. Technol.* **26**(5):851–858.

Axley, J., 1988, Progress toward a General Analytical Method for Predicting Indoor Air Pollution in Buildings, National Bureau of Standards, Washington, D.C.

Bardeschi, A., Colucci, A., Gianelle, V., Gnagnetti, M., Tamponi, M., and Tebaldi, G., 1991, Analysis of the impact on air quality of motor vehicle traffic in the Milan urban area, *Atmos. Environ.* **25B:** 415–428.

Biersteker, K., DeGraaf, H., and Nass, C. A. G., 1965, Indoor air pollution in Rotterdam homes, *Int. J. Air Water Pollut.* **9:**345–350.

CEQ, 1989, Risk Analysis. Council on Environmental Quality, Executive Office of the President of the United States, Washington, D.C.

Chuang, J. C., Cao, S. R., Xian, Y. L., Harris, D. B., and Mumford, J. L., 1992, Chemical characterizations of indoor air of homes from communes in Xuan Wei, China, with high lung cancer mortality rate, *Atmos. Environ.* **26A:**2193–2201.

Coghlin, J., Hammond, S. K., and Gann, P. H., 1989, Development of epidemiological tools for measuring environmental tobacco smoke, *Am. J. Epidemiol.* **130:**606–704.

Colome, S. D., Kado, N. Y., Jaques, P., and Kleinman, M., 1992, Indoor–outdoor air pollution's relations, *Atmos. Environ.* **26A:**2173–2178.

Cox, C. S., 1987, *The Aerobiological Pathway of Microorganisms*, Wiley-Interscience, Chichester, England.

Friedlander, S. K., 1977, *Smoke, Dust and Haze*, Wiley-Interscience, New York.

Garfinkel, L., 1980, Cancer mortality in nonsmokers, *J. Natl. Cancer Inst.* **65:**1169–1173.

Guerin, M. R., Jenkins, R. A., and Tomkins, B. A., 1992, *The Chemistry of Environmental Tobacco Smoke*, Lewis Publishers, Boca Raton, Florida.

Hileman, B., 1991, Multiple chemical sensitivity, *Chem. Eng. News* **1991**(July 22):26–42.

Hirayama, T., 1981, Passive smoking and lung cancer, *Br. Med. J.* **282:**1393–1394.

ICIAQ, 1992, Current Federal Indoor Air Quality Activities, Interagency Committee on Indoor Air Quality, United States Environmental Protection Agency.

Jarvis, M. J., and Russell, M. H., 1985, Passive exposure to tobacco smoke, *Br. Med. J.* **291:**1646.

Jenkins, P. L., Phillips, T. J., Mulberg, E. J., and Hui, S. P., 1992, Activity patterns of Californians: Use of and proximity to indoor pollutant sources, *Atmos. Environ.* **26A:**2141–2148.

Lappe, M., 1991, *Chemical Deception*, Sierra Club, San Francisco.

Ligoki, M. P., Salmon, L. G., Fall, T., Jones, M. C., Nazaroff, W. W., and Case, Glen R., 1993, Characteristics of airborne particles inside southern California museums, *Atmos. Environ.* **27A:**679–711.

Lowrey, A. H., Kantor, S., and Repace, J. L., 1993, Outdoor and indoor respirable suspended particulates in Budapest, Hungary, Indoor Air '93, Helsinki, Finland.

Nagda, N. L., Rector, H. E., and Koontz, M. D., 1987, *Guidelines for Monitoring Indoor Air Quality*, Hemisphere Publishing, Washington, D.C.

NAPCA, 1969, Air Quality Criteria for Particulate Matter, National Air Pollution Control Administration, Public Health Service, Washington, D.C.

NRC, 1981, *Indoor Pollutants*, National Research Council, National Academy of Sciences, Washington, D.C.

NRC, 1983, *Risk Assessment in the Federal Government: Managing the Process.* National Research Council, National Academy of Sciences, Washington, D.C.

NRC, 1989, *Improving Risk Communication*, National Research Council, National Academy of Sciences, Washington, D.C.

NRC, 1992, *Multiple Chemical Sensitivities*, National Research Council, National Academy of Sciences, Washington, D.C.

OAR, 1988, The Inside Story: A Guide to Indoor Air Quality, Office of Air and Radiation, U.S. Environmental Protection Agency.

OAR, 1989, Environmental Tobacco Smoke, Office of Air and Radiation, U.S. Environmental Protection Agency.

OHEA, 1992, Respiratory Health Effects of Passive Smoking: Lung Cancer and Other Disorders, Office of Health and Environmental Assessment, U.S. Environmental Protection Agency.

Ott, W., 1988, Human Activity Pattern, Research Planning Conference, Las Vegas, Nevada, U.S. Environmental Protection Agency.

Owen, M. K., Ensor, D. S., and Sparks, L. E., 1992, Airborne particle sizes and sources found in indoor air, *Atmos. Environ.* **26A:**2149–2162.

Parker, A., 1978, *Industrial Air Pollution Handbook*, McGraw-Hill, London.

Pastuszka, J. S., 1993, Fibrous and particulate pollution of indoor air in Upper Silesia, Indoor Air '93, Helsinki, Finland.

Pellizaari, E. D., Thomas, K. W., Clayton, C. A., Whitmore, R. W., Shores, R. C., Zelon, H. S., and Peritt, R. L., 1993, Particle Total Exposure Assessment Methodology (PTEAM): Riverside California Pilot Study, U.S. Environmental Protection Agency.

Raiyani, C. V., Shah, S. H., Desai, N. M., Venkaiah, K., Patel, J. S., Parikh, D. J., and Kashyap, S. K., 1993, Characterization and problems of indoor pollution due to cooking stove smoke, *Atmos. Environ.* **27A:**1643–1655.

Repace, J. L., 1987, Indoor concentrations of environmental tobacco smoke: Models dealing with the effects of ventilation and room size, in *Passive Smoking* (I. K. O'Neill, K. B. Brunnemann, B. Dodet, and D. Hoffmann, eds.), World Health Organization, Lyon.

Repace, J. L., and Lowrey, A. H., 1980, Indoor air pollution, tobacco smoke, and public health, *Science* **208:**464–474.

Repace, J. L., and Lowrey, A. H., 1982, Tobacco smoke, ventilation and indoor air quality, *ASHRAE Trans.* **88(**I**):**895–914.

Repace, J. L., and Lowrey, A. H., 1985a, Quantitative estimate of nonsmokers lung cancer risk from passive smoking, *Environ. Int.* **11:**3–22.

Repace, J. L., and Lowrey, A. H., 1985b, Indoor air quality standard for ambient tobacco smoke based on carcinogenic risk, *N.Y. State J. Med.* **85:**381–383.

Repace, J. L., and Lowrey, A. H., 1990, Risk assessment methodologies for passive smoking-induced lung cancer, *Risk Anal.* **10(**1**):**27–37.

Repace, J. L., and Lowrey, A. H., 1993, An enforceable indoor air quality standard for environmental tobacco smoke in the workplace, *Risk Anal.* **13(**4**):**443–455.

Repace, J. L., Ott, W. R., and Wallace, L. A., 1980, Total human exposure to air pollution, 73d Annual Meeting of the Air Pollution Control Association, Montreal, Canada.

Rodricks, J. V., 1992, *Calculated Risks*, Cambridge University Press, Cambridge.

Schulze, R. H., 1993, The 20-year history of the evolution of air pollution control legislation in the USA, *Atmos. Environ.* **27B:**15–22.

Sparks, L., 1988, *Indoor Air Quality Model Version 1.0*, U.S. Environmental Protection Agency.

Spengler, J. D., Dockery, D. W., Turner, W. A., Wolfson, J. M., and Ferris, B. C., Jr., 1981, Long term measurements of respirable sulfates and particles inside and outside homes, *Atmos. Environ.* **15:** 23–30.

Spengler, J. D., Reed, M. P., Lebret, E., Chang, B. H., Ware, J. H., Speizer, F. E., and Ferris, B. G., Jr., 1986, Harvard's indoor air pollution/health study, 79th Annual Meeting of the Air Pollution Control Association, Minneapolis, Minnesota.

Szalai, A., 1972, *The Use of Time: Daily Activities of Urban and Suburban Populations in Twelve Countries*, Moughton, The Hague.

Telliard, W. A., 1992, "EPA's Environmental Monitoring Methods Index." *Environ. Sci. Technol.* **27(**1**):**39–41.

U.S. EPA, 1982, Air Quality Criteria for Particulate Matter and Sulfur Oxides, Environmental Criteria and Assessment Office, U.S. Environmental Protection Agency.

U.S. EPA, 1986a, Assessment of Newly Available Health Effects Information, Environmental Criteria and Assessment Office, U.S. Environmental Protection Agency.

U.S. EPA, 1986b, Guidelines for carcinogen risk assessment, *Federal Register* **51:**33992–34003.

U.S. EPA, 1991, National Air Quality and Emissions Trends Report 1989, Washington, D.C., U.S. Environmental Protection Agency.

Wallace, L., Pellizzari, E., Sheldon, L., Whitmore, R., Zelon, H., Clayton, A., Shores, R., Thomas, K., Whitaker, D., Reading, P., Spengler, J., Ozkaynak, H., Froelich, S., Jenkins, P., Ota, L., and Westerdahl, D., 1991, The TEAM study of inhalable particulates, Paper presented at the

84th Annual Meeting of the Air and Waste Management Association, Vancouver, British Columbia.

Wick, C. H., Edmonds, R. L., and Blew, J., 1994, Rapid detection and identification of background levels of airborne biological particles, ERDEC Technical Report TR-155, U.S. Army Edgewood Research, Development and Engineering Center, Aberdeen Proving Ground, Maryland 21010.

Willeke, K. and Baron, P., 1993, *Aerosol Measurement*, van Nostrand Reinhold, New York.

Yocum, J. E., Clink, W. L., and Cote, W. A., 1971, Indoor/outdoor air quality relationships, *J. Air Pollut. Control Assoc.* **21**:251.

Chapter 11

Bulk and Surface Studies of Fly Ash Particles

Costantino Boni, Ezio Cereda, Grazia Maria Braga Marcazzan,
and Fulvio Parmigiani

1. INTRODUCTION

Coal and oil combustion produces fly ash particulates whose dispersion into the environment represents one of the major sources of pollution.

When released into the atmosphere, fly ashes have long residence times and slow settling rates, thus undergoing long-distance transport phenomena. This behavior together with the toxicity of some of the compounds present in fly ashes makes the question of the particulates released during combustion processes one of the most important topics of modern ecology.

Studies related to this particulate matter involve several steps of the combustion cycle, such as the formation of ash particles upon combustion itself, the collection mechanisms of control systems, the release of ash particles into the environment, and their disposal and utilization.

CostanTINO Boni AND Ezio Cereda • Materials Division, CISE Tecnologie Innovative S.p.A., 20134 Milan, Italy.   Grazia Maria Braga Marcazzan • Istituto di Fisica Generale ed Applicata, Università degli Studi di Milano, 20133 Milan, Italy.   Fulvio Parmigiani • Materials Division, CISE Tecnologie Innovative S.p.A., 20134 Milan, Italy and Facoltà di Scienze, Università di Como, Como, Italy.

Combustion Efficiency and Air Quality, edited by István Hargittai and Tamás Vidóczy. Plenum Press, New York, 1995.

In this light, the improvement of combustor designs and particulate control devices are subjects of great interest both for developing new technologies and for acquiring a better basic understanding of the formation of solid particulates during fossil fuel burning processes.

The assessment of the environmental impact of coal combustion based only on the particulate total mass released into the atmosphere is inadequate. The element emissions, the particle size distribution and its association with potentially toxic trace elements (Valkovic, 1983), and the surface structure and composition have to be taken into account (Caruso and Parmigiani, 1987).

Coal combustion results in the emission of several elements; the total amount emitted depends on the mineral matter content in the parent coal, on the combustor design and operating conditions, and on the efficiency of the control devices. The elements present in the parent coal can be divided into three groups: inert elements with a uniform distribution among all combustion products, elements that are partially volatilized and further recondense on the ash particle surfaces, and elements that are completely volatilized and emitted in the vapor phase. As a result of the particle formation mechanisms, fly ashes have high surface concentrations of trace elements which increase with decreasing particle size (Davison *et al.*, 1974; Lee *et al.*, 1975; Coles *et al.*, 1979; Smith *et al.*, 1979; Markowski and Filby, 1985; Tazaki *et al.*, 1989; Kauppinen and Pakkanen, 1990).

When electrostatic precipitators are used to reduce the amount of fly ashes released into the atmosphere, most of the emitted particulate matter is made up of fine particles since the efficiency of the electrostatic precipitators has a minimum in the 0.1–1.0-μm range. This particulate matter has important and dangerous health effects because of the high degree of retention in the human respiratory system and the high concentration of toxic trace elements (Hansen and Fisher, 1980; Valkovic *et al.*, 1984; Boni *et al.*, 1990a). A depletion of the surface concentration upon leaching has been observed for some elements (Pb, Tl, Mn, Cr, V) but not for others (S, Fe, K, Na, Li) although both groups show surface enrichment (Linton *et al.*, 1977). Because of the leachibility and the enhanced surface concentration of some potentially toxic trace elements, the environmental and health impact of fly ash is much more significant than indicated by bulk analysis (Linton *et al.*, 1976). The study of coal combustion-related phenomena can be carried out in both macroscopic and microscopic terms (Bellotto *et al.*, 1990). In the former approach, the chemical composition (in particular with regard to trace elements) and physical properties of fly ashes, such as particle size distribution and resistivity, are investigated and related to the operating conditions of the power plant in order to estimate the enrichment, penetration, and emission factors and the trace element behavior through the whole combustion cycle (Kaakinen *et al.*, 1975; Klein *et al.*, 1975; Braga Marcazzan *et al.*, 1990; Caridi *et al.*, 1992; Bellagamba *et al.*, 1993). Analytical results should be collected for the development of a data base for trace elements in three areas: their occurrence mode in fuels, their partitioning among the different combustion products, and their collection efficiency when electrostatic precipitators are used as control devices. These results could then be used to implement and validate semiempirical models for the prediction of trace element emission factors, assuming as input data only the coal composition, the fly ash particle size distribution, and the electrostatic precipitator penetration function (Cernuschi and Giuliano, 1987).

In the latter approach, the chemical, physical, and morphological properties of individual particles are considered. They are most important from the environmental point of view, in particular for the assessment of the environmental availability of potentially toxic trace elements, that is, their probability of interaction with the surrounding environment. There is evidence that some elements are mainly surface concentrated: in this case the environmental availability is much greater than estimated by ordinary bulk analysis. Another subject to be investigated is the possible different trace element concentrations in particles belonging to the same size range. This can be relevant to the study and optimization of control devices, through the identification of classes of particles which, despite being of the same size and morphology, can undergo different precipitation mechanisms.

Several reports in the literature deal with the characterization of individual fly ash particles. Linton et al. (1977) have obtained evidence for the surface enrichment of some trace elements by Auger electron spectroscopy and ion microprobe mass spectrometry. Cereda et al. (1988) have shown the possibility of analyzing the elementary radial composition of ash particles by Auger depth profiling, while X-ray photoelectron spectroscopy (XPS) can be employed to study the oxidation state of the elements present on the particle surfaces in detail. Fisher et al. (1978) have identified 11 morphological classes by light microscopy and have estimated their relative abundance with respect to particle size. Stinespring and Stewart (1981) as well as Hock and Lichtman (1982) gave evidence of the shell structure of fly ash particles. Ramsden and Shibaoka (1982) applied quantitative electron microprobe analysis to the study of the chemical composition of particles previously categorized in seven groups according to morphological properties determined by optical and electron microscopies. Vis et al. (1983) used proton microprobe analysis for the characterization of individual particles belonging to the magnetic and nonmagnetic fractions of fly ash. Mamane et al. (1986) have investigated the morphology, size, and chemical composition of individual coal and oil ash particles by electron microscopy combined with quantitative microprobe analysis. Secondary ion mass spectrometry has been used by Cox et al. (1987) to study the trace element distribution on single particles.

Proton microprobe analysis (Watt and Grime, 1987) can provide information about the mode of occurrence of major and trace elements in fly ash particles. In particular, it can be established whether trace elements are uniformly distributed over all particles or preferentially associated with specific morphologies, sizes, and major elements. In addition, areal scans and line scans along the particle diameter can be operated to investigate possible surface enrichment of trace elements (Jaksic *et al.*, 1991; Caridi *et al.*, 1993; Jaksic *et al.*, 1993).

The aim of the present chapter is to describe surface and bulk diagnostic techniques suitable for the study of fly ash particles produced during the combustion of fossil fuels in large power plants. The studies reported in this chapter supply rather complete information about the surface and bulk chemical composition of fly ash particles allowing the evaluation of parameters of paramount importance with respect to environmental impact, particle formation during combustion processes, efficiency of electrostatic precipitators, and fly ash disposal.

The morphology of the fly ash particles strongly depends on the combustion processes and on the nature of the fuel. Those produced by fossil coal have a shape quite close to spherical and are, in most cases, hollow. Figure 1 shows a picture

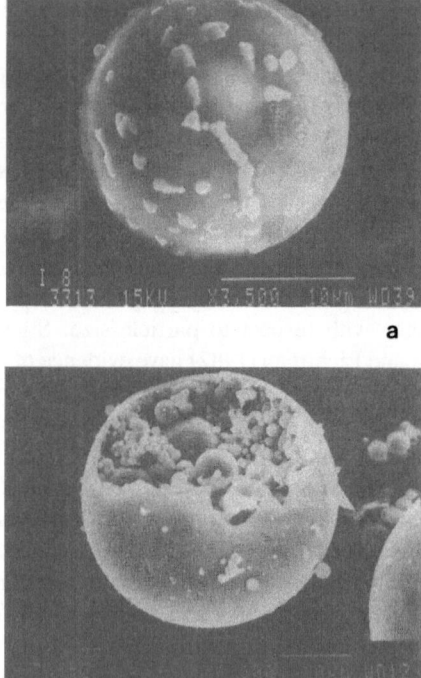

a

b

Figure 1. Scanning electron micrography of a fly ash particle produced by a coal-burning plant. Micrograph a was obtained from an integral particle, while micrograph b shows a broken particle filled with smaller spheres. Since fly ash particles are empty inside, they are also known by the neologism of "cenosphere."

obtained by scanning electron microscopy (SEM). The spherical shape of the integral particle may be noted, while the broken particle, filled with smaller particles, gives a clear and representative picture of the structure and morphology of fly ash particles, which are sometimes referred to by the neologism "cenosphere," derived from the ancient Greek words χενός (empty) and σφαιρα (sphere).

2. ANALYTICAL TECHNIQUES

2.1. Particle-Induced X-Ray Emission (PIXE)

The PIXE technique (Johansson and Campbell, 1988), based on the use of charged particles (generally protons) in the megaelectronvolt energy range, was initiated at the Lund Institute of Technology in 1970, and it is now extensively applied to the elemental chemical analysis of a large variety of samples in several fields of research such as biology, medicine, geology, environment, art, archaeology, and materials science. PIXE has multielemental capability (in ordinary experimental setups, elements with $Z \geqslant 11$ are simultaneously detectable), provides good sensitivity (of the order of nanograms per square centimeter or parts per million, even for samples available in small quantities) and rapidity of analysis, and is nondestructive. These features make it particularly suitable for routine analyses in fields such as atmospheric applications which require the analysis of a great number of samples (Braga Marcazzan *et al.*, 1987).

The physical process underlying the PIXE technique involves the excitation of an atom and its subsequent deexcitation by X-ray emission. The interaction between a charged particle (such as an energetic proton) and the electronic structure of an atom can lead to the creation of vacancies in the inner shells, which are then promptly (within about 10^{-16} s) filled by the outer electrons. The residual energy is released as X rays (K, L, or M X rays, depending on the shell where the vacancy has been created) or Auger electrons. The X-ray energy is characteristic of the atom so that an energy dispersive analysis provides the identification of the atomic species in the sample.

In Fig. 2 a typical PIXE spectrum of urban particulate matter is shown. The characteristic X rays are superimposed on a background due to the bremsstrahlung of the incident beam and of the secondary electrons released by the sample. Also, gamma rays following nuclear interactions can contribute to the background (Compton photons).

For quantitative analysis, different procedures are followed for thin, thick, and intermediate samples. A sample is considered thin if the average value of the ionization cross section through the sample depth is within 5% of the value at the sample surface and the X-ray self-absorption effects in the sample are negligible.

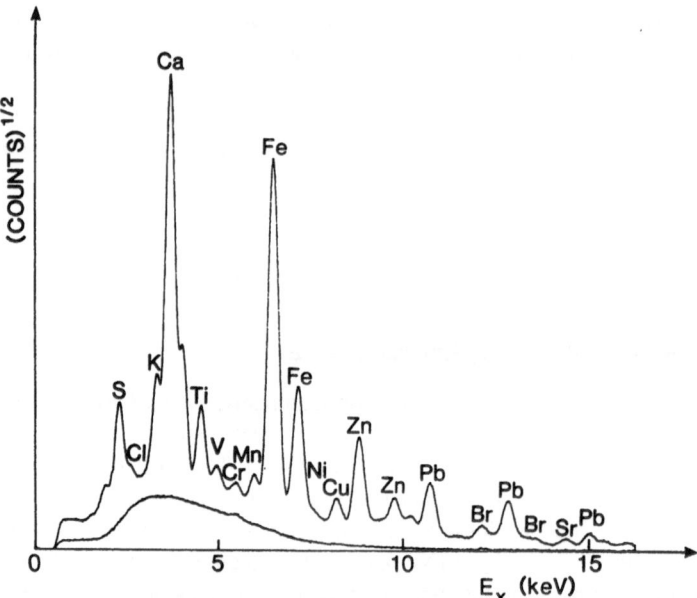

Figure 2. PIXE spectrum of urban particulate matter.

At the usual energies of proton beams (a few megaelectronvolts), a sample is assumed thin for a sample thickness $t \approx 0.1$ mg/cm^2. The quantitative analysis is based on the following equation:

$$Y = t \cdot N_0 \cdot N_p \cdot \Omega \cdot \epsilon \cdot \sigma_i \cdot \omega \cdot k/4\pi \cdot A \cdot \cos \theta_i \qquad (1)$$

where Y is the characteristic X-ray yield, N_0 is Avogadro's number, N_p is the number of incident particles, Ω is the detector solid angel, ϵ is the detection efficiency, σ_i is the ionization cross section, ω is the fluorescence yield, k is the relative intensity for the specific X ray, A is the atomic weight of the element, and θ_i is the angle between the incident beam and the normal to the sample. The quantitative analysis can be carried out also in a relative way, that is, with respect to standard reference samples such as thin films (20–50 μg/cm^2) of single elements. Calibration data (that is, counts/unit charge per μg/cm^2) are determined for each characteristic X ray.

A sample is considered thick when its thickness is greater than the incident beam range in the sample. For instance, a silicon sample is considered thick for a 3-MeV proton beam when its thickness is greater than about 100 μm. The yield for a characteristic X ray relative to an element with concentration C (μg/g) is

$$Y = \frac{CN_0 \Omega N_p \epsilon}{4\pi A} \int_{E_p}^{0} \frac{\sigma(E)T(E)}{S(E)} dE \qquad (2)$$

where $S(E)$ is the specific energy loss of the incident beam in the sample (MeV per g/cm^2), $T(E)$ is the photon transmission function through the sample, and E_p is the incident beam energy. In this equation, effects such as secondary fluorescence are neglected. To remove the uncertainties due to the matrix effects, the use of internal and external standards is effective. For samples with intermediate thickness (a few milligrams per square centimeter), corrections for the incident beam energy loss and X-ray self-absorption must be applied. The sensitivity of the technique can be expressed by the minimum detection limit (MDL) values (g/cm^2) of each element, defined as $(3/S)(N_b/I)^{1/2}$, where N_b is the number of counts per unit charge in the background within an energy interval of two full widths at half-maximum around the peak, S is the X-ray yield (counts per unit charge per unit mass per unit area), and I is the total irradiated charge. Figure 3 presents the detection limits for an irradiation of 60 μC for thin samples deposited on Nucleopore filters and polyvinylacetate films (Boni *et al.*, 1990a). The detection limits are worse in the former case because of the thickness of the substrate (10 μm). The reported values have been estimated for single element standards. Effects due to element inter-ferences are not taken into account. For thick samples the detection limits are in general worse because of the higher background.

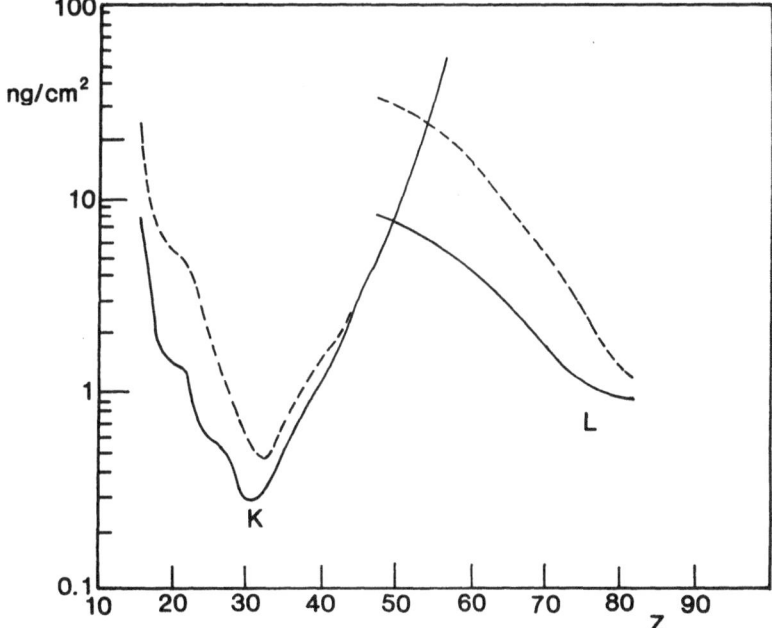

Figure 3. PIXE detection limits for an irradiation of 60 μC for thin samples deposited on nucleopore (---) and polyvinylacetate films (——).

The analytical precision is mainly dependent on the count statistics and on the beam and sample homogeneities. The accuracy depends on the standard reference sample accuracy (in general within 5%) or, when absolute calculations are performed, on the accuracies of certain parameters such as the ionization cross section, the florescence yields, and the X-ray relative intensities, which are in general within 10%.

2.2. Proton-Induced Gamma-Ray Emission (PIGE)

Because of the inadequate resolution and efficiency of Si(Li) detectors and the high attentuation and the poor signal-to-noise ratio for low energy X-rays, PIXE cannot detect light elements ($Z < 11$) unless a specific experimental apparatus is set up. Moreover, for light elements (in principle, detectable) such as Na, Mg, Al, Si, and P, the analytical accuracy is strongly dependent on the self-absorption corrections which, in general, can be applied provided the sample thickness, composition, and morphology are known. In particular, for particulate matter samples with a nonhomogeneous particle size distribution, such as fly ashes, these corrections are troublesome and unreliable.

PIGE technique, based on the detection of prompt gamma rays following nuclear reactions, is particularly suitable for the analysis of light elements so that the simultaneous and combined use of PIXE and PIGE techniques is an effective analytical tool for a complete multielemental analysis.

So far, PIGE has mainly been applied to the analysis of thick samples, while for thin ones there are only a few reports in the literature, concerning the detection of C and N (Maclas *et al.*, 1978), Na (Robaye *et al.*, 1985; Asking *et al.*, 1987), and S (Raisanen and Lapatto, 1988). The limited application of PIGE to thin samples is mainly due to the strongly resonant behavior of the gamma-ray nuclear cross sections (Boni *et al.*, 1988). To overcome this difficulty, a method based on the use of an energy-spread proton beam has been set up at CISE (Boni *et al.*, 1990b).

The nuclear reactions involved in the application of the PIGE technique are capture reactions (p,γ), inelastic scattering (p,p'γ), and transmutation reactions such as (p,nγ) and (p,$\alpha\gamma$). The energy analysis of gamma rays allows the identification not only of the element but also of the isotope which has taken part in the reaction. In Fig. 4 a typical PIGE spectrum of urban particulate matter is shown.

For quantitative analysis, the number N of gamma rays following nuclear reactions in a homogeneous sample of thickness t is

$$N = N_p n_0 \epsilon \int_0^t \sigma(E)\, dx = N_p n_0 \epsilon \int_{\Delta E} [\sigma(E)/S(E)]\, dx \qquad (3)$$

where N_p is the number of incident protons, n_0 is the number density of target atoms, ϵ is the detection efficiency, $\sigma(E)$ is the nuclear reaction cross section, and $S(E)$ is the stopping power.

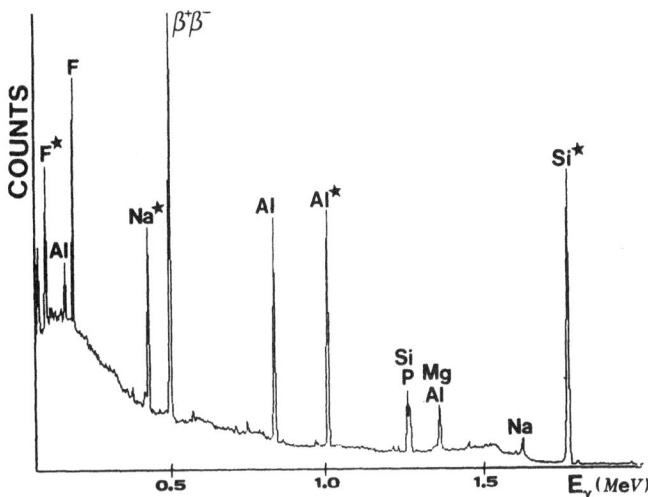

Figure 4. PIGE spectrum of urban particulate matter.

For thick samples the upper limit of integration is the beam range in the sample, and ΔE ranges from 0 to the incident beam energy value. When performing analyses relative to standard reference samples, assuming that $S(E)$ has the same trend for both the unknown and the standard samples, the following equation can be used:

$$N_x/N_s = n_{0x}S_s(E^*)/n_{0s}S_x(E^*) \tag{4}$$

where the subscripts x and s stand for unknown and standard reference samples, respectively, and E^* is the beam energy at which the gamma yield is half the value corresponding to the incident energy. For gamma ray yield values, plenty of data are available in the literature (Kiss et al., 1985). For thin samples the integral should be computed numerically, provided the sample thickness t is accurately known and an estimate of the sample composition is available in order to calculate $S(E)$ and ΔE. The calculation is then iteratively carried out according to the sample composition obtained at each step. However, the main difficulty is due to the strongly energy-dependent trend of the cross sections.

Only when the cross section is approximately constant over the energy interval ΔE, a simple equation is obtained

$$N = N_p \cdot n_0 \cdot \sigma \cdot \epsilon \cdot t \tag{5}$$

allowing the possibility of analyses with respect to standard reference samples:

$$N_x/N_s = n_{0x}t_x/n_{0s}t_s \tag{6}$$

A simple method aimed at smoothing the strongly energy-dependent cross section trends has been devised at CISE to overcome the above-mentioned difficulties for the analysis of thin samples (Boni *et al.*, 1990b). The method is based on the use of an energy-spread proton beam. A rotating circular Al diffuser foil, whose sectors have different thicknesses, is placed in front of the irradiation chamber so that the energy of the emerging proton beam is spread over a suitable range. Under these experimental conditions, the gamma ray yields for reactions on Li, B, F, Na, Mg, Al, Si, and P turn out to be smooth and constant enough over an energy range equivalent to the proton energy loss in a typical particulate matter sample up to 1 mg/cm^2 thick and made up of particles as large as 20 μm.

The sensitivity of the PIGE technique is strongly dependent on the specific target isotope and on the incident energy because of the nonsystematic trend of the nuclear reaction cross sections. Table I reports the detection limits estimated in real fly ash and coal samples for the CISE experimental setup (Boni *et al.*, 1990b). Columns A and B refer to thin (about 1 mg/cm^2 thick) fly ash and coal layers, respectively, deposited on polycarbonate substrates for an irradiation of 60 μC of 3.52-MeV protons. The detection limits are not low for all elements, but they prove to be adequate for ash and coal samples. For instance, in a typical 10-min irradiation, Li, F, Na, Al, and Si are detectable. Columns C and D refer to thick samples (pellets) of fly ash and coal, respectively, for an irradiation of 60 μC of 3.5-MeV protons. The detection limits for some elements are one order of magnitude better than those obtained with thin samples.

Table I. Nuclear Reactions, Gamma Ray Energies, and Detection Limits (DL) of PIGE Technique for Thin and Thick Fly Ash and Coal Samples[a]

			DL (μg/cm^2)		DL (μg/g)	
Element	Reaction	E_γ (keV)	A[a]	B	C	D
Li	^{7}Li(p,p$_{1\gamma}$)^{7}Li	478	0.030	0.015	25	10
B	^{11}B(p,p$_{1\gamma}$)^{11}B	2125	0.600	0.600	100	90
F	^{19}F(p,p$_{1\gamma}$)^{19}F	110	0.035	0.025	5	10
Na	^{23}Na(p,p$_{1\gamma}$)^{23}Na	440	0.080	0.050	10	20
Mg	^{25}Mg(p,p$_{1\gamma}$)^{25}Mg	585	2.000	1.200	300	550
Al	^{27}Al(p,p$_{2\gamma}$)^{27}Al	1014	0.300	0.150	40	70
Si	^{28}Si(p,p$_{1\gamma}$)^{28}Si	1779	0.400	0.350	250	320
P	^{31}P(p,p$_{1\gamma}$)^{31}P	1266	0.300	0.150	500	250

[a]Samples are designated as follows: A, Thin fly ash samples; B, thin coal samples, C, thick fly ash samples; D, thick coal samples.

2.3. Auger Analysis of Solid Particles

When an electron is removed from an inner shell of an atom by electron irradiation or other forms of energy transfer, i.e., energetic photons, the return to the ground state is not always accompanied by a radiative process. The nonradiative return to the ground state is accomplished by the Auger effect.

In the nonrelativistic limit an Auger process can be viewed in two ways: as a radiationless reorganization via a direct interaction of two electrons of an atom ionized in an inner shell or by a two-step mechanism where a quantum X ray is first produced and then absorbed by an outer electron which is ejected (internal conversion). Both these points of view lead to the same results. A detailed description of the Auger process is beyond the scope of the present work; in the literature there are excellent books and review articles on the subject (Agarwal, 1991; Briggs, 1983).

The Auger effect is represented schematically in Fig. 5. An electron is removed by an ionization process from the K shell. A second electron falls from an L_1 shell to the K shell with the consequent transfer on the hole from the K shell to the L_1 shell. The energy released by this process is used to remove a second electron from the $L_{2,3}$ shell, which leaves the atom with a defined kinetic energy E_k. This electron is called an Auger electron. The net result is the presence of two holes in the L shells. For atoms with a large nuclear charge, the two L shell holes can be filled with electrons from outer shells, and other Auger electrons can be ejected in competition with X-ray emission. The Auger process described above is formally defined as a $KL_1L_{2,3}$ Auger decay, where the first letter indicates the ionized shell, the second letter the shell from which an electron is removed to fill the K hole, and the third letter the shell from which the Auger electron is ejected.

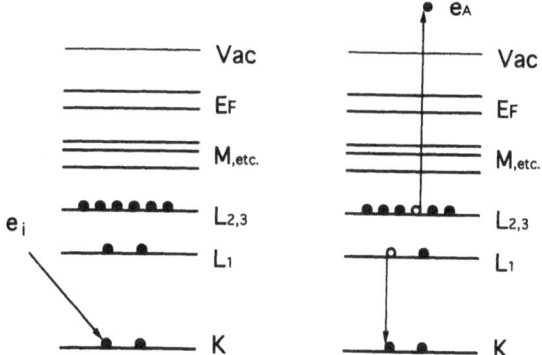

Figure 5. Schematic representation of a *KLL* and CVV Auger process.

To a first and simplified approximation, the kinetic energy of an Auger electron is given by

$$E_{kin} = h_\nu - E_L = E_K - E_L - E_L = E_K - 2E_L \qquad (7)$$

where E_L and E_K are the binding energy of the L and K shells, respectively. The E_{kin} for Auger electrons is a characteristic quantity, and so its determination can identify the presence of the elements that undergo Auger emission. Very important for the study of the electronic structure of a solid are the core–core–valence (CCV) and core–valence–valence (CVV) Auger transitions. In both these cases the Auger electron is ejected from the valence band, and the Auger spectrum results from the convolution of the singly or doubly ionized density of states.

Auger processes can be observed in atoms, molecules, and solids. In solids they are, in general, more complex because they reflect, besides the effects due to the local electronic properties, also those due to the nonlocal electronic behaviors. Auger line shapes, satellite emissions, and line widths are important information that reflect properties of the solid.

To fully interpret an Auger spectrum, it is necessary to distinguish the initial state from the final state effects. This represents the main problem for Auger spectroscopy, and it is a source of controversy and open questions. In the present context, it is important to highlight the fact that the Auger kinetic energy is influenced by the oxidation states of the emitting atoms, and an Auger spectrum contains also information about their oxidation states. For a more detailed descriptions, see, for example, Carlson (1976) and Briggs (1983).

Auger electrons escaping from a solid are subject to inelastic scattering processes that strongly reduce the mean free path, λ_m, of the unscattered Auger electrons, defined as

$$\lambda_m = 1/l \int_0^\infty dl = 1/\sigma \qquad (8)$$

where σ is the inelastic cross section per unit length. Also useful is the definition of escape depth at half intensity, $\lambda_{1/2}$, or depth below the surface from which half of the Auger electrons are observed:

$$\lambda_{1/2} = \ln 2/\sigma = 0.693/\sigma \qquad (9)$$

For the most common solids, the escape depth ranges from a few angstroms up to a few tens of angstroms, making Auger spectroscopy a technique sensitive to the topmost layers of a solid.

Auger spectroscopy applied to solids presents intrinsic limits. In particular, for nonconducting samples electrostatic charging effects of the sample can inhibit the proper collection of the Auger electrons. A second important source of problems arises from possible damage, in particular when the solid is excited by an electron beam accelerated at energies of several kiloelectronvolts. In this case the sample can be locally heated to an extent where stoichiometric and structural changes take place, thus modifying the intrinsic properties.

In a modern Auger spectrometer the Auger analysis is performed together with the detection of secondary electrons. High-quality electron micrographs can be obtained by performing a scanning of the sample surface. This technique is very powerful since it allows morphological information to be combined with Auger analysis.

So far, Auger spectroscopy allows the surface of a solid to be studied, while only limited information is acquired about the underneath layers. The use of ion sputtering to remove the topmost layer offers, when used properly, a possibility to analyze the bulk. The Auger profile of a solid, however, has to be used very carefully since the surface modifications due to the ion irradiation can introduce artifacts about the stoichiometry, while the chemical oxidation state of the elements is certainly modified.

The Auger spectrometers most commonly employed to analyze solid surfaces use gun sources to excite the Auger electrons. The electron beam in these sources can be focused up to a few hundreds of angstroms, allowing a comparable scanning spatial resolution. Later in this chapter, some examples of Auger profile and high-spatial-resolution measurements will be presented.

The electronic structure of a solid surface can be studied also by exciting the valence band and the core-level electrons using energetic X-ray or UV photons. The one step photoionization process is based on the measurement of the kinetic energy of an electron which is ejected from the solid after the absorption of a photon, $A + h\nu = A^+ + e^-$. X-ray and UV photoelectron spectroscopy (XPS and UPS) have become important tools for the study of the core-level and valence-band electronic structure of atoms, molecules, and solids after the work of Siegbahn *et al.* (1969).

An incident quantum $h_{\nu 0}$ of X-ray radiation can remove an inner-shell electron of an atom in a solid; the electron is ejected with a kinetic energy given by the energy conservation equation

$$E_{kin} = h\nu_0 - E_{BK} - \phi \tag{10}$$

where E_{BK} is the Koopmans' theorem frozen-orbital binding energy of the ionized core level, and ϕ is the solid work function. This one-electron picture used to describe the photoionization process has only a limited validity since the initial and final states represented are those strictly described by Koopmans' theorem, where the process of photoionization is approximated by a frozen transition from N-electron system to an $(N - 1)$-electron system plus an electron taken to infinity. In such a simple description, the relaxation of the passive electron orbitals to screen the hole induced by the photoionization is ignored as are the spin–orbit and spin–spin interactions. The relaxation of the valence electrons reduces the energy of the final state by E_R, and thus the measured binding energy is given by $E_{Bm} = E_{BK} - E_R$.

The measured E_B of the core levels as well as the XPS spectral line shapes of a given element is, therefore, related to the valence-band (valence orbital) electronic structure, and, accordingly, it is possible to obtain detailed information on the nature of the chemical bonds that the element forms in the molecule or in a condensed state.

In XPS the sample is irradiated by monochromatic soft X rays. The most common are the Mg and Al $K\alpha$ lines at $h\nu_0 = 1253.4$ eV and $h\nu_0 = 1486.6$ eV, respectively. Modern photoelectron spectrometers employ electrostatic analyzers. The most common are the hemispherical and cylindrical mirror analyzers. The photoelectrons are dispersed along an axial or radial coordinate after being deflected by a potential V applied to the outer hemisphere or cylinder according to their kinetic energy.

3. EXPERIMENTAL RESULTS

3.1. PIXE–PIGE Setup

Figure 6 shows a general view of the CISE PIXE–PIGE set up. The system has been described in detail elsewhere (Boni *et al.*, 1990b). The main features are summarized here. The proton beam, provided by a 3.5-MV tandem Van de Graaff accelerator, is analyzed by a bending magnet and switched into the PIXE–PIGE beam line. The beam crosses a 1.5-mg/cm^2 Al diffuser foil and four Ta diaphragms in order to obtain a uniform charge distribution. A rotating disk made up of five Al sectors of different thicknesses and a sixth sector transparent to the beam is placed 40 cm from the irradiation chamber. In this way, the protons get an energy spread in which six main components, partly superimposed by the straggling effect, are present. The resulting energy distribution can reasonably be assumed to have a 270-keV-wide rectangular shape. The main effect of using an energy-spread proton beam is the smoothing of the prompt gamma excitation functions for PIGE analyses. In this way, constant cross sections over a suitable energy interval are obtained, thus allowing a straightforward analysis relative to standard reference samples. The beam is finally collimated by a graphite collimator before entering the irradiation chamber, where the samples are mounted on a remotely controlled 100-position sample holder. The beam spot on the sample is about 1 cm^2. The beam current is measured by a Faraday cup. The fluorescence X rays are detected by a Si(Li) detector (80 mm^2 wide, 3 mm thick, 185-eV resolution at 5.9 keV) placed at 90° to the beam. The filter system between the irradiation chamber and the

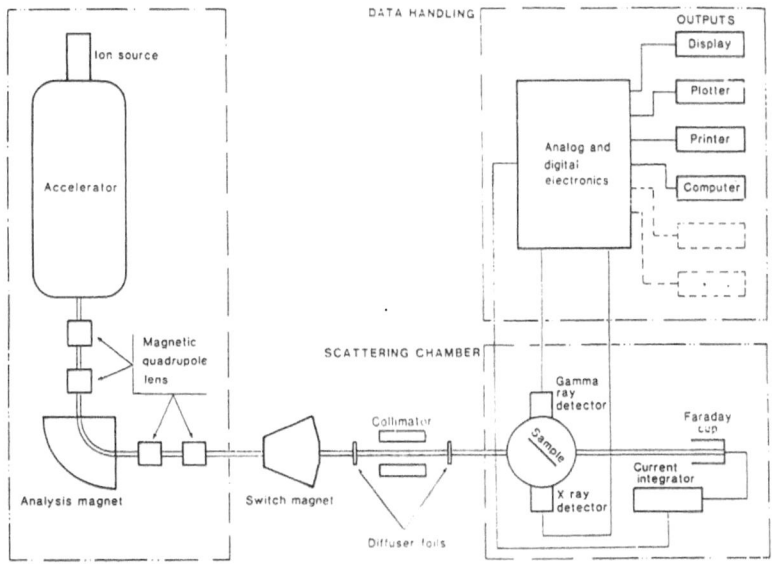

Figure 6. CISE PIXE–PIGE setup.

Si(Li) detector consists of a 25-μm-thick aluminized Mylar foil to withstand the pressure difference and to stop the scattered protons, a 0.5-mm-thick Be layer with a 6-mm-diameter hole, and a 0.25-mm-thick Teflon foil with a 3.5-mm diameter hole. This filter configuration reduces the count rate and pileup effects due to the main element X rays while allowing the detection of low-energy X rays due to light elements.

In principle, all elements with $Z \geqslant 16$ can be detected by PIXE with this detection configuration. The gamma rays following nuclear reactions are counted by a high-purity 4.93×6.28 cm Ge detector (1.71-keV resolution and 20% efficiency at 1.31 MeV) placed at 180° to the Si(Li) detector. Both detectors, accurately shielded by lead screens, are in the air and connected to a standard electronic chain with pileup rejectors. The data acquisition, storage, and reduction is supervised by a VAX microcomputer. For the reduction of PIXE spectra, the PIXAN computer package (Clayton *et al.*, 1987) is used. For quantitative analysis, the system has been calibrated using certified reference samples.

As for thin-sample preparation, for powders (coal and fly ash) a small quantity of previously ground and homogenized material is resuspended in cyclohexane and filtered through preweighed Nucleopore filters. After drying and conditioning, the samples are weighed to obtain the thickness of the powder layer.

The samples are finally coated with a thin layer of polyvinylacetate in order to prevent material loss. For thick samples, the powder is ground and blended in preground ultrapure graphite spiked with yttrium (as internal standard) and finally pressed into pellets. In general, coals can be pelletized with no graphite dilution, while for coal ashes 1:10 dilution is applied in order to reduce the count rate and enhancement effects and to simplify matrix corrections.

3.2. Nuclear Microscopy Facility

Proton microprobe analysis of fly ash samples has been carried out using the Oxford facility. A detailed description of the system is available in the paper by Grime *et al.* (1991).

Fly ash samples are first prepared by suspending the particles in a dilute solution of pioloform in analytical grade chloroform. Droplets of this suspension are allowed to flow over the surface of polished glass slides to form a thin layer (about 0.2 μm), which after drying can be peeled from the slides as a self-supporting film of pioloform with the dispersed particles embedded in it. The particles are analyzed using a 100-pA beam of 3-MeV protons, provided by a 1.7-MV Pelletron tandem accelerator, focused to submicron diameter (between 0.5 and 1 μm).

Particles for analysis are selected by mapping a 50 μm × 50 μm region of the film using off-axis scanning transmission ion microscopy. In this technique a surface barrier detector is mounted behind the sample at an angle of approximately 20° to the transmitted proton beam. The extra areal density of the particles causes enhanced small-angle scattering, thus providing an efficient method for mapping the particle distribution in the film.

Single particles, selected by shape and size, are analyzed by positioning the proton beam in the center of the particles and collecting simultaneously PIXE and Rutherford backscattering spectrometry (RBS; Chu *et al.*, 1978) spectra (Fig. 7). The X rays are detected using an 80-mm^2 Si(Li) detector with an 8-μm Be window. To avoid saturation of the spectrum by intense Al and Si X rays, a 114-μm Be filter is fitted. For the reduction of PIXE spectra, the matrix composition and thickness of the particles are first determined by the analysis of the RBS spectra. Quantification of the PIXE data is related to the total charge incident on the sample as collected by a graphite Faraday cup behind the film and corrected for detector and data acquisition system dead time. X-ray attenuation corrections are applied. The absolute accuracy of the determination is estimated at 15%, although relative concentrations at each point have greater accuracy. For the measurement of possible surface enrichment of element concentrations and of the complex structure of the particles, line scans across the particle diameter as well as areal scans over the whole particle can be performed.

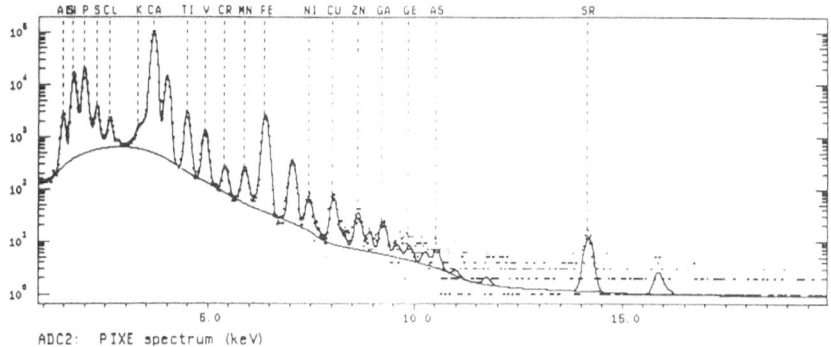

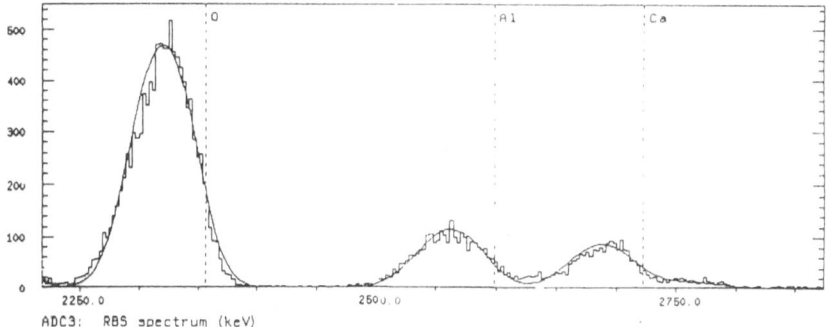

Figure 7. Proton microbeam PIXE and RBS spectra recorded from a fly ash particle.

3.3. PIGE–PIXE Results

As an example of the application of PIGE–PIXE techniques, the results of the analysis of coal fly ash sample collected at the inlet and outlet of the electrostatic precipitators (ESP) are reported. Table II reports the average element concentrations in fly ash samples collected at the ESP inlet and outlet in the conventional energization mode (collection efficiency < 99%).

The enrichment factors, that is (el. conc.)$_{out}$/(el. conc.)$_{in}$, are reported as well. The fly ash composition is different at the ESP inlet and outlet. In particular, for several elements such as F, S, V, Zn, Ga, Ge, As, and Pb, although the absolute concentrations decrease, the relative concentrations are enhanced. This trend can be related to the volatile nature of these elements, which are therefore likely to condense on the surface of the particles. Surface condensation, even though independent of particle size, yields a higher specific concentration for small

Table II.   Average Element Concentrations (ppm) in
Fly Ash Samples Collected at the Inlet and Outlet of
the Electrostatic Precipitators in Conventional
Energization Mode and Enrichment Factors (EF)

| Element | Average concentration (ppm) | | EF[a] |
	Inlet	Outlet	
F	260 ± 50	710 ± 65	2.73
Na	1,300 ± 450	1,230 ± 150	0.95
Li	300 ± 45	390 ± 15	1.30
Al	176,000 ± 1,000	187,000 ± 800	1.06
Si	226,000 ± 2,500	226,000 ± 1,500	1.00
S	1,140 ± 200	1,610 ± 110	1.41
K	4,600 ± 900	5,100 ± 100	1.10
Ca	40,000 ± 200	43,500 ± 900	1.09
Ti	9,000 ± 1,000	11,500 ± 500	1.28
V	150 ± 55	305 ± 30	2.03
Cr	160 ± 45	180 ± 15	1.12
Mn	320 ± 65	440 ± 30	1.37
Fe	17,000 ± 2,000	18,000 ± 400	1.06
Ni	150 ± 20	180 ± 40	1.20
Cu	160 ± 70	130 ± 20	0.81
Zn	105 ± 10	230 ± 25	2.19
Ga	75 ± 6	220 ± 25	2.93
Ge	23 ± 3	57 ± 7	2.47
As	14 ± 3	41 ± 6	2.93
Rb	30 ± 3	40 ± 15	1.33
Sr	2,300 ± 175	2,960 ± 450	1.18
Y	105 ± 25	120 ± 10	1.14
Zr	540 ± 40	555 ± 30	1.03
Ba	1,950 ± 400	2,400 ± 360	1.23
Pb	90 ± 10	245 ± 35	2.72

[a]$EF = (El.\ conc.)_{out}/(El.\ conc.)_{in}$.

particles due to the increased surface-to-volume ratio. Finally, the decrease in the ESP collection efficiency with decreasing particle size must be taken into account. In addition, fly ash particle resistivity, which depends on trace element concentrations, can contribute to these enrichment phenomena. For another group of elements, including Al, Si, Ca, and Fe, the enrichment factor is about unity. The uniform concentration versus particle size can account for this result.

3.4. Nuclear Microscopy Results

As an example, results are reported here from the analysis of ∼ 100 particles collected at the inlet of the electrostatic precipitator of a coal-fired power plant.

Particles have been analyzed in two size ranges, i.e., ≤ 2 μm and >2 μm, which are known to have different properties. In terms of the main element concentrations, different groups of particles have been found. Table III contains the average concentrations and standard deviations for groups 1 and 2, whose relative abundances account for about 60–70% of the analyzed particles. The particles in these groups are composed of an aluminosilicate matrix with differences due to higher concentrations of P, Ca, and Ti in group 2. As for trace elements, S, Cr, Cu, and Zn have higher concentrations in group 2. Moreover, trace element concentrations are higher in the fine range (≤ 2 μm). Standard deviations are low for Al and Si since the mineral matter in the parent coal is made up of aluminosilicates which undergo fusion during combustion. The high standard deviations for K, Ca, Ti, and Fe represent a small degree of coalescence upon combustion, taking into account that, although present in high concentrations, they are not the main constituents of the coal mineral matter. The high standard deviations for trace elements can be attributed to different particle formation mechanisms (segregation, evaporation–condensation).

Table III. Average Concentrations, with Their Standard Deviations, of Elements in Individual Fly Ash Particles in Two Size Ranges Collected at the Inlet of the Electrostatic Precipitators[a]

	Particle size ≤ 2 μm		Particle size > 2 μm	
	Group 1	Group 2	Group 1	Group 2
Abundance (%)	51	16	50	9
Al (%)	23.1 (5.3)	18.6 (3.0)	23.6 (6.2)	19.4 (9.7)
Si (%)	29.0 (6.3)	25.6 (7.0)	28.4 (5.6)	24.1 (8.7)
P (%)	0.18 (41)	0.49 (61)	0.28 (96)	0.78 (83)
S (ppm)	1500 (115)	4450 (86)	140 (105)	490 (68)
K (%)	0.97 (65)	1.1 (81)	0.72 (31)	0.64 (43)
Ca (%)	0.69 (74)	3.4 (55)	0.98 (73)	1.90 (53)
Ti (%)	0.68 (100)	1.16 (59)	1.05 (93)	1.29 (86)
Cr (ppm)	210 (76)	380 (23)	130 (87)	410 (82)
Mn (ppm)	280 (72)	590 (70)	130 (55)	—[b]
Fe (%)	0.79 (88)	1.57 (61)	0.55 (44)	0.45 (43)
Ni (ppm)	120 (50)	—[b]	85 (65)	110 (56)
Cu (ppm)	490 (103)	1480 (92)	110 (167)	250 (139)
Zn (ppm)	220 (76)	540 (53)	110 (48)	120 (17)
Ga (ppm)	160 (80)	—[b]	80 (38)	—[b]
Sr (ppm)	—[b]	2140 (62)	1630 (174)	790 (761)
Zr (ppm)	—[b]	—[b]	640 (128)	—[b]
Al/Si (ppm)	0.8 (85)	0.73 (8.2)	0.83 (7.8)	0.81 (15)

[a]Data are presented for two different matrix groups.
[b]Below the detection limit in most particles.

As for the other particle groups not reported in Table III, both groups 3 and 4 contain Ca as the main element and high concentrations of P and S, but they differ in Al and Si content. They have the highest concentrations of V, Cr, Mn, Ni, Ga, Ge, and As. Groups 5 and 6 are made up of Fe-rich and Ti-rich particles, while group 7 contains quartz particles with low trace element concentrations. Work on the analysis of particles collected at the ESP outlet is in progress.

3.5. Auger and XPS Results

For photoelectron spectroscopic measurements, fly ash samples were supported on pure In foils. This procedure allows for a reduction of the electrostatic charging effects sometimes induced by the photoemission process on dielectric compounds.

Figure 8 shows the electron micrograph of a fly ash particle obtained during the Auger measurements. The In foil on which the particle is supported is well visible in the picture. The particle has a diameter of about 10 μm while the surface is mainly composed of aluminosilicate, as shown by the Auger spectrum in Fig. 9. Rather interesting to note are the two components of the Si signal. This behavior has been observed also with other samples. The lower kinetic energy component is close to that reported for elementary silicon; however, an oxidation state close to elementary Si is very unlikely in fly ash materials, and the origin of the lower kinetic energy silicon line is still obscure. From the Auger spectrum of Fig. 9, it is possible to note the presence of K, C, Ca, and P. The presence of CaO is not surprising since CaO was used during the burning process to reduce SO_2 emission, while K and P belong to the original coal minerals. The In signal is due to the substrate used for the Auger measurements.

Rather important is the possibility of performing a depth profile of the fly ash particle along its radius. This kind of measurement can be achieved when it is possible to alternate Auger measurements with highly controlled sputtering

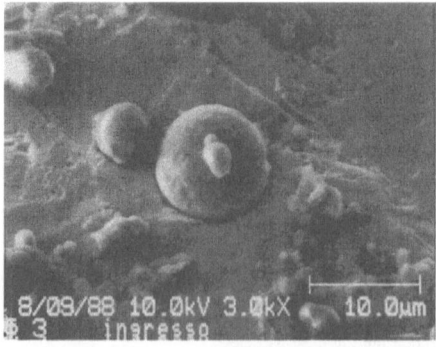

Figure 8.   Scanning Auger micrograph of a fly ash particle.

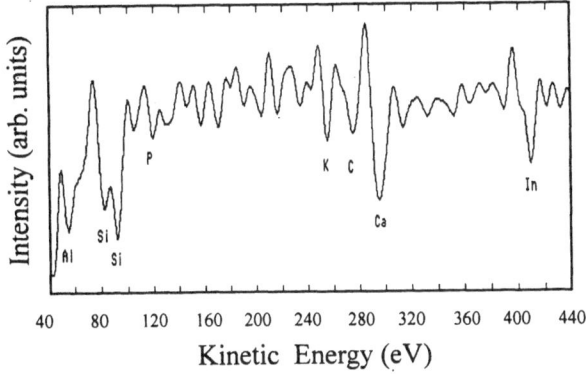

Figure 9. Auger spectrum measured on the particle shown in Fig. 8.

erosion of the particle surface. For this analysis the particle is pressed on an indium foil. In the present experiment the sputtering has been performed by Ar^+ ions accelerated at 1 keV using an ion current of $\sim$500 μA. The sputtering time was previously calibrated to obtain the proper conversion between the ion irradiation time and the thickness of the sputtered material. Figure 10 shows the dependence on the sputtering time of the atomic concentrations of Al, Si, and O as measured from the intensity of the Auger line and using the sensitivity factors reported in the literature (Davis *et al.*, 1976). This Auger depth profile is a typical one for fly ash aluminosilicate particles. Aluminum, silicon, and oxygen are the dominant elements. It is important to note that photoelectron spectroscopies, i.e., Auger spectroscopy and XPS, have a rather reduced sensitivity when used for qualitative analysis (0.5–1.0 at. %); the most important information provided by

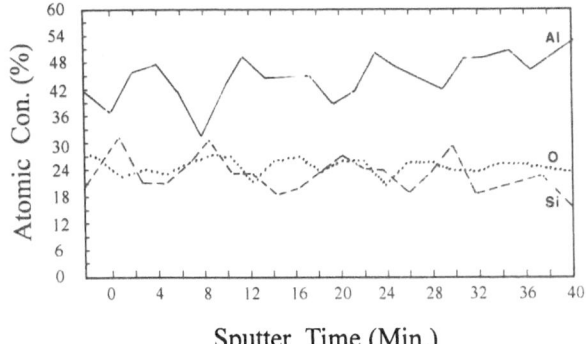

Figure 10. Auger depth profile of the particle shown in Fig. 8. It may be noted that the surface is mainly composed of aluminosilicates. The sputtering rate was 14 nm/min.

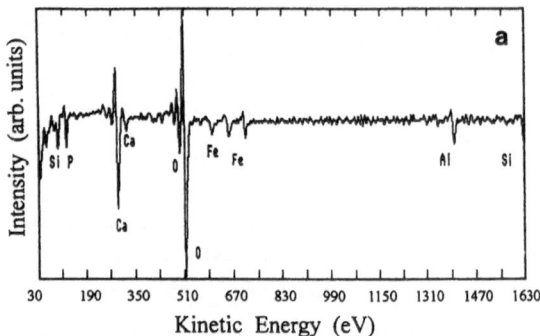

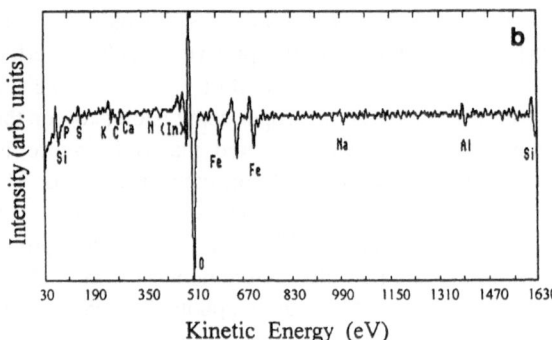

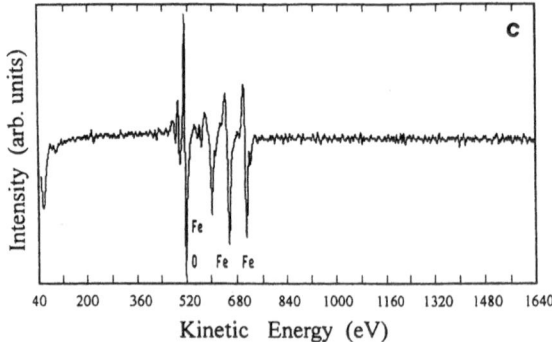

Figure 11. Auger spectra of a particle composed of shells, measured before sputtering (a) and with increasing sputtering times (b and c).

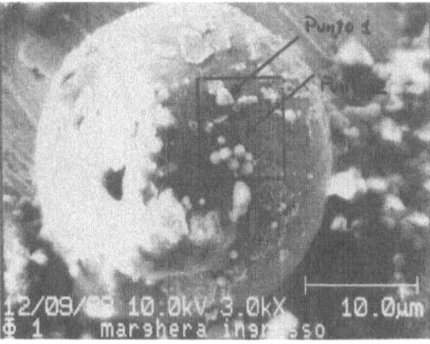

Figure 12.   SAM micrograph of the fly ash particle whose Auger spectra are given in Fig. 11.

these techniques is that related to the nature of the chemical bonds at the surface of a solid. In this respect, detailed information on the surface composition can be obtained by secondary ion mass spectroscopy (SIMS).

Fly ash particles formed by shells with different composition and structure are often observed. The Auger spectrum of Fig. 11a, recorded with a high spatial resolution (~1000 Å) from the particle shown in Fig. 12, shows a surface composition where, besides Al, Si, S, P, Ca, and O, Fe is present, as significant Fe *LMM* lines are detected. In Fig. 11b and c, the Auger spectra after increasing sputtering times are presented. The inner shell has mainly the composition Fe_xO_y. The shell structure of this particle is well evinced by the Auger depth profile presented in Fig. 13. This result is interesting in particular with regard to particle

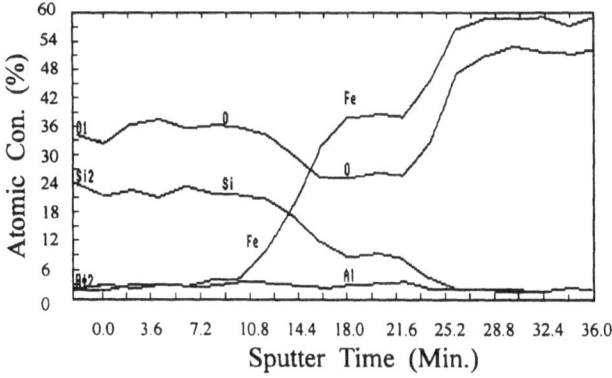

Figure 13.   Auger depth profile of the particle shown in Fig. 12 (sputtering rate 25 Å/min). The shell composition of this particle is unambiguously demonstrated. Three different shells can be observed. The topmost shell is mainly composed of aluminosilicates. Underneath this shell, an aluminosilicate and iron oxide shell is observed, while the inner shell is mainly composed of iron oxides.

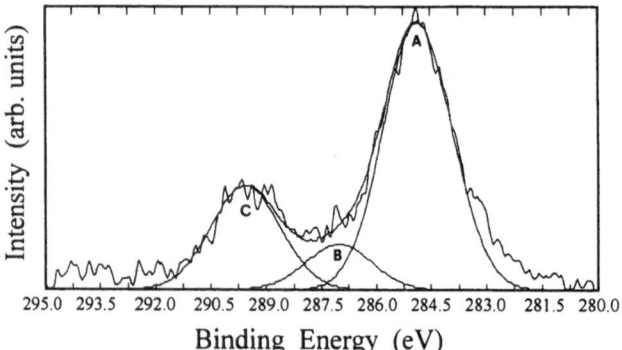

Figure 14. X-ray photoelectron spectrum measured in the C1s binding energy region of fly ash particles produced by a large coal-burning thermoelectric power plant. At least three different carbon compounds are detected.

nucleation processes and the structure of magnetic particles, and it deserves more detailed analysis and studies that are beyond the scope of the present work.

Detailed information about the oxidation states of the elements present on the particle surfaces is provided by XPS. Due to the relatively large area of detection XPS measurements have a statistical meaning since the surfaces of several hundred fly ash particles are measured contemporaneously. Figure 14 shows the XPS C1s spectral region of a fly ash sample from a coal-burning thermoelectric power plant. Several components (A–C in Fig. 14) are detected. This indicates that different kinds of carbon compounds are present on the surface, with the oxidation state of carbon increasing on going from component A to component C. In such a complex sample, an unambiguous identification of the carbon bond or carbon compounds is not possible since several carbon groups present similar C1s binding

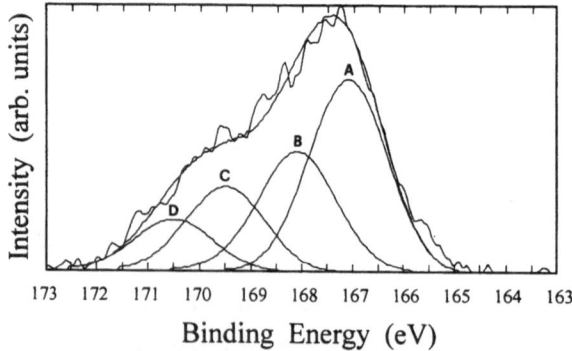

Figure 15. X-ray photoelectron spectrum measured in the S2p binding energy region of fly ash particles produced by a large coal-burning thermoelectric power plant. Peak A is due to SO_3^{2-} groups, while peaks B–D are due to SO_4^{2-} compounds.

energy values (Moulder *et al.*, 1992). A more complete characterization could be achieved by the use of other analytical techniques such as time-of-flight SIMS.

Figure 15 shows the XPS spectrum taken in the S2p binding energy region. Also in this case several oxidation states for the sulfur are detected. In particular, component A can be attributed to SO_3^{2-} groups, while components B–D are due to SO_4^{2-} compounds (Moulder *et al.*, 1992).

In Fig. 16a and b the C1s XPS spectra of fly ash samples, collected after and before the electrofilter, are presented. It is interesting to note the selective action of

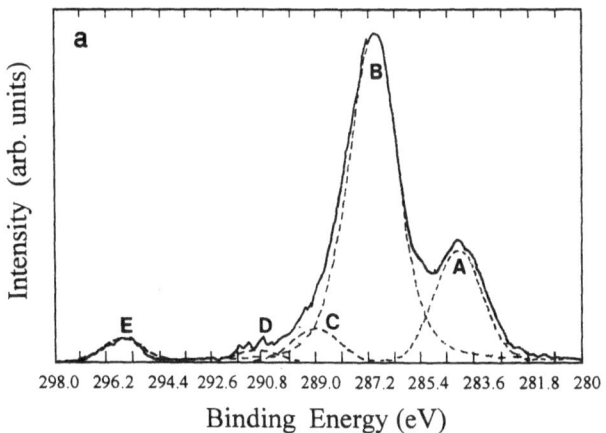

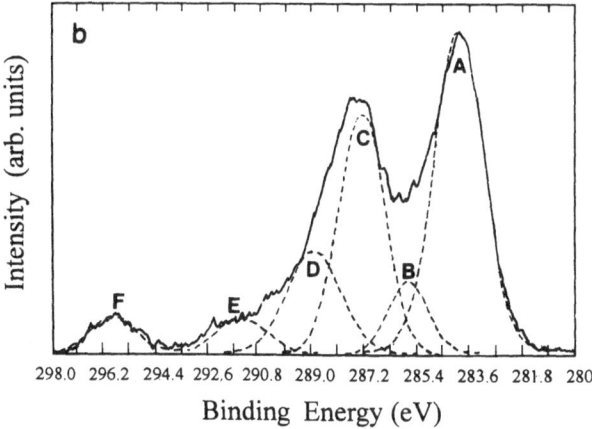

Figure 16. X-ray photoelectron spectra measured in the C1s binding energy region of fly ash collected after (a) and before (b) the electrofilter. It is interesting to note that component A, attributed to carbonaceous compounds is strongly reduced by the electrofiltering process, spectrum (b) versus spectrum (a). On the contrary, component C results enriched by the electrofilter.

the electrofilter on the different carbon components. Component A of the spectrum reported in Fig. 16b, which is attributed to carbonaceous compounds, is strongly reduced in the fly ash collected after the electrofilter, as shown by the spectrum of Fig. 16a. On the contrary, component C seems to be enriched by the electrofilter (Fig. 16a versus Fig. 16b). A possible explanation may be proposed if the different surface electrical conductivities of the fly ash particles are considered. Particles with carbonaceous compounds on the surface are more conductive, and they seem to be more gettered by the electrofilters.

4. CONCLUSIONS

In this work surface and bulk techniques used to study fly ash particles have been presented. In particular, the applicability of PIGE–PIXE analysis in the study of the bulk composition of the particulates produced during the combustion process of fossil fuels has been demonstrated. The results are of paramount importance for control of the burning process, for design of more efficient burner plants and electrofilter systems, and for the study of the environmental impact of such materials.

Complementary information can be obtained by surface analysis techniques such as Auger and XPS electron spectroscopies. In this case the surface composition and the chemical oxidation state of single particles (Auger) or of a large number of them (XPS) can be obtained. In addition, the elementary radial composition of these particles can be investigated by Auger depth profile measurements.

REFERENCES

Agarwal, B. K., 1991, *X-Ray Spectroscopy*, Springer-Verlag, New York.

Asking, L., Swietlick, E., and Garg, M. L., 1987, PIGE analysis of sodium in thin aerosol samples, *Nucl. Instrum. Methods* **B22**:368–371.

Bellagamba, B., Caridi, A., Cereda, E., Braga Marcazzan, G. M., and Valkovic, V., 1993, PIXE application to the study of trace element behaviour in coal combustion cycle, *Nucl. Instrum. Methods* **B75**:222–229.

Bellotto, M., Boni, C., Caridi, A., Cereda, E., Chemelli, C., Braga Marcazzan, G. M., Parmigiani, F., Scagliotti, M., and Bellagamba, B., 1990, Analysis of coal fly ash by bulk and surface characterisation techniques, *Mater. Res. Soc. Symp. Proc.* **178**:45–56.

Boni, C., Cereda, E., Braga Marcazzan, G. M., and De Tomasi, V., 1988, Prompt gamma emission excitation functions for PIGE analysis of Li, B, F, Mg, Al, Si and P in thin samples, *Nucl. Instrum. Methods* **B35**:80–86.

Boni, C., Caridi, A., Cereda, E., and Braga Marcazzan, G. M., 1990a, PIXE–PIGE analysis of thin fly-ash samples, *Nucl. Instrum. Methods* **B45**:352–355.

Boni, C., Caridi, A., Cereda, E., and Braga Marcazzan, G. M., 1990b, A PIXE–PIGE set-up for the analysis of thin samples, *Nucl. Instrum. Methods* **B47**:133–142.

Braga Marcazzan, G. M., Caruso, E., Cereda, E., and Redaelli, P., 1987, The CISE PIXE system for the aerosol characterization in several Italian sites, *Nucl. Instrum. Methods.* **B22**:305–314.

Braga Marcazzan, G. M., Bellagamba, B., Bellotto, M., Boni, C., Caridi, A., Cereda, E., and Chemelli, C., 1990, Fly ash from a coal power plant: Correlation of elemental and structural composition with electrostatic precipitator collection efficiency, *J. Aerosol Sci.* **21**:697–701.

Briggs, D., and Seah, M. P., 1983, *Practical Surface Analysis by Auger and X-Ray Photoelectron Spectroscopy*, John Wiley & Sons, Norwich.

Caridi, A., Cereda, E., Fazinic, S., Jaksic, M., Braga Marcazzan, G. M., Valkovic, O., and Valkovic, V., 1992, Fluorine enrichment phenomena in coal combustion cycle, *Nucl. Instrum. Methods* **B66**:298–301.

Caridi, A., Cereda, E., Grime, G. W., Jaksic, M., Braga Marcazzan, G. M., Valkovic, V., and Watt, F., 1993, Application of proton microprobe analysis to the study of electrostatic precipitation of single fly ash particles, *Nucl. Instrum. Methods* **B77**:524–529.

Carlson, T. A., 1976, *Photoelectron and Auger Spectroscopy*, Plenum Pres, New York.

Caruso, E., and Parmigiani, F., 1987, Studio dell' applicazione della Spettroscopia Elettronica alla Diagnostica delle ceneri di Combustione, CISE Report 4364, Segrate, Italy.

Cereda, E., Chemelli, C., Marcazzan, G. M., and Parmigiani, F., 1988, Spettroscopia Elettronica di Fotoemissione da Particolato Prodotto Durante Processi di Combustione in Caldaia, CISE Report 5153, Segrate, Italy.

Cernuschi, S., and Giuliano, M., 1987, Trace element emission factors from coal combustion, *Sci. Total Environ.* **65**:95–107.

Chu, W. K., Mayer, J. W., and Nicolet, M. A., 1978, *Backscattering Spectrometry*, Academic Press, New York.

Clayton, P., Duerden, P., and Cohen, D. D., 1987, A discussion of PIXAN and PIXANPC: The AAEC PIXE analysis computer packages, *Nucl Instrum. Methods* **B22**:64–67.

Coles, D. G., Ragaini, R. C., Ondov, J. M., Fisher, G. L., Silberman, D., and Prentice, B. A., 1979, Chemical studies of stack fly ash from a coal-fired power plant, *Environ. Sci. Technol.* **13**: 455–459.

Cox, X. B., Bryan, S. R., Linton, R. W., and Griffis, D. P., 1987, Microcharacterization of trace elemental distribution within individual coal combustion particles using secondary ion mass spectrometry and digital imaging, *Anal. Chem.* **59**:2018–2023.

Davis, L. E., MacDonald, N. C., Palmberg, P. W., Riach, G. E., and Weber, R. E., 1976, *Handbook of Auger Electron Spectroscopy*, Physical Electronic Industries, Inc., Eden Prairie, Minnesota.

Davison, R. L., Natusch, D. I. S., Wallace, J. R., and Evans, C. A. Jr., 1974, Trace elements in fly ash, dependence of concentration on particle size, *Environ. Sci. Technol.* **8**:1107–1113.

Fisher, G. L., Prentice, B. A., Silberman, D., Ondov, J. M., Biermann, A. H., Ragaini, R. C., and McFarland, A. R., 1978, Physical and morphological studies of size-classified coal fly ash, *Environ. Sci. Technol.* **12**:447–451.

Grime, G. W., Dawson, M., Marsh, M., McArthur, I. C., and Watt, F., 1991, The Oxford submicron nuclear microscopy facility, *Nucl. Instrum. Methods* **B54**:52–63.

Hansen, L. D., and Fisher, G. L., 1980, Elemental distribution in coal fly ash particles, *Environ. Sci. Technol.* **14**:1111–1117.

Hock, J. L., and Lichtman, D., 1982, Studies of surface layers on single particles of in-stack coal fly ash, *Environ. Sci. Technol.* **16**:423–427.

Jaksic, M., Watt, F., Grime, G. W., Cereda, E., Braga Marcazzan, G. M., and Valkovic, V., 1991, Proton microprobe analysis of the trace element distribution in fly ash particles, *Nucl. Instrum. Methods* **B56/57**:699–702.

Jaksic, M., Bogdanovic, I., Cereda, E., Fazinic, S., and Valkovic, V., 1993, Quantitative PIXE analysis of single fly ash particles by a proton microbeam, *Nucl. Instrum. Methods* **B77**:505–508.

Johansson, S. A. E., and Campbell, J. L., 1988, *PIXE: A Novel Technique for Elemental Analysis*, John Wiley & Sons, New York.

Kaakinen, J. W., Jordan, R. M., Lawasani, M. H., and West, R. E., 1975, Trace element behavior in coal-fired plant, *Environ. Sci. Technol.* **9**:862–869.

Kauppinen, E. I., and Pakkanen, T. A., 1990, Coal combustion aerosol: A field study, *Environ. Sci. Technol.* **24**:1811–1818.

Kiss, A. Z., Koltay, E., Nyako, B., Somorjai, E., Anttila, A., and Raisanen, J., 1985, Measurements of relative thick target yields for PIGE analysis of light elements in the proton energy interval 2.4–4.2 MeV, *J. Radioanal. Chem.* **89**:123–141.

Klein, D. H., Andren, A. W., Carter, J. A., Emery, J. F., Feldman, C., Fulkenon, W., Lyon, W. S., Ogle, J. C., Talmi, Y., Van Hook, R., and Bolton, N., 1975, Pathways of thirty-seven trace elements through coal-fired power plant, *Environ. Sci. Technol.* **9**:973–979.

Lee, R. E., Jr., Crist, H. L., Riley, A. E., and MacLeod, K. E., 1975, Concentration and size of trace metal emissions from a power plant, a steel plant and a cotton gin, *Environ. Sci. Technol.* **9**:643–647.

Linton, R. W., Loh, A., Natusch, D. F. S., Evans, C. A., and Williams, P., 1976, Surface predominance of trace elements in airborne particles, *Science* **191**:852–854.

Linton, R. W., Williams, P., Evans, C. A., and Natusch, D. F. S., 1977, Determination of the surface predominance of toxic elements in airborne particles by ion mass spectrometry and Auger electron spectrometry, *Anal. Chem.* **49**:1514–1521.

Maclas, E. S., Radcliffe, C. D., Lewis, C. W., and Sawicki, C. R., 1978, Proton induced γ-ray analysis of atmospheric aerosols for carbon, nitrogen, and sulfur composition, *Anal. Chem.* **50**:1120–1124.

Mamane, Y., Miller, J. L., and Dzubay, T. G., 1986, Characterization of individual fly ash particles emitted from coal- and oil-fired power plants, *Atmos. Environ.* **20**:2125–2135.

Markowski, G. R., and Filby, R., 1985, Trace element concentration as a function of particle size in fly ash from a pulverized coal utility boiler, *Environ. Sci. Technol.* **19**:796–804.

Moulder, J. F., Stickle, W. F., Sobol, P. E., and Bomben, K. D., 1992, *Handbook of X-ray Photoelectron Spectroscopy* (Jill Chastain, ed.), Perkin-Elmer Corporation Physical Electronics Division, Eden Prairie, Minnesota.

Raisanen, J., and Lapatto, R., 1988, Analysis of sulphur with external beam proton induced gamma-ray emission analysis, *Nucl. Instrum. Methods* **B30**:90–93.

Ramsden, A. R., and Shibaoka, M., 1982, Characterization and analysis of individual fly-ash particles from coal-fired power stations by a combination of optical microscopy, electron microscopy and quantitative electron microprobe analysis, *Atmos. Environ.* **16**:2191–2206.

Robaye, G., Delbrouck-Habaru, J. M., Roelandts, I., Weber, C., Girard-Reydet, L., Morelli, J., and Qiusefit, J. P., 1985, PIGE coupled with PIXE for sodium determination in atmospheric aerosol samples, *Nucl. Instrum. Methods.* **B6**:558–561.

Siegbahn, K., *et al.*, 1969, *ESCA Applied to Free Molecules*, North-Holland, Amsterdam.

Smith, R. D., Campbell, J. S., and Nielson, K. K., 1979, Characterization and formation of submicron particles in coal-fired plants, *Atmos. Environ.* **13**:607–617.

Stinespring, C. D., and Stewart, G. W., 1981, Surface enrichment of aluminosilicate minerals and coal combustion ash particles, *Atmos. Environ.* **15**:307–313.

Tazaki, K., Fyfe, W. S., Sahn, K. C., and Powell, M., 1989, Observations on the nature of fly ash particles, *Fuel* **68**:727–734.

Valkovic, V., 1983, *Trace Elements in Coal*, CRC Press, Boca Raton, Florida.

Valkovic, V., Makjanic, J., Jaksic, M., Popovic, S., Bos, A., Vis, R. D., Wiederspahn, K., and Verheul, H., 1984, Analysis of fly ash by X-ray emission spectroscopy and proton microbeam analysis, *Fuel* **63**:1357–1363.

Vis, R. D., Bos, A., Valkovic, V., and Verheul, H., 1983, The analysis of fly ash particles with a proton microbeam, *IEEE Trans. Nucl. Sci.* **NS-30**:1236–1239.

Watt, F., and Grime, G. W., 1987, *Principles and Applications of High-Energy Ion Microbeams*, Adam Hilger, Bristol.

Chapter 12

Polycyclic Aromatic Hydrocarbons: From External to Internal Environments

Arthur Greenberg

1. INTRODUCTION AND SCOPE

1.1. Introduction

Ambient airborne particulate matter has associated with it thousands of organic compounds (Graedel *et al.*, 1986). These compounds are derived from primary emissions, secondary reactions from gas-phase components to yield particulate-bound species, and reactions on the particulates themselves. In attempting to assess the health impacts of particulate-bound organic matter, one might consider acute or chronic human illness as end points. Our studies have largely focused upon carcinogenic substances thought to be associated with the chronic disease cancer.

Of the numerous potential carcinogens on airborne particulate matter, one class has received the greatest amount of attention. These are the polycyclic aromatic hydrocarbons (PAHs) (National Academy of Sciences, 1983), which are exemplified by naphthalene (**1**), anthracene (**2**), phenanthrene (**3**), pyrene (**4**),

ARTHUR GREENBERG • Department of Chemistry, University of North Carolina at Charlotte, Charlotte, North Carolina 28223.

Combustion Efficiency and Air Quality, edited by István Hargittai and Tamás Vidóczy. Plenum Press, New York, 1995.

fluoranthene (**5**), cyclopenteno[*cd*]pyrene (**6**), tetracene (**7**), benzo[*a*]pyrene (**8**), benzo[*e*]pyrene (**9**), pentacene (**10**), coronene (**11**), and dibenzo[*al*]pyrene (**12**).

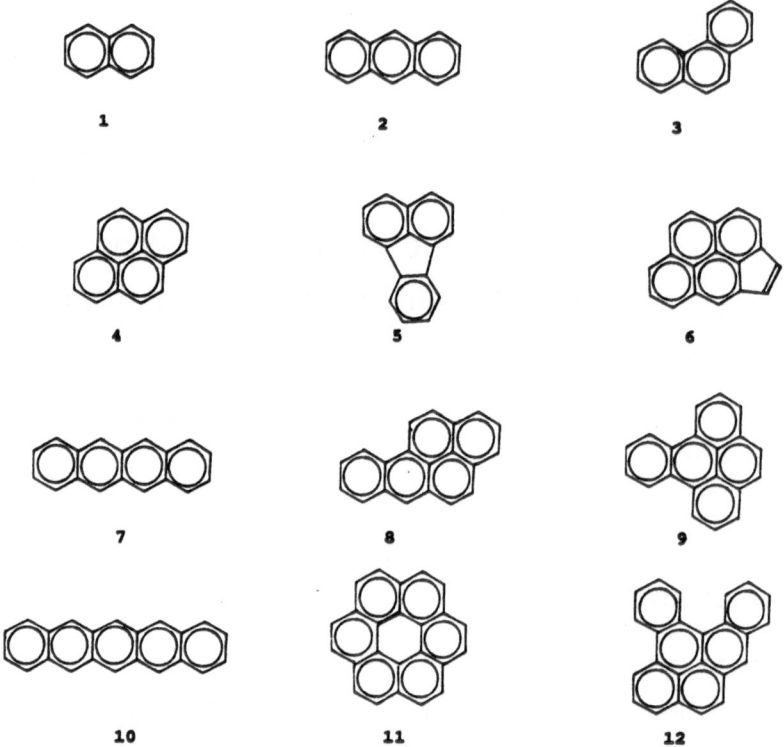

Indeed, the original study of occupation-related scrotal cancer among chimney sweeps by Percival Pott in 1775 identified chimney soot as the causative agent, and the active components were shown, in the 20th century, to be PAHs (National Academy of Sciences, 1983). The specific compounds depicted represent 12 of perhaps 200 or more compounds detected at measurable concentrations (ca. 0.01– 0.1 ng/m^3 in air or roughly 0.01–0.1 ppt since 1 m^3 of air has a mass of 1 kg). In general, as each ring is added (e.g., naphthalene to anthracene), the concentrations are reduced by about an order of magnitude due to the accretion nature of their formation (Badger, 1965).

PAHs are ubiquitous in the environment. They are trace-level by-products of the combustion of organic matter including gas, gasoline, oil, wood, coal, vegetable matter, food, and natural and synthetic polymers (National Academy of Sciences, 1983). Background levels that predate man have been assessed in remote regions (Blumer, 1976; La Flamme and Hites, 1978) and are probably mostly due to natural forest fires. Deep-sea hydrothermal vents appear to create PAHs *de novo* under conditions in which organic material is exposed to hyperbaric superheated water (Simoneit, 1988). Nevertheless, most of the PAH mass in modern populated areas is attributable to human activities (National Academy of Sciences, 1983).

Although the classical work on scrotal cancer involved primarily the dermal exposure route, most modern studies of human exposure to PAHs concentrate on inhalation and ingestion routes. The inhalation route has long been studied, particularly in occupational settings such as aluminum smelting operations, motor vehicle tunnels, and highly urban/industrial settings (Santodonato *et al.*, 1981; Agency for Toxic Substances and Disease Registry, 1990). More recent studies that have focused on total human exposure in ambient environments usually examine combined inhalation and ingestion routes. However, it is worth noting that certain occupations such as asphalt and roofing work may give rise to significant dermal exposure and that a very small population treats psoriasis with dermal applications of coal tar that are rich in PAHs.

In discussing exposure to airborne PAHs, one should note that most work focuses on particulate matter. Total suspended particulate (TSP) is considered to be <45 μ aerodynamic diameter (AD). Inhalable particulate matter smaller than 10 μ AD ("PM10") is most often collected in environmental studies, but the PAHs are really associated with the lung-damaging fine particulate matter (usually classified as <2.5 μ AD) due to their genesis during combustion. The smaller PAHs such as naphthalene, anthracene and phenanthrene are volatile solids which have low equilibrium concentrations on the particulate matter and may be re-volatilized during particulate collections on filters. This has largely not been considered to be a problem in the past since these small PAHs are not carcinogenic. However, they may be collected by backing the filter sampling train with a tube packed with an adsorbent such as Tenax. Intermediate-size PAHs such as pyrene and fluoranthene are found in significant concentrations in both the gas and particulate phases, and the balance between these phases varies seasonally (Greenberg *et al.*, 1985). Larger PAHs, including the carcinogenic benzo[*a*]pyrene, have little volatility and are found almost quantitatively in the particulate phase under ambient conditions, regardless of season. Analysis of PAHs in food is somewhat easier in that exposure is defined by what is extracted from the bulk food without much concern about condensed-phase versus gas-phase distribution as long as the food is stored frozen and unexposed to UV light, which can decompose some PAHs.

Although the primary topic of this chapter is the PAHs, it is worthwhile to note that they are a relatively small percentage of the particulate-bound organic matter and that there are other bioactive substances present. Indeed, if mutagenic activity employing the Ames assay is used as an end point, most of the bioactivity of organic matter extracted from ambient airborne particulate matter is associated with compounds much more polar than PAHs (Nishioka *et al.*, 1988). We will return to this issue later.

1.2. Scope of This Chapter

This chapter will present an outline of our odyssey over a 15-year period in environmental PAH studies in the context of the evolution of the field. Initial studies involved outdoor measurements of a broad group of PAHs as well as one specific PAH—benzo[a]pyrene (BaP). BaP, a highly carcinogenic PAH [its isomer BeP (9) is noncarcinogenic], has often been used as a surrogate for the class. In large part, this is an accident of discovery in that BaP is highly fluorescent under UV light and has unusual retention properties under thin-layer chromatography (TLC) and thus called attention to itself early. There are two complementary strategies in our studies. The first, analysis of a full battery of PAHs (up to 25 compounds in our hands), is extremely work-intensive and strongly limits the number of environmental samples. However, this approach yields information on a variety of carcinogenic and noncarcinogenic PAHs as well as reactive in addition to chemically stable PAHs. The second approach, involving measurement of BaP alone, provides much more limited information and involves an implicit assumption that the fraction of BaP in the PAH mixture is relatively constant but allows almost an order of magnitude more measurements of this significant carcinogen.

From the initial studies of outdoor ambient samples, we became concerned with sources of PAHs and their reactivity and how these relate to environmental measurements. Following the lead of Nishioka et al. (1988), we undertook studies of other fractions beyond the PAHs. More recently, we have addressed the question of human exposure to PAHs and the potential for biomarkers of exposure.

2. RESULTS AND DISCUSSION OF AMBIENT AIRBORNE PAHs

2.1. Studies of Multiple PAHs in Airborne Particulates

An early study (Greenberg et al., 1985) involved comparisons of airborne PAH concentrations at four sites in New Jersey: three urban locations (Newark, Elizabeth, and Camden) and one rural location (Ringwood). Not surprisingly, the urban PAH concentrations were typically about five times the rural concentrations, and the average winter concentrations were typically about fivefold higher than the summer concentrations.

However, if one examines these ratios a bit more carefully, some interesting data come to light. If the winter/summer ratio for BaP is normalized to 1.0 and other PAH ratios are expressed similarly as W/S (really a relative winter/summer ratio), then there is an interesting diversity in values. This is indicated in Table I (Greenberg, 1989). Thus, the high W/S value for pyrene (1.85), which is relatively unreactive [higher electrophile bond localization parameter L_r (Greenberg and Darack, 1987) and medium reactivity class (Nielsen, 1984)], is probably primarily attributable to the aforementioned summer volatilization losses for this tetracyclic

Table I. Ratios (Relative to BaP, Which Is Normalized to 1.00), of Relative Winter/Summer Ratios (W/S) Averaged for Four Periods Along with Selected Chemical Parameters[a]

Compound	Avg W/S	Class	L_R^+
Pyrene	1.85	III	2.19
Benz[a]anthracene	1.36	III	2.05
Cyclopenteno[cd]pyrene	2.30	II	2.08
Benzo[k]fluoranthene	0.73	V	2.13
Benzo[b]fluoranthene	0.79	V	2.28
Indeno[1,2,3-cd]pyrene	0.76	V	2.22

[a]Adapted from Greenberg (1989).

PAH (Greenberg, 1989). In contrast, the even higher W/S value (2.30) for cyclopenteno[cd]pyrene (6), which is much less volatile than pyrene, is strongly suggestive of the high chemical reactivity known for this PAH (Greenberg, 1989). If one defines W/S relative to the very unreactive PAH benzo[b]fluoranthene (L_r, 2.28; class V) as is done in Table II, then the value for the nonvolatile BaP (1.26) indicates reactivity losses, but these are much less than for cyclopenteno[cd]pyrene, with a W/S value in Table II of 3.35 (Greenberg, 1989).

2.2. Studies of BaP in Ambient Airborne Particulates

During the early 1980s, we "piggybacked" a study of airborne BaP on the New Jersey Department of Environmental Protection and Energy (NJDEPE) Air Monitoring Network. Samples were obtained once every six days from 27 of the NJDEPE sites throughout the state (Harkov and Greenberg, 1985) for a 13-month

Table II. Ratios (Relative to Benzo[b]fluoranthene, Which Is Normalized to 1.00), of Relative Winter/Summer Ratios (W/S) Averaged for Four Periods Along with Selected Chemical Parameters[a]

Compound	Avg W/S	Class	L_R^+
Pyrene	2.75	III	2.19
Benz[a]anthracene	1.64	III	2.05
Cyclopenteno[cd]pyrene	3.35	II	2.08
Benzo[k]fluoranthene	0.90	V	2.13
Benzo[a]pyrene	1.26	II	1.96
Indeno[1,2,3-cd]pyrene	0.90	V	2.22

[a]Adapted from Greenberg (1989).

period, yielding about 1700 samples. The averages at each site during 1982 are listed in Table III. Again, one sees an urban/rural ratio as high as 5.

Figure 1 depicts the daily variation in the average value for the BaP concentrations of the 27 sites. This furnishes a higher degree of time resolution of winter/summer differences than provided by the previous study, and the highest winter/summer ratio was over 30 (Harkov and Greenberg, 1985).

2.3. Sources of Seasonal Differences in PAH Concentrations

Why are summer concentrations of BaP much lower than winter concentrations? It is clear from the aforementioned study by Greenberg (1989) that there are

Table III. Mean and Maximum Airborne BaP Concentrations (ng/m^3) at Coastal (C), Northeast Industrial (NEI), Southwest Industrial (SWI), and Rural Interior (RI) Sites in New Jersey during 1982[a]

Site	Zone	Mean BaP (ng/m^3)	Max BaP (ng/m^3)
1. Asbury Park	C	0.32	2.24
2. Atlantic City	C	0.22	1.29
3. Bayonne	NEI	0.44	3.20
4. Bridgeton	C	0.28	1.63
5. Burlington	SWI	0.40	1.51
6. Camden	SWI	0.47	2.44
7. Carteret	NEI	0.33	1.43
8. Cheesequake	C	0.26	1.50
9. Fairlawn	NEI	0.64	3.47
10. Frenchtown	RI	0.37	2.68
11. Hackensack	NEI	0.58	2.65
12. Hackettstown	RI	0.40	2.40
13. Hoboken	NEI	0.73	5.36
14. Jackson	RI	0.27	1.23
15. Jersey City	NEI	0.91	6.28
16. Kean College	NEI	0.76	4.62
17. Metuchen	NEI	0.62	3.85
18. Newark	NEI	0.71	4.09
19. Perth Amboy	NEI	0.50	2.09
20. Phillipsburg	RI	0.81	7.93
21. Pilesgrove	C	0.21	2.24
22. Red Bank	C	0.45	2.77
23. Ringwood	RI	0.19	1.06
24. Tom's River	C	0.27	2.13
25. Trenton	SWI	0.74	3.35
26. Waretown	C	0.27	1.64
27. Woodbury	C	0.54	3.13

[a]Adapted from Harkov and Greenberg (1985).

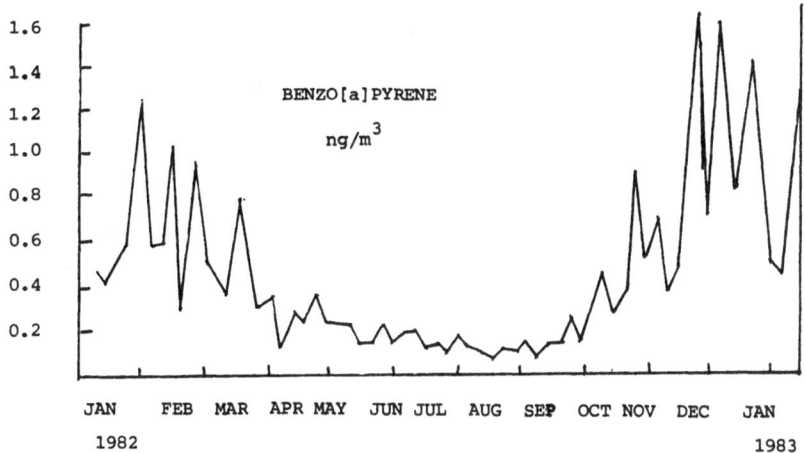

Figure 1. Plot of average BaP concentrations (ng/m³) for 27 New Jersey sites sampled once every six days from January 1982 through January 1983. Adapted from Harkov and Greenberg (1985).

some reactivity losses of BaP during summer when ambient levels of ozone and hydroxyl radical are highest. These losses are not high relative to those for benzo[*b*]fluoranthene. The answer seems to be in the sources of PAH emissions. Table IV lists the results of our study of average BaP emission rates based upon an examination of the literature. Clearly, there is much uncertainty in the actual values, but, surprisingly, it has little effect upon the conclusions. Noteworthy is the relatively low value for coal incineration at a power facility where emissions and operation efficiency are controlled and the extremely high numbers for residential

Table IV. BaP Emission Rates (ng/Btu) for Various Combustion Sources[a]

Fuel	User	Rate (ng/Btu)
Coal	Utilities	5.6×10^{-2}–7.0×10^{-2}
	Residential	0.12–61
Wood	Residential	27–6300
Oil		
Heating	Residential	$<2.6 \times 10^{-3}$
	Residential	$<4 \times 10^{-2}$–5×10^{-2}
Miscellaneous distillates	Commercial/industrial	$<2 \times 10^{-2}$–4.7×10^{-2}
Residual oil	Utility/commercial/industrial	4.3×10^{-4}
Natural gas	Residential	$<2 \times 10^{-2}$
Gasoline	Autos/trucks	0.6
Diesel fuel	Trucks/buses	2.3

[a]Adapted from Harkov and Greenberg (1985).

wood combustion where emissions and operation efficiency are uncontrolled or crudely controlled.

Table V lists total emissions from each source on an annual basis and on a presumed seasonal basis upon multiplication of emission rates by fuel usage in New Jersey. Some of the usage numbers, such as gasoline consumption, are highly accurate. In contrast, the quantity of residential wood combustion seems to be uncertain by about an order of magnitude. Again, these uncertainties do not affect the conclusions, which are: (1) wood combustion accounts for the vast majority of BaP emissions during winter (perhaps over 95%); (2) motor vehicles account for the vast majority of BaP emissions during summer; and (3) the total winter emissions of BaP are more than an order of magnitude greater than the summer BaP emissions (Harkov and Greenberg, 1985). Thus, it would appear that atmospheric sources of PAH rather than summertime reactivity losses are the major reason for the seasonal variation in atmospheric PAH concentrations. The moral, incidentally, would not be to abandon fireplaces and wood stoves, if this were a health issue, but to control source emissions.

2.4. Polar Derivatives of PAHs

The excellent work by Nishioka *et al.* (1988) led our group to the use of bioassay-directed fractionation in order to investigate the relative bioactivity of the PAHs compared to that of other classes of substances on airborne particulates. In

Table V.   Estimated Annual Heating Season (November–March) and Nonheating Season BaP Emissions[a]

Fuel	User	Rate (ng/Btu)	BaP emission (kg) Heating season	Nonheating season
Coal	Utilities	6.1×10^{-2}	1.9	2.6
Coal	Residential	37.7	3.8	
Wood	Residential	227	6129	
Oil				
Heating	Residential	2.6×10^{-3}	0.4	
Miscellaneous distillates	Commercial/industrial	2×10^{-4}	0.1	0.1
Residual oil	Utility/commercial/ industrial	4.3×10^{-4}	<0.1	
Gas (heating)	Residential	2.0×10^{-4}	0.1	
Gasoline	Autos/trucks	0.6	95	133
Diesel fuel	Trucks/buses	2.3	35	50
		Total	6266	186

[a]Adapted from Harkov and Greenberg (1985).

addition, we hoped to isolate and identify fractions, classes, and specific compounds. The biological end point employed was the Ames mutagenicity assay (Ames *et al.*, 1975), performed by Dr. Thomas Atherholt, then of the Coriell Institute for Medical Research. The assay employs special mutant strains of *Salmonella* in order to test their abilities to genetically revert upon exposure to chemicals that cause mutations. This rapid, inexpensive test is often used as a screening tool and an indicator of *potential* carcinogenic activity. The use of the Ames assay on complex mixtures is commonly performed in environmental research although the potential for synergism in the mixture may cloud interpretation.

It is known that airborne PAHs react under ambient conditions to produce nitro-PAHs. Many of these nitro-PAHs are "direct mutagens" in the sense that mammalian microsomal enzyme extracts need not be added in order to observe mutagenic activity. In contrast, mutagenic PAHs are almost universally inactive in the Ames assay without enzymatic activation. Furthermore, nonmutagenic PAHs such as pyrene are converted to mutagenic nitro-PAHs under ambient conditions, and, indeed, some dinitropyrenes are termed "supermutagens." The chemistry leading to the nitro-PAHs is quite interesting.

Nitration of pyrene with nitric acid yields 1-nitropyrene (**13**), consistent with all simple expectations of pyrene reactivity. 1-Nitropyrene is found in considerable abundance in freshly collected diesel particulate (Siak *et al.*, 1985). However, in ambient airborne particulate the dominant species is the isomer 2-nitropyrene (**14**). The explanation for this seeming anomaly is that initial attack on pyrene is via hydroxyl radical, generally considered to be the dominant reaction sink for most PAHs, followed by reaction with NO_2 and subsequent elimination of water (Scheme 1) (Pitts, 1987). A similar explanation is invoked to rationalize the

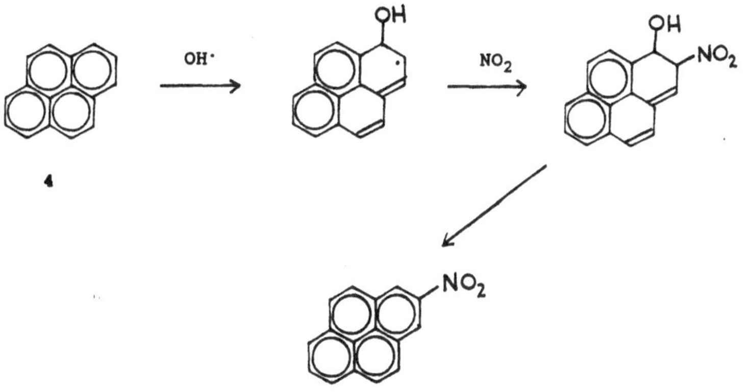

Scheme 1

observation of 2-nitrofluoranthene (15) as a significant component in air as opposed to 3-nitrofluoranthene, which is produced in nitric acid (Pitts, 1987). The most fascinating aspect is that it is the significant volatility of pyrene and fluoranthene that permits apparent gas-phase chemistry to yield 14, for example, rather than condensed-phase chemistry that might yield 13. In addition, N_2O_5 also produces 14 and 15 in the gas phase and in nonpolar media (in polar media and in the crystalline state N_2O_5 exists as $NO_2^+NO_3^-$ and produces 13) and may also be a factor in nighttime generation of 14 and 15 (Pitts, 1987).

13   14   15

Another class of mutagenic PAH derivatives produced in the ambient air are quinones such as 16 (Pierce and Katz, 1976). Another interesting observation is the formation of 17 via photochemical reaction of BaP with singlet oxygen (Lee-Ruff et al., 1990). The study of Nishioka et al. (1988) discovered the importance of mutagenic hydroxynitro-PAHs such as 18; more recently, the discovery of extremely mutagenic species such as 2-nitro-6H-dibenzo[b,d]pyran-6-one (19) on

16   17   18   19

airborne particulate matter (Helmig et al., 1992) has supported the importance of compounds more polar than the PAHs and nitro-PAHs, and this is consistent with the findings of our own studies (Greenberg et al., 1993a).

3. ROUTES OF HUMAN EXPOSURE TO PAHs

In an occupational environment such as aluminum smelting, inhalation is considered to be the dominant human exposure route to PAHs. However, in ambient environments, even in polluted urban areas, ingestion may play a significant role in human exposure. Table VI provides some concentrations of BaP found in foods under different conditions of preparation (Greenberg et al., 1990). It is apparent that some foods, particularly barbecued and smoked meats, can have fairly significant concentrations of BaP (and other PAHs). Interestingly enough,

Table VI.   Levels (wt/wt) of BaP in Various
Foodstuffs[a]

	BaP concentration (ppb)
A. Smoked food products	
Herring (dried)	1.0
Salami	2.0
Frankfurters	2.0
Barbecued beef	3.3
Smoked whiting	6.6
B. Beverages	
Coffee powder	0.8
Coffee infusion	0.01
Tea leaves	9.5
Tea infusion	0.02
Milk	ND
C. Char-broiled food	
Porterhouse steak	3.0
Hamburger	20.0
Beef	21.5
Pork	29.3
Lamb	10.5
D. Vegetables, fruits	
Cereals	0.2–4.1
Spinach	6.9
Apples	0.1–0.5
Tomatoes	0.2
Soybean	3.1

[a]Adapted from Greenberg et al. (1990).

we found relatively high levels of the reactive PAH cyclopenteno[cd]pyrene (6), and other PAHs that react rapidly in the atmosphere, in food (Greenberg et al., 1993b).

In a collaborative study in Phillipsburg, New Jersey, we found that exposure to BaP via the ingestion route was highly variable and diet dependent. In a small, nonstatistical study, we found that more people obtained larger BaP dose from their diet than through inhalation and that, if one eliminates smokers, workers in industries such as smelting, or people with special activities (e.g., soldering), high BaP exposures were usually associated with diet (Butler et al., 1993) (Table VII).

It is wise to place PAHs in food which arise from environmental contamination, processing, and cooking in the context of other carcinogens present in food (National Institute of Environmental Health Sciences, 1986). Specifically, foods

Table VII. Estimate of Mean Daily Inhalation and Food Ingestion
of BaP for Each of Ten Homes during the First Phase of the Total
Human Environmental Exposure (THEES) Project[a]

Home ID	Week no.	Inhaled BaP (ng/day)	Ingested BaP (ng/day)
010	1	19.6	205
	2	13.8	65
020	1	31.0	177
	2	20.7	63
030	1	18.6	9
	2	16.0	10
040	1	16.7	2
	2	16.7	17
050	1	11.1	1
	2	9.3	55
060	1	12.7	7
	2	11.2	120
070	1	13.7	3
	2	15.0	15
080	1	36.9	21
	2	35.4	89
090	1	24.0	241
	2	31.9	572
100	1	17.4	25
	2	45.8	36

[a]Adapted from Butler et al. (1993).

are contaminated with natural carcinogens such as aflatoxins and safrole (a
flavoring ingredient) and anticarcinogens (seemingly important in garlic and green
tea among many food products). The cooking of proteins involves formation of
highly mutagenic heterocyclic amines. The processing of foods is likely to add or
form carcinogens, and it is not clear how much of a risk all of this represents nor
whether PAH ingestion is a significant health risk. The general guidelines offered
by the experts are the Greek credo of "all things in moderation" accompanied by
diets low in fat, high in fruits and vegetables, and buttressed by antioxidants such
as vitamins C and E and, of course, no smoking or exposure to sidestream smoke.

4. BIOMARKERS OF HUMAN EXPOSURE TO PAHs

When foreign organic substances (xenobiotics) are absorbed by humans, they
are eliminated directly or through metabolism. Humans have evolved to metabo-
lize nonpolar hydrocarbons to more water-soluble oxygenated species (phase I
metabolism) which are often conjugated (phase II metabolism) to form highly

water-soluble species prior to elimination. Most of the metabolic activation occurs in liver microsomes. The mechanistic details are extremely complex. *Some* of the major metabolites from BaP metabolism are depicted in Scheme 2. There are

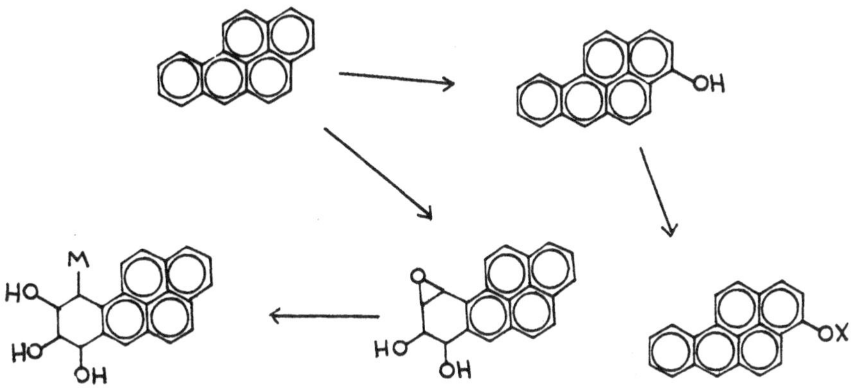

Scheme 2

perhaps 20 or so metabolites which may arise from a single PAH. Obviously, the typical environmental "gemisch" of ca. 200 *measurable* PAHs may give rise to thousands of urinary metabolites at trace levels.

If one wants to measure total human exposure to PAHs, effective biological dose is the key issue. Perhaps the best measure at present is the determination of DNA adducts of PAH metabolites (Santella *et al.*, 1984). However, these methods, which include ^{32}P post-labeling and ELISA are still in the development stage, with PAH specificity representing a significant limitation at present. Measurements of PAH metabolites in feces are invasive. Thus, measurements of urinary metabolites are highly desirable. The identification of a biomarker is highly desirable. The compound 1-hydroxypyrene (20) has found considerable application in the past (Jongeneelen *et al.*, 1985) and may well be useful as one of a battery of urinary biomarkers (Grimmer *et al.*, 1994). The nice aspect of 20 is that it is the dominant metabolite of pyrene, its glucuronide and sulfate conjugates are readily en-zymatically deconjugated to yield 20, and, as noted earlier, concentrations of this tetracyclic PAH are usually considerably higher than those of the pentacyclic BaP. Thus, 20 is readily detected in urines. Unfortunately, the corresponding derivative of BaP, the 3-hydroxy species considered most abundant in this group, is not usually detected in urines except for those derived from individuals having massive exposures. This is because 3-hydroxy-BaP is one of perhaps 20 metabolites derived from a PAH present at considerably lower levels than pyrene. However, our group (Ouyang *et al.*, 1994) as well as Weston *et al.* (1994) have discovered that acidification of urine with HCl yields 7,8,9,10-tetrahydroxy-7,8,9,10-tetrahydrobenzo[*a*]pyrene (21), which shows promise as a biomarker for

BaP exposure. We feel that **21** is likely to be derived from highly polar mercapturate conjugates derived from the 7,8-dihydrodiol-9,10-epoxide (**22**), which is considered to be the ultimate carcinogen derived from metabolism of BaP. Mer-

capturates are not enzymatically deconjugated by the glucuronidase–sulfatase mixture that is effective for producing 1-hydroxypyrene.

In conclusion, more work has to be accomplished in identifying biomarkers for PAH exposure. However, this must be done in the broader context of understanding the total picture of human exposure to carcinogens present in primary emissions, formed by environmental reactions, or present in foods.

REFERENCES

Agency for Toxic Substances and Disease Registry, 1990, Toxicological Profile for Benzo(*a*)pyrene, ATSDR/TP-88/05, U.S. Department of Health & Human Services, Public Health Service, Atlanta, Georgia.

Ames, B. N., McCann, J., and Yamasaki, E., 1975, Methods for detecting carcinogens with the *Salmonella*/mammalian-microsome mutagenicity test, *Mutat. Res.* **31**:347–364.

Badger, G. M., 1965, Formation of polycyclic aromatic hydrocarbons, *Prog. Phys. Org. Chem.* **3**: 1–40.

Blumer, M., 1976, Polycyclic aromatic compounds in nature, *Sci. Am.* **234**(3):34–45.

Butler, J. P., Post, G. B., Lioy, P. J., Waldman, J. M., and Greenberg, A., 1993, Assessment of carcinogenic risk from personal exposure to benzo(*a*)pyrene in the Total Human Environmental Exposure Study (THEES), *J. Air Waste Manage. Assoc.* **43**:970–977.

Graedel, T. E., Hawkins, D. T., and Claxton, L. D., 1986, *Atmospheric Chemical Compounds: Sources, Occurrence and Bioassay*, Academic Press, New York.

Greenberg, A., 1989, Phenomenological study of benzo[*a*]pyrene and cyclopenteno[*cd*]pyrene decay in ambient air using winter/summer comparisons, *Atmos. Environ.* **23**:2797–2799.

Greenberg, A., and Darack, F., 1987, Atmospheric reactions and reactivity indices of polycyclic aromatic hydrocarbons, in *Molecular Structure and Energetics*, Vol. 4, *Biophysical Aspects* (J. F. Liebman and A. Greenberg, eds.), pp. 1–47, VCH Publishers, New York.

Greenberg, A., Darack, F., Harkov, R., Lioy, P., and Daisey, J., 1985, Polycyclic aromatic hydrocarbons in New Jersey: A comparison of winter and summer concentrations over a two-year period, *Atmos. Environ.* **19**:1325–1339.

Greenberg, A., Luo, S., Hsu, C. H., Creighton, P., Waldman, J., and Lioy, P. J., 1990, Benzo[*a*]pyrene in composite prepared meals: Results from the THEES (Total Human Exposure to Environmental Substances) study, *Polycycl. Aromat. Cmpds.* **1**:221–231.

Greenberg, A., Lwo, J. H., Atherholt, T. B., Rosen, R., Hartman, T., Butler, J., and Louis, J., 1993a,

Bioassay-directed fractionation of organic compounds associated with airborne particulate matter: An interseasonal study, *Atmos. Environ.* **27**A:1609–1626.

Greenberg, A., Hsu, C. H., Rothman, H., and Strickland, P. T., 1993b, PAH profiles of charbroiled hamburgers: Pyrene/B[a]P ratios and presence of reactive PAH, *Polycycl. Aromat. Cmpds.* **3**: 101–110.

Grimmer, G., Dettbarn, G., Naujack, K. W., and Jacob, J., 1994, Relationship between inhaled PAH and urinary excretion of phenanthrene, pyrene and benzo[a]pyrene metabolites in coke plant workers, *Polycycl. Aromat. Cmpds.* **5**:269–277.

Harkov, R., and Greenberg, A., 1985, Benzo(a)pyrene in New Jersey—results from a twenty-seven site study, *J. Air Pollut. Control Assoc.* **35**:238–243.

Helmig, D., Arey, J., Harger, W. P., Atkinson, R., and Lopez-Cancio, J., 1992, Formation of mutagenic nitrodibenzopyranones and their occurrence in ambient air, *Environ. Sci. Technol.* **26**:622–624.

Jongeneelen, F. J., Anzion, R. B. M., Leijdekkers, C. M., Box, R. P., and Henderson, P. T., 1985, 1-Hydroxypyrene in human urine after exposure to coal tar and coal tar derived product, *Int. Arch. Occup. Environ. Health.* **57**:47–55.

La Flamme, R. E., and Hites, R. A., 1978, The global distribution of polycyclic aromatic hydrocarbons in recent sediments, *Geochim. Cosmochim. Acta* **42**:289–304.

Lee-Ruff, E., Kazarians-Moghaddam, H., and Katz, M., 1990, in *Polynuclear Aromatic Hydrocarbons: The Tenth International Symposium*, Battelle Press, Columbus, Ohio, pp. 519–534.

National Academy of Sciences, 1983, *Polycyclic Aromatic Hydrocarbons*, National Academy Press, Washington, D.C.

National Institute of Environmental Health Sciences, 1986, *Environmental Health Perspectives*, Vol. 67, Research Triangle Park, North Carolina.

Nielsen, T., 1984, Reactivity of polycyclic aromatic hydrocarbons toward nitrating species, *Environ. Sci. Technol.* **18**:157–163.

Nishioka, M. G., Howard, C. C., Contos, D. A., and Ball, L. M., 1988, Detection of hydroxylated nitro aromatic and hydroxylated nitro polycyclic aromatic compounds in an ambient air particulate extract using bioassay-directed fractionation, *Environ. Sci. Technol.* **22**:908–915.

Ouyang, Z., Greenberg, A., Kwei, G. Y., Kauffman, F. C., and Faria, E., 1994, A rapid assay for urinary metabolites of B[a]P, *Polycycl. Aromat. Cmpds.* **5**:259–268.

Pierce, R. C., and Katz, M., 1976, Chromatographic isolation and spectral analysis of polycyclic quinones. Application to air pollution analysis, *Environ. Sci. Technol.* **10**:45–51.

Pitts, J. N., Jr., 1987, Nitration of gaseous polycyclic aromatic hydrocarbons in simulated and ambient urban atmospheres: A source of mutagenic nitroarenes, *Atmos. Environ.* **21**:2531–2547.

Santella, R. M., Lin, C. D., Cleveland, W. L., and Weinstein, I. B., 1984, Monoclonal antibodies to DNA modified by benzo[a]pyrene diol epoxide, *Carcinogenesis* **5**:373–377.

Santodonato, J., Howard, P., and Basu, D., 1981, Health and ecological assessment of polynuclear aromatic hydrocarbons, *J. Environ. Pathol. Toxicol.* **5**:1–77.

Siak, J. S., Chan, T. L., Gibson, T. L., and Wolff, G. T., 1985, Contribution to bacterial mutagenicity from nitro-PAH compounds in ambient aerosols, *Atmos. Environ.* **19**:369–376.

Simoneit, B. R. T., 1988, Petroleum generation in submarine hydrothermal systems: An update, *Can. Mineral.* **26**:827–840.

Weston, A., Santella, R. M., and Bowman, E. D., 1994, Detection of polycyclic aromatic hydrocarbon metabolites in urine from coal tar treated psoriasis patients and controls, *Polycycl. Aromat. Cmpds.* **5**:241–247.

Chapter 13

Quality of Indoor Air

Istvan L. Gebefuegi

1. INTRODUCTION

The quality of indoor air is influenced by chemical contamination from several sources. The main sources of indoor air pollution are infiltration of the outdoor air, volatile compounds from contaminated ground, and emissions from building materials. Naturally, the indoor use and generation of chemicals are relevant sources too. The research activities of the Institute for Ecological Chemistry have been focused on the behavior of semivolatile organic compounds (SVOCs) coming from wood preservation paints used indoors, case studies following problems dealing with indoor air quality, and the occurrence of indoor volatile organic compounds (VOCs) in the Bavarian region.

Istvan L. Gebefuegi • Institute for Ecological Chemistry, GSF Research Center, Neuherberg, D-85758 Oberschleissheim, Germany.

Combustion Efficiency and Air Quality, edited by István Hargittai and Tamás Vidóczy. Plenum Press, New York, 1995.

2. INDOOR BEHAVIOR OF SVOCs

Measurements of semivolatile chemicals in real buildings indicate large variations in the indoor air concentrations (Gebefuegi and Korte, 1988a,b). Systematic studies show that semivolatiles do not build up steady-state air concentrations but that great variations in concentration are possible because of the influence of outdoor–indoor temperature variations and the change of the relative air humidity.

Small-chamber studies show accumulation phenomena of the semivolatile organic compounds pentachlorophenol (PCP) and Lindane. The SVOCs coming from prepared wood samples contaminated the adsorptive surfaces of textile fibers via air transport. The average air concentration of PCP was 0.6 $\mu g/m^3$, resulting in a contamination on the cotton fiber surface of up to 32 mg/kg within 48 hours. For Lindane, a concentration in air of 1.6 $\mu g/m^3$ resulted in up to 8 mg/kg on the fiber surface. According to these small-chamber studies, reduction of the average air concentration does not lead to a decrease of the contamination of the fiber surfaces. The observed phenomenon can therefore be classified as accumulation (Gebefuegi, 1989).

With this knowledge, we began to collect sedimented indoor particles for analysis of SVOCs. Electron microscopic investigations show that the larger part of the particles comes from textile fibers. Therefore, a good accumulation behavior was suspected for indoor particles, that is, the contamination of the indoor air and the indoor surfaces should lead to measurable residues on the particles. The particles were collected with commercial vacuum cleaners by our laboratory or by the inhabitants.

The samples were worked up for biocide analysis by liquid extraction and analyzed by capillary gas chromatography (GC) with electron capture and mass spectrometric detection. The results showed the ubiquitous presence of PCP and Lindane on sedimented indoor particles (Gebefuegi, 1993). After treatment of a home by professional pest control with insecticide agents such as permethrine, residues were found both in the various materials and on the collected particles. Table I shows the permethrine residue values measured three years after the treatment.

The relatively high residue values three years after the treatment was very surprising. The stability of permethrine under outdoor conditions would not allow for such high residues. Permethrine is well known as a nonpersistent compound under outdoor conditions. The results listed in Table I suggest a high persistence under indoor conditions, however.

3. VOCs IN AN OFFICE BUILDING—A CASE STUDY

A ten-year-old office building in downtown Munich was investigated regarding the occurrence of organics in the air inside the building in view of the

Table I. Results of Residue Analysis of Permethrine, Three Years after Professional Treatment

	Permethrine concentration (mg/kg)	
	cis	trans
Wallpaper, living room	120	390
Wallpaper, bedroom	118	400
Furniture, bedroom	8	22
Wood sample, wardrobe	15	44
Carpet, living room	2	6
Sedimented particle (average for the house)	46	130
Sedimented particle (living room, children's room, cellar)	25	60
Plaster (1 cm under wallpaper)	2	3

possibility of human reactions to low levels of VOCs (Molhave et al., 1986). An inventory was carried out in the presence of VOCs from indoor and outdoor natural and man-made sources by collecting VOCs on active charcoal tubes and analyzing the samples by capillary gas chromatography–mass spectrometry (GC-MS). All commercial cleaning products used in this office building were analyzed for volatile organic compounds by a simple laboratory off-gassing experiment.

3.1. Materials and Methods

The building had been in use for ten years and was a typical office building with computer terminals, photocopy equipment, and laser printers as well as classic typewriters and desks in compartmented open-plan office style. On the other hand, smaller separated conference rooms were also available. The building was equipped with a very suitable air conditioning system with an air exchange rate of 7.5 h^{-1} in the compartments.

Indoor air samples were collected using an air sampling pump at the working desk in any of the compartments. Air was pumped at 1 liter per minute through sorbent tubes containing active charcoal for the adsorption of VOCs. Collected samples were analyzed for VOCs by desorbing the VOCs with 1 ml of CS_2 and analyzing them by capillary GC with a flame ionization detector. The identification of the chemicals was carried out by capillary GC-MS, and the quantification by the use of authentic reference substances.

The investigation of the chemical products used in the office building (cleaning agents) was made in a simple laboratory experiment. An undiluted 100-µl aliquot of each product was placed in a 3-liter glass bulb. The headspace was sampled at 22°C with 5 liters of air on active charcoal adsorption tubes, in a similar manner to that described by Kreuzig *et al.* (1988), and analyzed like air samples for VOCs.

3.2. Results and Discussion

3.2.1. VOC Analysis

The inventory of the identified volatile compounds with the range values of 132 indoor samples and the means of 28 outdoor samples is summarized in Table II. The pathway of aromatics was found to be the "fresh air" intake near the parking garage entrance at the street level. The time-dependent concentrations of the VOCs in the indoor air show clearly maximums in the main arrival and departure intervals of the company cars (fuel-powered, without catalytic converters).

3.2.2. Product Analysis

All chemical products (cleansers) used in the office area were collected and analyzed for volatile compounds. Table III shows an overview of the materials and their emittable volatile organic components. It should be mentioned that a number of chemicals detected in the indoor air may be influenced by the use of the cleansers. These cleansers contain the terpenes limonene, *p*-cymene, and α-pinene, typical hydrocarbons and some aromatics. The recommendation to replace the cleansers designated A, C, D, and F in Table III led to a significant reduction of terpene concentrations in the indoor air of the building. The concentrations of 1,4-diethylbenzene/butylbenzene, undecane, and decane were reduced too.

Thus, the concentrations of the observed volatile organic compounds in the indoor air were found to be influenced by contaminated fresh air and by the use of chemical products containing volatile organic compounds. Changing the cleansers and moving the location of the fresh air intake resulted in a significantly lower concentration of VOCs.

Although the chemical products (cleansers) are responsible for high indoor concentration values, we suspect that VOCs adsorbed on surfaces of materials such as textiles or carpets may show temporary higher indoor concentrations than outdoors, analogously to earlier results regarding semivolatiles (Korte and Gebefuegi, 1987). Another publication reporting higher indoor concentrations of VOCs suggests the importance of indoor or attached sources, but with unexplained factors (Cohen *et al.*, 1989).

4. OCCURRENCE OF VOCs IN RESIDENTIAL BUILDINGS

The energy crisis led to a trend toward airtight buildings and reduced ventilation in the upper Bavarian region, as elsewhere. The climatic situation in

Table II.   Range of VOC Concentrations Measured inside the Office
Building and Simultaneous Outdoor Measurements

	Concentration (mg/m³)		
		Outdoors	
Compound	Indoors	8:00 A.M.	11:00 A.M.
Hexadecane	0.4–2.7	n.d.[a]	1.60
Dodecane	n.d. –9.5	6.60	7.49
Camphene	n.d. –1.4	0.36	0.35
1,2,3,5-Tetramethylbenzene	n.d. –1.7	0.01	n.d.
1,2,4,5-Tetramethylbenzene	n.d. –3.1	n.d.	n.d.
Undecane	n.d. –2.4	0.58	0.62
1-Undecene	n.d. –0.8	n.d.	n.d.
1,2-Diethylbenzene	0.2–3.5	n.d.	0.27
γ-Terpinene	n.d. –0.5	n.d.	n.d.
1,4-Diethylbenzene/butylbenzene	0.5–1.7	n.d.	0.51
1,3-Diethylbenzene	n.d. –0.5	n.d.	0.67
Limonene	0.40–84.0	n.d.	0.44
p-Cymene	0.01–4.5	n.d.	0.01
1,2,3-Trimethylbenzene	1.1–70.0	0.89	1.17
α-Terpinene	n.d. –1.0	n.d.	n.d.
Decane	2.5–52.0	3.35	3.48
Myrcene/1-decene	1.0–75.8	n.d.	74.81
1,2,4-Trimethylbenzene	n.d. –65.36	44.82	n.d.
2-Ethyltoluene	0.8–1.3	1.69	0.99
β-Pinene	0.1–1.23	n.d.	0.21
1,3,5-Trimethylbenzene	n.d. –14.0	34.95	n.d.
4-Ethyltoluene	7.3–17.54	n.d.	17.32
3-Ethyltoluene	12.0–25.5	49.88	24.72
Propylbenzene	1.2–1.66	2.80	1.66
δ-3-Carene	n.d. –0.4	n.d.	n.d.
α-Pinene	n.d. –10.9	n.d.	0.74
Nonane	3.5–24.5	4.90	4.13
1-Nonene	n.d. –0.24	n.d.	n.d.
o-Xylene	9.8–34.51	48.92	25.89
m/p-Xylene	2.5–22.94	15.09	7.96
Ethylbenzene	13.4–28.6	44.56	24.61
Octane	1.0–4.0	4.62	3.04
1-Octene	n.d. –2.09	n.d.	n.d.
Toluene	14.5–60.0	96.23	54.95
Heptane	0.9–5.0	4.68	3.65
1-Heptene	3.2–15.5	17.58	14.85

[a]n.d., Not detectable; detection limit (average): 0.01 mg/m³.

Table III.   List of Chemical Products Used in the Office and the Main Analyzed Volatile Components

Product	Main components
A. Floor wax	1,4-Diethylbenzene, butylbenzene, decane, 1,2,5-trimethylbenzene, 1-nonene, ethylbenzene, 2-*m/p*-xylene, limonene (5500 mg/m³)
B. Carpet cleanser	Aqueous solution with ammonia, no volatile organics
C. Ceiling cleanser	Limonene (20500 mg/m³), *p*-cymene, (3300 mg/m³), undecane, α-pinene (fragrance?)
D. Ceramic-floor cleanser	Limonene (20500 mg/m³), *p*-cymene (3300 mg/m³)
E. Stone-floor cleanser	Heptene, undecane, nonane, and decane (120–1270 mg/m³)
F. Desk cleanser	Strong smell of limonene, undecane, and *p*-cymene emissions
G. Glass plate cleanser	Heptane, ammonia but no fragrance

this part of Europe results in a long heating period. Poor indoor air quality is often associated with low air exchange rates and higher concentrations of volatile organic chemicals and sensory irritation.

The presence of volatile chemicals in the indoor environment is often discussed since residues can be caused by the building technology, e.g., by glues and coats, or by the use of household chemicals. Infiltration from the outdoor air is an additional source. Natural ventilation occurs in the typical Bavarian housing technology with very low air exchange rates, typically between 0.1 and $0.5h^{-1}$.

Exposure to VOCs has often been suspected to be the reason for complaints and health effects, but the VOC level can also be an indication of poor indoor air quality, resulting from air tight buildings and insufficient ventilation.

In epidemiological studies related to complaints of the respiratory tract versus outdoor air quality, it is necessary to know the extent of exposure to VOCs in indoor environments. In this study an inventory of charcoal-trapped organics was made in 180 private houses of Bavarian families (all with school children) with the goal of obtaining information about the indoor air quality. All examined objects were older than six years. The sampling technique employed was developed for outdoor background level measurements of VOCs and is useful for long-term and higher volume measurements also. The indoor results were compared with outdoor values analyzed by the same procedures.

4.1. Materials and Methods

Charcoal tubes were prepared in our laboratory in the following manner. Borosilicate glass tubes approximately 12 cm in length with an i.d. of 0.6 cm were each filled with 1 g of Merck charcoal (Merck, Darmstadt, no. 9631.0100, 0.2–0.5 mm). The filled tubes were conditioned under vacuum at 240°C for 12 h and then ventilated with purified nitrogen. Using 1 ml of CS_2 for desorption and high-resolution capillary GC with flame ionization detection (GC-FID), the tubes did not give any interferences. These high-quality and high-capacity charcoal tubes were successfully used for outdoor background concentrations studies (Kreuzig et al., 1986, 1988). The breakthrough efficiency and suitability for long-term sampling periods were verified.

The VOCs collected in the charcoal tube were desorbed with 1 ml of CS_2 (Merck Uvasol no. 2213.0500, max. 1 ppm benzene; Aldrich HPLC no. 27.066-0; Promochem no. 9056, free from aromatic compounds, max. 0.5 ppm) with 10 ng of the cyclododecane internal standard.

The triple high-resolution capillary gas chromatography was carried out on a Hewlett-Packard 5890 A gas chromatograph with the following parameters: Capillary column: DB 5, 30-m column; i.d. 0.32 mm, film thickness 0.25 mm (J & W Scientific, Inc., Folsom, California, USA); carrier: H_2, 3 ml/min; oven temperature program: 35°C, 5 min, 2°C/min to 50°C, 4°C/min to 120°C, 8°C/min to 240°C. Data registration and processing were performed with the 3350 Hewlett-Packard laboratory automation system (LAS). Detection limits were, on average, 0.005 mg/m³. The gas-chromatographic results were verified by GC-MS (Finnigan ITS 40, same chromatographic parameters as for GC-FID).

Active sampling was carried out by using constant-flow volumetric pumps (Desaga GS 312 or SKC 224-PCXR7KB) with a sampling rate of 1 liter/min for 2–3 days. The typical sample volume was 2000 liters. The sampling site was the living room or the "center of the home activity," as defined by the volunteers.

4.2. Results and Discussions

The results are shown in Table IV. The levels of each individual compound were quantified by the use of authentic standard mixtures for aliphatic, aromatic,

Table IV. Average, Median, and Maximum Values of Total Volatile Organic Compounds (TVOC) and the Subsums of Aliphatic, Aromatic, and Natural Compounds[a,b]

	Concentration (μg/m³)		
	Mean	Median	Maximum
Σ Aliphatics	19	4.6	284
Σ Aromatics	53	28	626
Σ Terpenes	18	7	205
TVOC[c]	90	45	886

[a]n = 180.
[b]Detection limit (average), 0.005 μg/m³.
[c]Sum of individual measured and calibrated components.

and natural terpene compounds. We used individually calibrated values for every compound to develop sum values from each of the sampled objects. The sums of the individually calibrated and calculated values are significantly lower than the values for total volatile organic compounds (TVOC), determined as the total integrated area from the FID signal (Wolkoff *et al.*, 1992).

The highest detected concentrations are those of the aromatic hydrocarbons. Table V shows the mean, median, and maximum values of the individual aromatic hydrocarbon components measured in the study objects.

Aliphatic hydrocarbons can come from indoor sources. They can also be due to infiltration from the outdoor air since they are components of diesel car engine emissions. The levels of aliphatic hydrocarbons are, however, significantly lower than those of aromatics, as shown in Table VI.

Volatile organic chemicals of natural origin are ubiquitous. They are mainly monoterpene type hydrocarbons, emitted by coniferous trees or by industrial products containing turpentine-like solvents. In particular, limonene is very often used as a fragrance in household chemicals. Table VII summarizes the concentration values for the monoterpene hydrocarbons α-pinene, δ-3-carene, *p*-cymol, and limonene; these compounds were always present in all air samples.

The main compounds detected in the study objects are listed in Tables IV–VII. Active sampling followed by solvent extraction was chosen for VOC measurements in order to allow for greater flexibility regarding time resolution of sampling and accurate analytical procedures. GC-FID quantification and GC-MS control provided correct analytical findings. The use of solvent extracts allowed multiple gas-chromatographic analyses with high precision for each individual compound. The relatively large amounts of sample provided the opportunity for

Table V.   Mean, Median, and Maximum Concentrations of the Aromatic Hydrocarbons[a,b]

	Concentration ($\mu g/m^3$)		
	Mean	Median	Maximum
Benzene	11.39	2.21	445.04
Toluene	31.61	13.38	372.76
Xylene	5.88	3.25	55.09
Alkyl aromatics[c]	1.49	0.61	13.35
Others	3.76	0.01	55.26

[a] $n = 180$.
[b] Detection limit (average), 0.005 $\mu g/m^3$.
[c] Propylbenzene, 4-Ethyltoluene, 1,3,5-trimethylbenzene, 2-ethyltoluene, 1,2,3-trimethylbenzene, 1,3-diethylbenzene, 1,4-diethylebenzene, 1,2-diethylbenzene, 1,2,3,5-tetramethylbenzene, and 1,2,4,5-tetramethylbenzene.

Table VI.   Mean, Median, and Maximum
Concentrations of the Aliphatic Hydrocarbons[a,b]

	Concentration (μg/m^3)		
	Mean	Median	Maximum
1-Heptene + heptane	2.51	0.71	33.96
1-Octene + octane	1.64	0.51	12.21
1-Nonene + nonane	1.45	0.01	18.45
1-Decene + decane	3.96	0.41	47.97
1-Undecene + undecane	5.57	0.27	141.66
Dodecane	3.45	0.01	96.49
Hexadecane	1.20	0.15	13.29

[a]n = 180.
[b]Detection limit (average), 0.005 μg/m^3.

quantification of all the 30 main compounds. GC-MS controls were carried out for every fifth sample.

Outdoor background concentration measurements were performed in the late eighties in urban and rural regions (Kreuzig *et al.*, 1986, 1988). The occurrence of anthropogenic and natural volatile chemicals in the vertical profile and their seasonal variation in forest and urban regions were studied. The results showed that the concentrations of volatile anthropogenic hydrocarbons were influenced by the distance between the forest regions and urban centers with heavy traffic. Moreover, in summer, the concentrations of α-pinene exceeded the concentrations of toluene in the forest regions. Table VIII summarizes the outdoor concentration values of aromatic, aliphatic, and monoterpene hydrocarbons in rural (forest) and urban (center of the city of Munich) regions.

Table VII.   Mean, Median, and
Maximum Concentrations of the
Monoterpene Hydrocarbons[a,b]

	Concentration (μg/m^3)		
	Mean	Median	Maximum
α-Pinene	4.67	1.61	57.01
δ-3-Carene	4.38	0.01	195.10
p-Cymol	0.41	0.00	4.14
Limonene	8.12	2.86	62.38

[a]n = 180.
[b]Detection limit (average), 0.005 μg/m^3.

**Table VIII.   Outdoor Air Concentration Values of Aromatic,
Aliphatic, and Monoterpene Hydrocarbons**[a]

| | Concentration (μg/m^3) | | | |
| | Forest sampling site | | Urban sampling site | |
	Range	Mean	Range	Mean
Σ Monoterpenes	0.09–3.80	1.21	0.17–1.21	0.57
Σ Aliphatics	0.38–2.62	1.31	0.07–4.49	2.27
Σ Aromatics	0.32–10.11	3.93	8.04–47.20	22.21

[a]Results from 1987–88.

A comparison of the values in Table VIII with the indoor concentration values measured in 1992 shows that all indoor values exceed the outdoor mean values. The aromatics and alkylbenzenes are well known as anthropogenic chemicals, coming from vehicle emissions (i.e., from gasoline burning in a spark-ignition engine). The main components of these emissions are benzene, alkylbenzene, toluene, ethylbenzene, m/p-xylene, o-xylene, 3-ethyltoluene, and 1,2,4-trimethylbenzene (Oelert, 1974). The highest maximum value for benzene, 445 mg/m^3, was identified as evaporation from a car fuel reservoir without a charcoal filter in a garage ventilating directly into the indoor air. Benzene, a carcinogenic compound, is nowadays only present in lead-free car fuel. In paint solvents, benzene has been substituted by other aromatics during the past ten years. It was generally found that the main source of aromatic compounds in indoor air is the infiltration of polluted outdoor air.

The constant findings of higher indoor than outdoor concentrations can be explained by indoor sources or accumulation phenomena from infiltrated chemicals. The retention of infiltrated chemicals such as by adsorption on indoor surfaces can cause an enrichment of VOCs as a result of poor air exchange rates. While monoterpene compounds can be additives in household chemicals and fragrance (limonene), aromatics are present mostly for technical reasons (such as solvents in paints and glues). Although the sources of residues in new buildings are evident, alternative sources of solvent residues in older houses have to be discussed. These may include infiltrates from the outdoor air which accumulate on indoor absorptive surfaces, causing elevated indoor air concentrations.

The potential sources of alkane hydrocarbons are more numerous: both residues caused by building technologies and consumer goods and infiltration from diesel engine emissions are possible. The measurement of indoor VOC levels is useful to detect high individual exposures, residues in new or renovated buildings, and the important role of infiltrating air as a marker of the general indoor air quality. To obtain comparable results, the exact analysis of all main components, perhaps by GC-MS (Wallace *et al.*, 1990), and the calculation of a TVOC value by summing the values obtained seems preferable. TVOC values are therefore suitable for epidemiological studies, in order to measure the relative indoor air quality, since indoor sources, residues, infiltration, and insufficient ventilation result in higher values. However, further investigations or measurements are necessary to find the causes or sources.

The mean indoor/outdoor TVOC ratio can be developed into a standardized general indoor air quality attribute. We believe that the indoor TVOC level cannot be lower than the outdoor fresh air level. Sampling periods of three days would be preferred to obtain representative mean values.

5. SUMMARY

During the past two decades, the Institute for Ecological Chemistry has investigated the behavior of volatile organic compounds (VOCs) and semivolatiles (SVOCs) and has conducted case studies in buildings with indoor air quality problems. The results that have been obtained demonstrate the importance of ventilation to control VOCs and the essential role of accumulation phenomena of SVOCs on material surfaces. Semivolatile biocide compounds were found to accumulate on the surface of textile fibers; concentrations in the range of micrograms per square meter in the air led to accumulations of up to milligrams per kilogram on textile fibers in 48 hours. This result points out that in the indoor environment dermal penetration of SVOCs could be more important than inhalation. Accumulation was observed indoors on sedimented particles with textile fiber contents, too. These results suggest that sedimented particles may be used for monitoring purposes as indicators or for passive sampling of SVOCs.

ACKNOWLEDGMENTS.   Parts of this study were financially supported by the Bavarian Environmental Authority. We would especially like to thank the 180 volunteers from the upper Bavarian cities for their participation and patience.

REFERENCES

Cohen, M. A., Ryan, B. R., Yanagisawa, Y., Spengler, J. D., Oezkaynak, H., and Epstein, P. S., 1989, Indoor/outdoor measurements of volatile organic compounds in the Kanawha Valley of West Virginia, *J. Air Pollut. Control Assoc.* **39:**1086–1093.

Gebefuegi, I., 1989, Chemical exposure in enclosed environments, *Toxicol. Environ. Chem.* **20–21:**121–127.

Gebefuegi, I., 1993, Biologisch aktive Chemikalien in Innenraeumen—Kontamination durch Baustoffe, in *Sick Building Syndrome* (W. Bischof *et al.*, eds.), pp. 78–82, Verlag C. F. Mueller, Karlsruhe.

Gebefuegi, I. L., and Korte, F., 1988a, Pathways and behaviour of semi-volatile chemicals in enclosed spaces, in *Indoor and Ambient Air Quality* (R. Perry and P. W. Kirk, eds.), pp. 393–398, Selper Ltd., London.

Gebefuegi, I. L., and Korte, F., 1988b, Chemicals in indoor air, in *Progress in Environmental Specimen Banking*, National Bureau of Standards Special Publication 740 (S. A. Wiese *et al.*, eds.), U.S. Government Printing Office, Washington, D.C.

Korte, F., and Gebefuegi, I., 1987, Avoidable and unavoidable exposition of indoor-chemicals, in *Indoor Air '87—Proceedings of the 4th International Conference on Indoor Air Quality and Climate* (B. Seifert *et al.*, eds.), pp. 239–241, Institute for Water, Soil and Air Hygiene, Berlin.

Kreuzig, R., Gebefügi, I., and Korte, F., 1986, Leichtflüchtige Kohlenwasserstoffe biogenen und anthropogenen Ursprungs in der Luft von Waldgebieten. *Forstwiss. Centralbl.* **105:**435–441.

Kreuzig, R., Gebefügi, I., Bahadir, M., and Korte, F., 1988, Jahresverlauf der Luftkonzentrationen anthropogener und biogener Kohlenwasserstoffe an drei unterschiedlich belasteten bayerischen Waldgebieten, *Centralbl. Gesamte Forstwes.* **107:**342–347.

Molhave, L., Bach, B., and Pedersen, O. F., 1986, Human reactions to low concentrations of volatile organic compounds, *Environ. Int.* **12**(1–4):167–175.

Oelert, H. H., Mayer-Gürr, W., and Zajontz, J., 1974, Zur motorischen Verbrennung von Benzin-kohlenwasserstoffen, *Erdöl Kohle, Erdgas, Petrochem. Brennst. Chem.* **27**(3):146–152.

Wallace, L., Pellizzari, E., and Wendel, C., 1990, Total organic concentrations in 2500 personal, indoor and outdoor air samples collected in the US EPA TEAM studies, in *Indoor Air '90, Proceedings of the 5th International Conference on Indoor Air Quality and Climate*, Toronto, Vol. 2, pp. 639–644, International Conference on Indoor Air Quality and Climate, Inc., Ottawa, Ontario.

Wolkoff, P., Johnsen, C. R., and Franck, C., 1992, A study of human reactions to office machines in a climate chamber, *J. Exposure Anal. Environ. Epidemiol.* **1992**(1):71–96.

Chapter 14

A Model Framework for Ranking of Measures to Reduce Air Pollution with a Focus on Damage Assessment

Kristin Aunan, Hans Martin Seip, and Hans Asbjørn Aaheim

1. INTRODUCTION

Both nationally and internationally, there is a growing interest in decision-support tools to assist decision makers in defining optimal combinations of abatement measures concerning a wide range of pollutants and their adverse effects. The importance given to cost-effectiveness and optimization has increased as the size of the investments needed to reduce environmental deterioration has been realized. Usually, each abatement measure influences the emissions of a number of pollutants, and some components are closely related in terms of abatement measures. This fact requires that different pollutants and abatement measures are considered as far as possible in an integrated way.

The project presented here is part of a comprehensive study planned to be carried out during 1994 and 1995: "Climate, Air Pollution and Energy: Cost-Effective Strategies for Reduction of Emissions," or the CAPE project for short.

KRISTIN AUNAN, HANS MARTIN SEIP, AND HANS ASBJØRN AAHEIM • Center for International Climate and Energy Research—Oslo (CICERO), University of Oslo, 0317 Oslo, Norway.

Combustion Efficiency and Air Quality, edited by István Hargittai and Tamás Vidóczy. Plenum Press, New York, 1995.

Two areas in Hungary, Budapest and Miskolc, will serve as examples of large cities and industrial areas, respectively, for application of the model. The damage types that probably are the main problems due to air pollution in Hungary and that will be the focus of the study are effects on human health, deterioration of forests and forest soils, and corrosion of materials. Global warming will be dealt with by estimating the reductions in greenhouse gases (GHGs) given in global warming potential (GWP) units as defined by the Intergovernmental Panel on Climate Change (IPCC) (Houghton *et al.*, 1990).

The project will be directed by the Center for International Climate and Energy Research—Oslo (CICERO) and will be performed by researchers at a number of Norwegian and Hungarian institutions. An introductory workshop with the Norwegian and Hungarian counterparts was held in Budapest in September 1993.

The overall objective of the CAPE project is to establish a comprehensive methodology and a decision-support tool for ranking of abatement measures regarding air pollutants and GHGs. Global and regional as well as local environmental effects will be addressed. The approach to be taken will be based on cost/benefit analysis (CBA). CBA proceeds by assessing, as far as possible, the social costs and benefits of a policy or action (e.g., an abatement measure). Theoretically, all measures with a positive present value of the net benefit should be implemented. However, a ranking is usually desirable, mainly because the budgetary resources may be lower than what is necessary for implementing all these measures and because some measures may be interdependent. The basic value judgment of CBA is that people act according to their individual preferences, and the goal is, in economic terms, to maximize social utility (OECD, 1989).

A number of questions have been raised as to whether monetary reductionism is feasible for all environmental attributes (for discussions, see Schrader-Frechette, 1991, and OECD, 1992). In our view, it is crucial that a physical description is also presented in connection with the monetized environmental attributes. We will, however, point out that monetization of health and environmental attributes, when pursued with integrity, has the advantage that implicit judgments are made explicit and subject to analysis and open discussion. As will be described in the following, we intend to integrate some features of other decision frameworks into our CBA approach. (For a systematized description of decision frameworks, see OECD, 1989.)

An important element of the project will be performed by political scientists and involves analyses of the socioeconomic and institutional environment within which alternative response measures may be chosen. In addition to contributing to development of methodology, the case studies in this project will hopefully also contribute to the development of cost-effective strategies concerning environmental and energy policy in Hungary.

2. THE OSLO AIR STUDY

The CAPE project will partly draw upon experiences from a project carried out in Norway—the Oslo Air Study. In this study, initiated and organized by the Norwegian State Pollution Control Authority, an analysis of air pollution abatement measures related to the city of Oslo was conducted (SFT, 1987; Trønnes and Seip, 1988). The principal steps in studies of this kind (see Fig. 1) include models for the fate of the emitted pollutants and the resulting exposure of humans, materials, and environment. Combining this information with quantitative knowledge of effects of the pollutants, the damages may be estimated. One control measure may affect a number of pollutants, and each compound may affect the population/recipient in several ways.

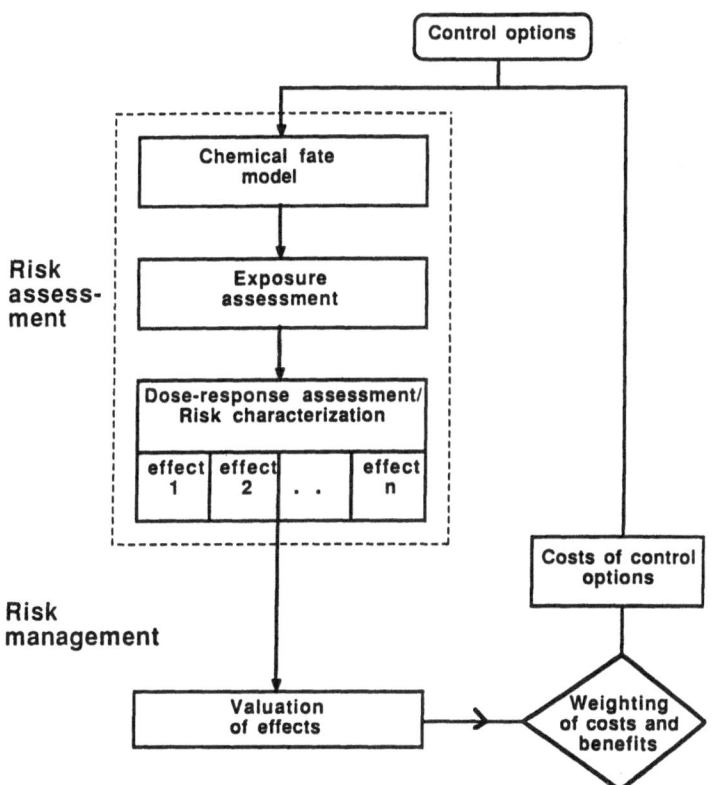

Figure 1. Components of an analysis of pollution control (from Trønnes and Seip, 1988).

Altogether, 38 measures (control options) were considered, including, for example, emission standards for buses, better maintenance of diesel vehicles, and substitution of oil by electricity in domestic heating. While most of the measures were applicable on a local scale, some were inherently national scale measures (i.e., emission standards for road vehicles).

The analysis was restricted to comparing the benefits (monetized effects of the measures) to the direct costs of implementation. Implementation costs for the state, the municipality, and the private sector were estimated. Other social effects were calculated as "side effects."

The primary motivation for the work on measures to curtail air pollution in Oslo was considerations of human exposure to air pollution and effects on health and well-being. Additionally, benefits concerning some other impacts were also included in the analysis. The objectives were structured in the following way:

1. Reduce human exposure to SO_2, NO_x, CO, and suspended particulate matter.
2. Increase well-being in the sense of reducing affliction by dust and/or odor.
3. Reduce contributions to acidification by reducing emissions of SO_2 and NO_x in Oslo.
4. Increase positive and reduce negative "side effects" (e.g., changes in number of traffic accidents, changes in travel times due to changed traffic flow on the roads, reduced corrosion of buildings and metals, and changes in heating expenses).
5. Reduce implementation, operation, and maintenance costs of abatement measures.

The attributes (the variables in the analysis) according to each of these objectives were estimated in different units:

- "Number of persons exposed to concentration levels above the guideline" (for SO_2, NO_2, CO, and particles, respectively). In considering effects on human health, the attributes were thus not based on dose–response functions, due to the lack of satisfactory models in the literature at that time (4 attributes, X_1–X_4).
- "Number of persons afflicted" (by dust and odor, respectively) (2 attributes, X_5 and X_6).
- "Tons emitted per year" (of SO_2 and NO_x, respectively) (2 attributes X_7 and X_8).
- "Annualized present value of net change in side effects" (X_9).
- "Annualized present value of implementation costs" (X_{10}).

Only the two last attributes were given directly in monetary units. In order to bring the different effects onto a common scale for internal comparison and be able to make trade-offs with abatement costs, thereby ranking the measures, a simplified multi-criteria decision analysis (MCDA) was applied (Trønnes and Seip, 1988). This method will also be applied in the CAPE study (see later).

The benefit/cost ratio for each measure was calculated simply as

$$\frac{B}{C} = \sum_{i=1}^{9} \frac{k_i \cdot X_i}{k_{10} \cdot X_{10}}$$

where the k_is ($i = 1$–10) are the weights bringing the attributes onto a common scale, and the X_i are the attributes.

When putting together a ranked selection of measures, dependency between measures was accounted for. An example of mutual dependency between measures is the relationship between the following measures: (a) reducing the sulfur content in oils used in domestic heating, and (b) partly substituting oil in domestic heating by electricity. The environmental effect of implementing both is lower than the sum of the isolated effects of implementing each of them. The benefit/cost ratio of the measures will therefore depend on the order in which they were entered.

The uncertainties in the attribute values for the measures are often large. The effect was studied by subjective estimation of probability distributions for the attributes and performing "Monte Carlo simulations": A large number of calculations were carried out by drawing values for the uncertain variables according to their probability distributions (stochastic simulation). The uncertainties are thus carried through the calculations and displayed in the result as a probability distribution of the B/C ratio.

The methodology developed in the Oslo Air Study has been further elaborated in other studies conducted by the State Pollution Control Authority in Norway, *inter alia* the analysis of measures to fulfill the commitments according to the North Sea Declaration on discharges of nutrients to the North Sea basin (SFT, 1992a). Although it is usually difficult to state with certainty what is actually decisive for a policy being chosen, it is a fact that in Norway these kinds of analyses are used in the official arguments for political decisions (Strand, 1990).

3. FURTHER METHODOLOGICAL DEVELOPMENT

In this project we will attempt to take some further steps toward a more comprehensive modeling tool regarding the ranking of abatement measures. The basic philosophy is that one must take an integrated approach comprising various environmental problems in order to sort out cost-effective measures. An underlying motive is also that the incentive for responding to the climate change issue may be strengthened by connecting it more closely to regional local problems.

The approach taken in the CAPE study includes some interrelations between emissions, environmental impacts, and economic activities that are vital to the evaluation of measures. However, many relations that indeed exist between these variables will have to be excluded. For instance, the social benefits obtained from a more healthy population cannot be derived from the reductions in the costs of health care alone. Productivity will increase, for instance, because of reductions in days off from work due to illness. A lower need for health care means that one can release some of the economic resources to other economic activities and so on. Indirect effects of this type will be excluded from the study. We do not either intend to include all types of pollutants but rather will focus, at least as a starting point, on some of the main air pollutants in addition to GHGs. Thus, the model outcome will only partly deserve the term "integrated assessment" as it will not include, for example, effects of hazardous waste and effects of discharges to water. If our concept turns out to be successful, the model might be extended to include such issues.

Methodological development is desirable for several reasons, and some of the items that will be dealt with in our study are outlined in the following sections.

3.1. Extension of the "Bottom-Up" Approach

The approach applied in the Oslo study and related studies is based on information on the microlevel, often referred to as the "bottom-up approach" (B/U). This implies that a number of abatement measures considered appropriate for fulfilling the objectives of concern are explored in detail. The emission reduction potentials (based on empirical evidence) and potentials for damage reductions (based partly on empirical studies, partly on theoretical assumptions) are estimated. Assessments of the values of the costs and benefits are then made according to observed or estimated market prices. The social net benefit from each alternative, found by adding up all costs and benefits, provides the basis for the evaluation of measures.

An alternative to the "bottom-up" approach is the "top-down approach" (T/D), where the effects of an abatement strategy are assessed by the use of a macroeconomic model. These models enable, in principle, analysis of direct and indirect economic effects of abatement measures on main macroeconomic variables. From the predicted changes in economic activity, the effects on emissions are deduced. In principle, the damage caused by the effects on emissions could be assessed similarly to the estimation of damage in the CAPE study and then be implemented in the macroeconomic model in order to make a simultaneous assessment of economic and environmental effects. However, this requires that the model distinguishes between different sources of emissions and different recipients of importance. For instance, if a given level of pollution affects health according to population density, the model will have to include a geographical dimension. This turns out to be a major problem in establishing integrative macroeconomic model tools.

A major weakness of B/U calculations is, on the other hand, that macro-economic effects are generally left out. If the measures under consideration require a vast amount of economic resources, for instance, large investments, they will displace activities which would have been carried out if the measure in question was not implemented. The short-term effect on this might be unemployment or increased pressure in the labor market. The long-term effect might be reallocation of resources and changes in relative prices. Thus, the reliance on observed market prices would no longer apply. However, some macroeconomic effects, e.g., unemployment, may be approximated as "side effects."

The properties of the T/D and B/U approaches are also crucial for their usefulness in making an evaluation of alternative policy instruments. The general framework of macroeconomic models makes them particularly suitable for analyzing general measures, such as taxes, and less suitable for the analysis of individual physical measures such as investments in abatement activities, for which a B/U approach would be preferable. Taxes appear to have some optimizing properties if there is a close correlation between economic activity and environmental effects. They do not, however, always provide the most cost-effective alternative. There may be many factors that complicate the relation between the taxed commodity (e.g., fossil fuel) and the environmental deterioration, such as patterns of dispersion, the mechanisms of formation of the pollutant, and atmospheric interactions with other pollutants. A tax may thus "overpunish" parts of the consumption that actually have only a small environmental impact.

In our view, the T/D approach is especially difficult to apply when effects of a number of components with different chemical features are considered. When the focus is on the complexity of relations between emissions and damage, macro-economic models may not provide enough relevant information for the ranking of measures.

In the CAPE study the intention is to take the B/U approach as the point of departure but to extend the approach with some new elements that widen the scope. This will be done by analyzing the socioeconomic and institutional environment, which is decisive for defining feasible measures to be analyzed. However, a complete macroeconomic analysis will not be performed. In addition to the problems mentioned above, the special transient economic situation in Hungary and other post-Communist countries at present would be extremely difficult to model.

Our primary goal is to contribute some novel ideas about how to perform damage assessment and how to evaluate the related costs and benefits. For the ranking of measures, special attention will be paid to the evaluation of uncertain outcomes of the alternatives. The damage assessments will be indicated by probability distributions. Accordingly, the net benefit of measures will also have to be expressed in terms of distributions. However, because of nonlinear cost functions and/or risk aversion, the relation between the marginal damage and the marginal net benefit is unlikely to be linear. If it is nonlinear, the net benefit will depend partly on the distribution of the assessment of the damage, not only on its expected level. Since the interval of probable damage may be large and may vary significantly between alternative measures, it is vital to take into account the uncertainty of the outcomes when evaluating the measures. As mentioned above, the probability distribution of damage will depend on several factors taken as assumptions. Sensitivity analysis on these assumptions enables evaluation of cost-effectiveness not only according to parameters of a given distribution, but also according to alternative distributions.

3.2. Application of Dose–Response Functions

A dose–response function gives the relationship between the concentration level and the prevalence of a certain effect in the recipient. The scientific basis for applying the dose–response approach has improved significantly since the Oslo Air Study was conducted, although there are still many unknowns and uncertainties as illustrated by examples given below.

In the CAPE study, we will draw upon other studies on damage assessment. A major study has been conducted jointly by the Commission of the European Community (CEC) and the U.S. Department of Energy on external costs of different fuel cycles (e.g., Markandya and Rhodes, 1992). International expert groups were convened to discuss methodologies and make recommendations concerning the most applicable dose–response functions and models and the basis for monetizing damages. Although the study made an effort to estimate damages on a marginal basis, it was realized that often average figures have to be used and that these are an approximation to the marginal damage. When it came to impacts on forestry, for instance, it was not considered possible to use the dose–response approach, and it was recommended to estimate average damage in a country per ton of SO_2 deposited. Concerning global warming, estimates of contribution to total global warming potential (GWP) were calculated.

A literature study with regard to present methods and knowledge concerning damage assessment will be the basis for selecting the dose–response functions and other types of assessment tools to be included in our model. The attributes which will be subject to monetization should be easily understood by decision makers (the user), and this objective must be pragmatically weighted against consideration of, for instance, precision.

Concerning *health effects*, a problem is that most studies report only the immediate consequences and not the possible delayed effects. It appears, however, that cumulative exposure to air pollutants also provides a significant risk. For instance, experiments with animals indicate that prolonged exposure to ozone may induce effects which seem to be irreversible, while most human studies cover relatively short periods (up to a few weeks in epidemiological studies), reporting impaired pulmonary function, cough, and other reversible effects (for an overview, see SFT, 1992b). Epidemiological studies show that elderly people are more vulnerable to air pollution in general, but it is hard to say whether this is due to increased vulnerability to short-term rises in the concentration level or if it is caused by a cumulative effect of long-term exposure (Gottinger, 1983).

Concerning *effects on forests*, a large number of variables, in the air, water, and soil, will be affected by emissions of air pollutants. The effects may be caused either directly when gases are taken up through the stomata or indirectly when deposition alters the soil chemistry. Resilience of trees is recognized through the estimation of critical levels and loads. A number of models of tree response to acidic deposition have been developed, e.g., the forest module of the RAINS model and a growth response model by Sverdrup and co-workers [both assessed in the CEC/U.S. Department of Energy (1993) study]. The first one identifies areas of increased risk of forest decline due to excess of a critical dose of SO_2, which, in principle, is the same as comparing the concentration levels with *critical levels* (although the model additionally includes considerations of the effects of altitude on tree health). The second one predicts growth response as a function of soil acidification expressed by the $(Ca+Mg)/Al$ ratio (*critical load* approach).

The main problem for the modelers is the lack of dose–response functions for acidic deposition and forest decline, which is due to a number of problems including lack of knowledge of basic growth processes and technical difficulties in experimental studies of long-term response of mature trees. The fact that forest decline probably is caused by different pollutants and different mechanisms in various regions makes it extremely difficult to make models for larger areas.

Impacts of sulfur dioxide and ozone on *materials* provide some of the clearest examples of damage related to air pollution. The detrimental effect depends on the material properties, concentration levels of the agents, and, for acid precursors, the presence of water. According to a survey undertaken in the CEC/U.S. Department of Energy study (Markandya and Rhodes, 1992), reliable estimates of damage from acidic deposition (based on dose–response functions) can be made for stone, mortar, concrete, paint, steel, zinc, and aluminum. In Norway several attempts have been made to estimate maintenance and replacement costs associated with material damage and the total social costs of these damages (see, e.g., Henriksen *et al.*, 1993; Glomsrød and Rosland, 1988). In addition to costs of maintenance, monetization can also involve estimates of existence value (for instance, for historical monuments).

Concerning *global warming*, there are some formidable problems with the cost/benefit approach. Quantitative insights (for example, response functions) concerning physical impact are still rudimentary, monetary assessment is accordingly difficult, the uncertainties are huge, and worst case scenarios predict extensive irreversible damage, which has made the precautionary principle a governing rule in global environmental policy processes. Advocates of the cost/

benefit approach point out that the drastic measures that are needed to curb greenhouse gas emissions should not be implemented without any serious attempt to weight the economic costs and benefits of alternative control strategies (for a discussion, see Pearce *et al.*, 1992). Since the potential effect of global warming will affect different regions in the world very differently, national or regional cost/ benefit analyses might, however, tend to have a too narrow perspective for properly taking care of the climate change issue.

3.3. The Process of Weighting and Evaluating Health and Environmental Attributes

The purpose of this part of the study is the monetization of damage to health and environment and other side effects that are considered in the analysis. These are external costs, defined as costs which fall on one group of people due to the social or economic activities of another group, and where the latter group does not take these costs into account.

When trying to monetize these effects, it is necessary to distinguish between costs in terms of a change in the trade of goods and services with observed market prices and attempts to relate a market price to effects on variables that are not traded in the market but indeed count for the social welfare, such as environmental amenities. The first category relates to so-called indirect economic effects, which include loss of worker productivity due to illness, medical expenses, and repair and replacement costs for material damage. The secondary category relates to a broader welfare concept than is usually applied in cost/benefit analysis. Nevertheless, such elements may have a significant influence on the choice of policy, especially when evaluating environmental measures (see, e.g., Sen, 1979). However, in monetizing these effects, several methodological problems arise. In the CAPE project, attempts will be made to include the value of environmental amenities, but a clear distinction between "real" economic effects and estimated value of environmental amenities will be made in presenting the results.

There are different methods for monetization of environmental amenities. The issue is subject to extensive research and implies considerations of fundamental questions of values. In the Oslo Air Study an expert panel arrived at values by using results from other studies, based on different methods, including contingent valuation assessments.[1] Here we will restrict ourselves to giving a brief introduction to one method that we intend to apply to bring attributes that are not directly commensurable onto a common scale, the multi-criteria decision analysis

---

[1]Contingent valuation includes asking for willingness to pay or willingness to accept—a measure of personal valuation of the respondent for increases or decreases in quantity of some good, contingent upon a hypothetical market. The method is only meaningful if the respondent is properly aware of the effects of concern (OECD, 1989).

(MCDA). Recently, this method has been applied for the purpose of estimating the Norwegian society's willingness to pay for environmental goods (Wenstøp et al., 1994).

The theoretical foundation for MCDA was laid down already in 1944 by von Neumann and Morgenstern (1944) through their work on the basis for utility theory. The theory was generalized by Keeney and Raiffa (1976) and made suitable for treating several conflicting objectives. Application of MCDA has increased significantly during recent years, specifically in areas where it is necessary to make trade-offs between health and environment versus economic criteria (Carlsen and Wenstøp, 1992).

In an MCDA the criteria, or attributes, are given weights by expert panels. These panels may consist of persons who have special knowledge of the topics of concern and/or have a specific profession or belong to a particular type of organization. The members of the panels are encouraged to aim at acting as general members of the society and not act tactically in order to promote sectional interests. The process of weighting different attributes is heuristic and helps the members of the panels, through discussions and assisted by interactive data software, to reveal their preferences in a consistent way (Carlsen and Wenstøp, 1992). The results from different panels will differ to varying extents.

3.4. Time Limits and Transferability

Questions of time limits and comparability of events at different points of time are essential for estimations of long-term effects, because of varying time profiles of different effects, among other reasons. Discounting monetary estimates of future environmental effects has been extensively discussed (Cline, 1992), but several problems remain to be solved before one can say that there is a consensus about how to treat the problem of discounting (see, e.g., Machina, 1989).

In studies where long-term environmental effects with uncertain outcomes are as important as they are in the CAPE study, a comparison between different ways of discounting is of particular interest. The problem relates partly to that of welfare comparisons mentioned above. In a long-term perspective, one may encounter considerable changes in the distribution of welfare between countries and between welfare components within one country. Generally, there is little knowledge of the income elasticity of demand for different environmental goods (Markandya, 1993). From damage estimates in a U.S. study on environmental costs of electricity (Ottinger et al., 1990), the income elasticities of health damages and material damages were estimated as 1.15 and 0.72, respectively. This implies that if income increases over time, the importance of health increases relative to that of material damages (see Markanyda, 1993). Since the value of health is expected to comprise more than its direct economic value, these noneconomic aspects of a future change in income should be reflected in the discount rate. The longer the time perspective, the more important is the inclusion of noneconomic aspects.

Concerning global warming, the output of our model will probably be limited to GWP units. GWPs for the important greenhouse gases have been estimated for various time horizons. However, this does not, of course, describe the time profiles of environmental effects.

Similar problems are related to transferability of dose–response functions from one region to another. National boundaries are in themselves not of any relevance to how people and the environment respond to pollution, but cultural differences may be. Easier to deal with, but equally important, are differences in, for instance, age distribution and overall health status in the population and vulnerability of vegetation due to, for example, differing climatic conditions.

4. OUTLINE OF A DAMAGE ASSESSMENT MODEL

In the CEC/U.S. Department of Energy study mentioned above, well over 200 impacts (considered as externalities) were identified for the coal fuel cycle. However, detailed analysis was restricted to those believed to be most important. This indicates the importance of identifying *priority impacts* at an early stage. The damage types that have been pointed out as probably the main problems due to air pollution in Hungary are effects on human health, deterioration of forests and forest soils, and corrosion of materials (Gajzago, 1992; Pais and Horváth, 1990; KTM, 1992).

Interdependency between measures calls for arranging the damage assessment model in such a way that for every measure that is fed into the model, the exposure status is adjusted before entering a new measure. The cost/benefit ratio of the measures will therefore depend on the order in which they were entered.

Four main modules have been outlined as necessary elements in the damage assessment model. These may be classified as follows:

A. Data base modules:
  • Emission data (source specification)
  • Recipient data (people, crops, forests, materials/buildings)
B. Model base modules:
  • Atmospheric chemistry, dispersion, and deposition
  • Damage assessment (based on dose–response functions and other approaches)

1. *Emission data base.* The input to this database is the results of the inventories of GHGs and other air pollutants according to different source categories (such as stationary combustion, industrial processes, and mobile

combustion). It will include CO_2 SO_2, NO_x, suspended particulates, CO, organic compounds, and metals.

2. *"Chemical fate" model(s)*. The input to this module will be emissions from different sources, and the output should give estimates of how dispersion and deposition of the different chemical species influence concentration levels and deposition loads in different areas. The different measures that are analyzed will result in changes in these quantities, and the status should be recalculated for every measure introduced. The format of the output has to be in accordance with the damage attributes in module 4. For instance, one has to decide whether the concentration levels should be given as hourly, daily, and/or yearly means.

3. *Population data base*. This data base will contain data concerning the geographical distribution of people, agriculture production, and forests within the areas of concern. We will also include data on buildings and materials (including historical/cultural objects) that are vulnerable to corrosion due to air pollution. This data base will be built up in close association with the following module (damage assessment) in order to ensure accordance between the two.

4. *Damage assessment module*. The output of the two former modules will be an assessment of the exposure of people, vegetation, and materials, and this will be input to the damage assessment module, which should provide estimates of potential damage (risk) to the populations/recipients of concern. When possible, dose–response functions will be used to calculate changed damage due to reductions in the emissions of air pollutants. Concerning global warming, estimates of indirect effects of other gases [NO_x, CO, and volatile organic compounds (VOCs)] on climate, as calculated by Fuglestvedt *et al.* (1994), will also be included in the GWP figures.

5. CONCLUSIONS

An integrated approach considering several pollution problems is generally necessary in cost/benefit analyses of abatement measures. When the focus is on the complexity of relations between emissions of different components and their associated damages (local, regional, and global), the microeconomic approach (often referred to as the "bottom-up approach") seems to have advantages compared to macroeconomic methods ("top-down"), which are often applied for evaluation of environmental policy instruments.

The assessment of impacts on health and environment, a crucial step in the analysis, requires knowledge concerning exposure and the mechanisms responsible for chemical and biological effects. Dose–response relationships give information which is very useful when it comes to analyzing the effects of reducing emissions.

Inherently, there will be uncertainties in every step in the calculations, and these should be reflected in the results.

The process of carrying out an analysis may in itself be valuable in terms of clarifying the main objectives of an abatement policy and explicitly describing adverse impacts and their assumed relative importances. This gives excellent opportunities for communication with those affected by the decisions. It also reveals the areas where the need for more insight is strongest. In our view, it is crucial that a physical description be also presented in connection with monetized environmental damages.

REFERENCES

Carlsen, A., and Wenstøp, F., 1992, Valuation of Health and Environmental Amenities by Use of Expert Panels, Norwegian School of Management, Research Note 1992 [in Norwegian].

CEC/U.S. Department of Energy, 1993, Joint Study on Fuel Cycle Costs. Assessment of the External Costs of the Coal Fuel Cycle, draft position paper prepared for DG XII of the Commission of the European Community, February 1993.

Cline, W., 1992, *The Economics of Global Warming*, Institute for International Economics, Washington, D.C.

Fuglestvedt, J.S., Wang, W-C., and Isaksen, I. S. A., 1994, Direct and Indirect Global Warming Potentials of Source Gases, Center for International Climate and Energy Research (CICERO) report, Oslo.

Gajzago, L., 1992, Air-pollution problems and priorities in Hungary, in *Coping with Crisis in Eastern Europe's Environment* (I. Alcamo, ed.), pp. 167–174, The Parthenon Publishing Group, Carnforth, United Kingdom.

Glomsrød, S., and Rosland, A., 1988, Air Pollution and Material Damages: Social Costs, Central Bureau of Statistics of Norway Report 88/31 [in Norwegian].

Gottinger, H. W., 1983, Air pollution health effects in the Munich metropolitan area: Preliminary results based on a statistical model, *Environ. Int.* **9**:207–220.

Henriksen, J. F., Bartonova, A., Støre, M., and Haagenrud, S. E., 1993, External Building Materials in a Norwegian Town, Sarpsborg—Quantities, Degradation and Costs Caused by Air Pollution, Norwegian Institute for Air Research Report F, November 1993.

Houghton, J. T., Jenkins, G. J., and Ephraums, J. J. (eds.), 1990, *Climate Change: The IPCC Scientific Assessment*, Intergovernmental Panel on Climate Change, Cambridge University Press, Cambridge.

Keeney, R., and Raiffa, H., 1976, *Decisions with Multiple Objectives*, John Wiley & Sons, New York.

KTM, Ministry for Environment and Regional Policy, 1992, State of the Environment in Hungary. The Short and Medium Term Environment Protection Action Plan of the Government, Budapest.

Machina, M., 1989, Dynamic consistency and non-expected utility of choice under uncertainty, *J. Econ. Lit.* **XXVII**:1622–1668.

Markandya, A., 1993, Air pollution and energy policies: The role of environmental damage estimation, Paper presented at the Workshop on "The International Dimension of Environmental Policy," Fondazione ENI Enrico Mattei, Milan, October 22–24, 1992.

Markandya, A., and Rhodes, B., 1992, External Costs of Fuel Cycles—An Impact Pathway Approach, Report prepared for the CEC/U.S. Fuel Cycle Study, Metroeconomica Ltd., draft, June 1992.

OECD, 1989, Environmental Policy Benefits: Monetary Valuation, Study prepared by D. W. Pearce and A. Markandya, Paris.

OECD, 1992, Benefits Estimates and Environmental Decision-Making, Study prepared by D. W. Pearce and K. Turner, Paris.

Ottinger, R., Wooley, D. R., Robinson, N. A., Hodas, D. R., and Babb, S. E., 1990, Environmental Costs of Electricity, Pace University Centre for Environmental Legal Studies, prepared for the New York State Energy Research and Development Authority and the United States Department of Energy.

Pais, I., and Horváth, L., 1990, Atmospheric acidic deposition and its environmental effects in Hungary, in *Acid Precipitation*, Vol. 5, *International Overview and Assessment* (A. H. M. Bresser and W. Salomons, eds.), pp. 193–214, Springer-Verlag, New York.

Pearce, D. W., Turner, R. K., and O'Riordan, T., 1992, Energy and social health: Integrating quantity and quality in energy planning, *World Energy Council J.* **1992**(December):76–88.

Schrader-Frechette, K. S., 1991, *Risk and Rationality*, University of California Press, Berkeley.

Sen, A., 1979, Personal utilities and public judgments: Or what's wrong with welfare economics?, *The Economic Journal* **89**:537–558.

SFT, Norwegian State Pollution Control Authority, 1987, Further Reductions of Air Pollution in Oslo, Norwegian State Pollution Control Authority Report, Oslo, August 1987 [in Norwegian].

SFT, Norwegian State Pollution Control Authority, 1992a, The North Sea Declaration—Measures to Reduce the Discharges of Nutrients, Norwegian State Pollution Control Authority Report 92:14, Oslo [in Norwegian].

SFT, Norwegian State Pollution Control Authority, 1992b, Effects of Ambient Air Pollution on Health and the Environment—Air Quality Guidelines, Norwegian State Pollution Control Authority Report 93:18, Oslo.

Strand, J., 1990, Principles and methods for valuation of environmental amenities, *Nordisk Tidsskrift för Politisk Ekonomi* **24**:9–25 [in Norwegian].

Trønnes, D. H., and Seip, H. M., 1988, Decision making in control of air pollutants posing health risks, in *Risk Assessment of Chemicals in the Environment* (M. L. Richardson, ed.), pp. 265–297, The Royal Society of Chemistry, London.

von Neumann, J., and Morgenstern, O., 1944, *The Theory of Games and Economic Behaviour*, Princeton University Press, Princeton, New Jersey.

Wenstøp, F., Carlsen, A. J., Bergland, O., and Magnus, P., 1994, Valuation of Environmental Goods with Expert Panels, The Norwegian School of Management Research Report.

Index